T0227823

Industrial Wireless Sensor Networks

APPLICATIONS, PROTOCOLS, AND STANDARDS

INDUSTRIAL ELECTRONICS SERIES
Series Editors:
Bogdan M. Wilamowski & J. David Irwin

PUBLISHED TITLES

Industrial Wireless Sensor Networks: Applications, Protocols, Standards, and Products, *V. V. Çağrı Güngör and Gerhard P. Hancke*

Power Electronics and Control Techniques for Maximum Energy Harvesting in Photovoltaic Systems, *Nicola Femia, Giovanni Petrone, Giovanni Spagnuolo, and Massimo Vitelli*

Extreme Environment Electronics, *John D. Cressler and H. Alan Mantooth*

Renewable Energy Systems: Advanced Conversion Technologies and Applications, *Fang Lin Luo and Hong Ye*

Multiobjective Optimization Methodology: A Jumping Gene Approach, *K.S. Tang, T.M. Chan, R.J. Yin, and K.F. Man*

The Industrial Information Technology Handbook, *Richard Zurawski*

The Power Electronics Handbook, *Timothy L. Skvarenina*

Supervised and Unsupervised Pattern Recognition: Feature Extraction and Computational Intelligence, *Evangelia Micheli-Tzanakou*

Switched Reluctance Motor Drives: Modeling, Simulation, Analysis, Design, and Applications, *R. Krishnan*

FORTHCOMING TITLES

Smart Grid Technologies: Applications, Architectures, Protocols, and Standards, *Vehbi Cagri Gungor, Carlo Cecati, Gerhard P. Hancke, Concettina Buccella, and Pierluigi Siano*

Multilevel Converters for Industrial Applications, *Sergio Alberto Gonzalez, Santiago Andres Verne, and Maria Ines Valla*

Data Mining: Theory and Practice, *Milos Manic*

Granular Computing: Analysis and Design of Intelligent Systems, *Witold Pedrycz*

Electric Multiphase Motor Drives: Modeling and Control, *Emil Levi, Martin Jones, and Drazen Dujic*

Sensorless Control Systems for AC Machines: A Multiscalar Model-Based Approach, *Zbigniew Krzeminski*

Next-Generation Optical Networks: QoS for Industry, *Janusz Korniak and Pawel Rozycki*

Signal Integrity in Digital Systems: Principles and Practice, *Jianjian Song and Edward Wheeler*

FPGAs: Fundamentals, Advanced Features, and Applications in Industrial Electronics, *Juan Jose Rodriguez Andina and Eduardo de la Torre*

Dynamics of Electrical Machines: Practical Examples in Energy and Transportation Systems, *M. Kemal Saioglu, Bulent Bilir, Metin Gokasan, and Seta Bogosyan*

Industrial Wireless Sensor Networks

APPLICATIONS, PROTOCOLS, AND STANDARDS

EDITED BY

V. Çağri Güngör and Gerhard P. Hancke

CRC Press
Taylor & Francis Group
Boca Raton   London   New York

CRC Press is an imprint of the
Taylor & Francis Group, an **informa** business

CRC Press
Taylor & Francis Group
6000 Broken Sound Parkway NW, Suite 300
Boca Raton, FL 33487-2742

First issued in paperback 2017

© 2013 by Taylor & Francis Group, LLC
CRC Press is an imprint of Taylor & Francis Group, an Informa business

No claim to original U.S. Government works
Version Date: 20130123

ISBN 13: 978-1-138-07620-4 (pbk)
ISBN 13: 978-1-4665-0051-8 (hbk)

This book contains information obtained from authentic and highly regarded sources. Reasonable efforts have been made to publish reliable data and information, but the author and publisher cannot assume responsibility for the validity of all materials or the consequences of their use. The authors and publishers have attempted to trace the copyright holders of all material reproduced in this publication and apologize to copyright holders if permission to publish in this form has not been obtained. If any copyright material has not been acknowledged please write and let us know so we may rectify in any future reprint.

Except as permitted under U.S. Copyright Law, no part of this book may be reprinted, reproduced, transmitted, or utilized in any form by any electronic, mechanical, or other means, now known or hereafter invented, including photocopying, microfilming, and recording, or in any information storage or retrieval system, without written permission from the publishers.

For permission to photocopy or use material electronically from this work, please access www.copyright.com (http://www.copyright.com/) or contact the Copyright Clearance Center, Inc. (CCC), 222 Rosewood Drive, Danvers, MA 01923, 978-750-8400. CCC is a not-for-profit organization that provides licenses and registration for a variety of users. For organizations that have been granted a photocopy license by the CCC, a separate system of payment has been arranged.

Trademark Notice: Product or corporate names may be trademarks or registered trademarks, and are used only for identification and explanation without intent to infringe.

Library of Congress Cataloging-in-Publication Data

Industrial wireless sensor networks : applications, protocols, and standards / editors, Vehbi Cagri Gungor, Gerhard P. Hancke.
     pages cm. -- (Industrial electronics series)
     Summary: "Preface In today's competitive industry marketplace, the companies face growing demands to improve process efficiencies, comply with environmental regulations, and meet corporate financial objectives. Given the increasing age of many industrial systems and the dynamic industrial manufacturing market, intelligent and low-cost industrial automation systems are required to improve the productivity and efficiency of such systems. The collaborative nature of industrial wireless sensor networks (IWSNs) brings several advantages over traditional wired industrial monitoring and control systems, including self-organization, rapid deployment, flexibility, and inherent intelligent-processing capability. In this regard, IWSN plays a vital role in creating a highly reliable and self-healing industrial system that rapidly responds to real-time events with appropriate actions. In this book, detailed reviews about the emerging and already employed industrial wireless sensor network applications and technologies are presented. In addition, technical challenges and design objectives are introduced. Specifically, radio technologies, energy harvesting techniques and network and resource management for IWSNs are discussed. Furthermore, industrial wireless sensor network standards are presented in detail. Overall, this book covers the current state of the art in IWSNs and discuss future research directions in this field. The remainder of this book is organized as follows. Applications of Industrial Wireless Sensor Networks Chapter X analyzes and evaluates Industrial Wireless Sensor Networks (IWSN) applications classified into three main groups, that is environmental sensing, condition monitoring and process"-- Provided by publisher.
     Includes bibliographical references and index.
     ISBN 978-1-4665-0051-8 (hardback)
     1. Digital control systems. 2. Wireless sensor networks--Industrial applications. 3. Automation. I. Gungor, Vehbi Cagri.

TJ223.M531536 2013
681'.2--dc23   2012048676

Visit the Taylor & Francis Web site at
http://www.taylorandfrancis.com

and the CRC Press Web site at
http://www.crcpress.com

To my wife Burcu, my other Ayten, my son Selim for their continuous love and support, and to the loving memory of my Dad, Ahmet.

Dr.V. Çağrı Güngör

Contents

16 Industrial WSN Standards

Tomas Lennvall, Krister Landernäs, Mikael Gidlund, and Johan Åkerberg

List of Figures

List of Tables

Preface

In today's competitive industry marketplace, companies face growing demands to improve process efficiencies, comply with environmental regulations, and meet corporate financial objectives. Given the increasing age of many industrial systems and the dynamic industrial manufacturing market, intelligent and low-cost industrial automation systems are required to improve the productivity and efficiency of such systems. The collaborative nature of industrial wireless sensor networks (IWSNs) brings several advantages over traditional wired industrial monitoring and control systems, including self-organization, rapid deployment, flexibility, and inherent intelligent-processing capability. In this regard, IWSNs play a vital role in creating a highly reliable and self-healing industrial system that rapidly responds to real-time events with appropriate actions.

In this book, detailed reviews about the emerging and already employed industrial wireless sensor network applications and technologies are presented. In addition, technical challenges and design objectives are introduced. Specifically, radio technologies, energy harvesting techniques, and network and resource management for IWSNs are discussed. Furthermore, industrial wireless sensor network standards are presented in detail. Overall, this book covers the current state of the art in IWSNs and discusses future research directions in this field. The remainder of this book is organized as follows.

Applications of Industrial Wireless Sensor Networks

Chapter 1 analyzes and evaluates Industrial Wireless Sensor Networks (IWSN) applications classified into three main groups, that is environmental sensing, condition monitoring, and process automation applications, which are based on the specific requirements of the groups of applications. It points out the technological challenges of deploying WSNs in the industrial environment. An extensive list of IWSN commercial solutions and service providers are provided and future trends in the field of IWSNs are summarized.

Machine Condition Monitoring with Industrial Wireless Sensor Networks

Chapter 2 discusses the system requirements for Industrial Wireless Sensor Networks (IWSNs) monitoring systems. The state-of-the-art wireless sensor platforms are also introduced and compared. The trade-off between the higher system requirements of industrial monitoring and the resource constrained characteristics of IWSN nodes are demonstrated through the example of induction motor condition monitoring. Three industrial wireless network standards, i.e., ZigBee, WirelessHART, ISA100.11a, and several protocols developed by individual researchers for IWSNs are also reviewed.

Wireless Sensor Networks for Intelligent Transportation Applications

Chapter 3 focuses on the use of wireless sensor networks for sensing and monitoring in intelligent transportation applications. It also describes three key application areas, that is traffic and car park management, intra-vehicle monitoring systems, and road safety. In addition, the application requirements for different automotive segments are discussed and promising research approaches are mentioned. The issue of privacy, especially in intra-vehicle monitoring systems, is pointed out.

Design Challenges and Objectives in Industrial Wireless Sensor Networks

Chapter 4 covers Industrial Wireless Sensor Network (IWSN) requirements and objectives common to many applications. It also summarizes the potential advantages offered by IWSNs. Research challenges in the area of IWSNs, such as safety, security, availability, real-time performance, system integration and deployment, coexistence and interference avoidance, and energy consumption, are discussed in detail. Important properties of industrial communication systems are also presented.

Resource Management and Scheduling in WSNs Powered by Ambient Energy Harvesting

Chapter 5 points out that resource (energy) management in WSNs equipped with energy harvesting capabilities is substantially and qualitatively different from resource management in traditional (battery-powered) WSNs. It summarizes a selection of state-of-the-art resource management and scheduling algorithms, developed for energy harvesting wireless sensor networks (WSNs), selected in particular with respect to their suitability to the industrial WSN environment. The drawbacks, advantages, and possible application areas of these algorithms are also discussed to inform a system designer about which one to implement for a particular application or in a particular industrial setting.

Energy Harvesting Techniques for Industrial Wireless Sensor Networks

Chapter 6 discusses different energy harvesting techniques including solar, thermal, vibration, air flow, acoustic, magnetic field, electromagnetic wave, radio frequency-based energy harvesting, and envisaged methods in terms of efficiency and applicability. These techniques are also compared in terms of power density in indoor and outdoor environments. In addition, node and network-level adaptations to improve energy harvesting schemes are explained in detail. Open research issues are discussed with future research directions proposed.

Fault Tolerant Industrial Wireless Sensor Networks

Chapter 7 introduces the main sources of faults in industrial wireless sensor networks (IWSNs). In particular, specific types of network level faults along with fault tolerance mechanisms for managing network faults in three major industrial wireless sensor network (WSN) standards (ZigBee, WirelessHART and ISA 100.11a) are discussed. Finally, some promising fault tolerant IWSN design approaches, including fault tolerant routing and fault tolerant node placement and clustering are presented to improve the fault tolerance capability at the network level of an IWSN.

Network Architectures for Delay Critical Industrial Wireless Sensor Networks

Chapter 8 focuses on communication challenges caused by the harsh deployment environments that are common in industrial settings, and the strict application requirements imposed by the real-time nature of the industrial processes. To address delay requirements of industrial applications, three cross-layer mechanisms on PHY and MAC layers have been proposed. These mechanisms utilize the deadline information associated with each packet for further transmission, backoff decision, and redundancy optimization. The authors also show comparative performance evaluations of their proposed solutions.

Network Synchronization in Industrial Wireless Sensor Networks

Chapter 9 emphasizes the importance of network synchronization for current and future Industrial Wireless Sensor Networks (IWSNs) and summarizes some functionalities having this requirement. It classifies and presents an overview of network synchronization approaches for IWSNs. Specifically, mutual network synchronization, random time source, and diffusion algorithms that improve on synchronization reliability and fault-tolerance, are introduced. The clock parameter estimation is also described briefly.

Wireless Control Networks with Real-Time Constraints

Chapter 10 focuses on the problem of operating an industrial system with real-time constraints over a wireless network, employing simple, off-the-shelf, IEEE 802.11 (WiFi) and IEEE 802.15.4 (ZigBee) radio technologies. Three case studies have also been presented under various channel conditions over different wireless technologies. Through extensive experiments, it is also shown that significant performance gains can be achieved by the integration of wireless model based predictive networked control system with various considerations in the network level and challenges of the wireless networked control problem can be well addressed with such multi-disciplinary approaches.

Medium Access Control and Routing in Industrial Wireless Sensor Networks

Chapter 11 presents existing protocol solutions for Industrial Wireless Sensor Networks (IWSNs) at the medium access control (MAC) and network layers. A classification of the protocols and a summary of the standardization activities at both layers as well as the bridging layers are provided. Testing the performance of the proposed protocols in real industrial environments with harsh conditions, cross-layer solutions that involve interactions between MAC and routing layers, protocols supporting mobile nodes and sinks, supporting real-time services, energy harvesting-aware power management and protocols, co-existence issues, and error control techniques are identified as some research challenges in the domain of IWSNs.

QoS-Aware Routing for Industrial Wireless Sensor Networks

Chapter 12 focuses on Quality of Service (QoS) provisioning at the routing layer for Industrial Wireless Sensor Networks (IWSNs). It also provides a general overview of industrial application requirements and the challenges imposed by the limitations of wireless sensor devices that influence the design and performance of QoS-aware routing protocols for IWSNs. In addition, it reviews current approaches to routing in industrial environments together with the mechanisms used to provide QoS at the Network Layer.

Reliable and Robust Communications in Industrial Wireless Sensor Networks

Chapter 13 introduces a cross-layer design of channel coding, resource allocation, and distributed source coding. In addition, the relationship between source coding rates and the source nodes correlation is presented. The trade-off between packet error rate and energy-latency performance is analyzed to provide unequal error protection among sensor nodes. Moreover, the channel coding technique is investigated as a promising technique to improve the overall network reliability and robustness.

Network Security in Industrial Wireless Sensor Networks

Chapter 14 describes the security solutions proposed in the context of Industrial Wireless Sensor Networks (IWSNs). A particular focus is put on network authentication and access control, key management, firmware update, and privacy protection solutions. The chapter also discusses their applicability in the industrial domain based on a number of derived security challenges associated with industrial applications.

Cognitive Radio Sensor Networks for Industrial Applications

Chapter 15 presents potential advantages and architecture configurations of Cognitive Radio Sensor Networks (CRSNs) for industrial applications to overcome challenges of wireless monitoring and control in industrial environments. In addition, spectrum management functionalities, including spectrum sensing, decision and mobility, are discussed, and communication protocol requirements of CRSN are investigated from the perspective of industrial applications. The open research issues on communication protocol development for CRSN in industrial applications are also presented.

Industrial Wireless Sensor Network Standards

Chapter 16 highlights the need for specific wireless communication standards for industrial application as opposed to the consumer industry, which has very different requirements. It gives an overview of main Industrial Wireless Sensor Network (IWSN) standards, these being IEEE 802.15.4, WirelessHART, ISA100.11a, and WIA-PA. It discusses the strengths and weaknesses of each standard. A future outlook on industrial WSNs is also given.

Chapter Outline:

1) Applications of Industrial Wireless Sensor Networks

2) Machine Condition Monitoring with Industrial Wireless Sensor Networks

3) Wireless Sensor Networks for Intelligent Transportation Applications

4) Design Challenges and Objectives in Industrial Wireless Sensor Networks

5) Resource Management and Scheduling in WSNs Powered by Ambient Energy Harvesting

6) Energy Harvesting Techniques for Industrial Wireless Sensor Networks

7) Fault Tolerant Industrial Wireless Sensor Networks

8) Network Architectures for Delay Critical Industrial Wireless Sensor Networks

9) Network Synchronization in Industrial Wireless Sensor Networks

10) Wireless Control Networks with Real-Time Constraints

11) Medium Access Control and Routing in Industrial Wireless Sensor Networks

12) QoS-Aware Routing for Industrial Wireless Sensor Networks

13) Reliable and Robust Communications in Industrial Wireless Sensor Networks

14) Network Security in Industrial Wireless Sensor Networks

15) Cognitive Radio Sensor Networks for Industrial Applications

16) Industrial Wireless Sensor Network Standards

V. Çağrı Güngör
Gerhard P. Hancke

Author Biographies

V. Çağrı Güngör received his B.S. and M.S. degrees in electrical and electronics engineering from Middle East Technical University, Ankara, Turkey, in 2001 and 2003, respectively. He received his Ph.D. degree in electrical and computer engineering from the Broadband and Wireless Networking Laboratory, Georgia Institute of Technology, Atlanta, GA, USA, in 2007. Currently, he is an Assistant Professor and Co-Director of the Computer Networks and Mobile Communications Lab at the Department of Computer Engineering, Bahcesehir University, Istanbul, Turkey. Before joining Bahcesehir University, he was working at Eaton Corporation, Innovation Center, WI, USA as a Project Leader. His current research interests are in wireless ad hoc and sensor networks, industrial communications, smart grid communications, next-generation wireless networks, cognitive radio networks, and IP networks. Dr. Güngör has authored several papers in refereed journals and international conference proceedings, and has been serving as a reviewer and program committee member to numerous journals and conferences in these areas. He is also the recipient of the IEEE Transactions on Industrial Informatics 2012 Best Paper Award, IEEE ISCN 2006 Best Paper Award, the European Union FP7 Marie Curie IRG Award in 2009, and the San-Tez Project Awards issued by Alcatel-Lucent and the Turkish Ministry of Science, Industry, and Technology in 2010. He is also the Principal Investigator of the Smart Grid Communications R&D project funded by Turk Telekom between 2010 and 2012.

Gerhard P. Hancke received B.Sc., B.Eng., and M.Eng. degrees from the University of Stellenbosch, Stellenbosch, South Africa, and the D.Eng. degree from the University of Pretoria, Pretoria, South Africa. He has been engaged in engineering education, research, and management for more than 36 years. He is the Program Coordinator for Computer Engineering at the University of Pretoria and has played a major role in developing this program. He is the founder and Head of the Advanced Sensor Networks Group, a collaborative intiniative with the Meraka Institute at the Council for Scientific and Industrial Research (CSIR) and is also collaborating in research projects internationally. He has published more than 150 papers in refereed journals and international conference proceedings. Amongst others he has been or is: refereeing papers for many journals and conferences; Area Editor of Elsevier's Ad Hoc Networks Journal; Guest Co-Editor of a Special Section on "Industrial Wireless Sensor Networks" in the IEEE Transactions on Industrial Electronics, 2009; Guest Co-Editor of a Special Section on "Information Technologies in Smart Grids" in the IEEE Transactions on Industrial Informatics, 2012/13; General (Co-)Chair of International IEEE Conferences, e.g., IEEE Industrial Informatics Conference (INDIN 2007), Vienna, and IEEE International Conference on Industrial Technology (ICIT 2013). He is a co-recipient of the IEEE Transactions on Industrial Informatics 2012 Best Paper Award and the recipient of two THRIP Technology Awards from the South African Department of Trade and Industry (DTI): in the SMME Development Category (2007) and the Advanced Hi-Tech Category (2011). In 2007 he received the IEEE Larry K. Wilson Award "For inspiring membership development and services as a member of several regional and technical conferences worldwide." Before joining academia, he was a Senior Engineer with the Atomic Energy Board.

1

Applications of Industrial Wireless Sensor Networks

Milan Erdelj and Nathalie Mitton

FUN Research Group, Inria Lille - Nord Europe, France

Enrico Natalizio

University of Technology of Compiègne, Heudiasyc Laboratory, France

CONTENTS

1.1   Introduction

Recent advances in wireless sensing technology encourage the further optimization and improvement of the product development and service provision processes. Industrial Wireless Sensor Networks (IWSNs) are an emerging class of WSN that face specific constraints linked to the particularities of the industrial production. In these terms, IWSNs face several challenges such as the reliability and robustness in harsh environments, as well as the ability to properly execute and achieve the goal in parallel with all the other industrial processes. Furthermore, IWSN solutions should be versatile, simple to use and install, long lifetime, and low-cost devices – indeed, the combination of requirements hard to meet.

In this chapter we discuss the applications of WSNs in industrial environments. Based on the specific requirements of the industrial production, the IWSN applications can be classified into three groups (Figure 1.1):

1. **Environmental sensing.** This group generally represents the widest field of WSN application nowadays. IWSN applications for environmental sensing cover the problems of air, water (together with waste water) *pollution*, but also the production material pollution monitoring. Furthermore, in *hazardous* environments, there are numerous needs for fire, flood, or landslide sensing. Finally, the *security* issues arise in markets with competing product and service providers, where IWSNs are used for point of interest, area, and barrier monitoring.

2. **Condition monitoring.** This group generally covers the problems of structure and human condition monitoring. The first and second subgroups provide both the structure health information (the condition of the buildings, constructions, bridges, supply routes, etc.) and the machine condition monitoring including possible factory automation. The third subgroup takes into consideration healthcare applications of IWSN.

3. **Process automation.** The last group of applications provides the users with the information regarding the resources for the production and service provision, including the materials, current stock and supply chain status, as well as the manpower included in the industrial process and building automation. Finally, one of the most important issues from the user perspective is the production performance monitoring, evaluation, and improvement that are achieved through IWSNs.

Table 1.1 provides the list of major industry branches and sectors with the specific group of applications applicable to the particular branch. Here we consider 4 industry sectors: primary (the extraction of resources directly from

Earth, includes farming, mining and logging with no product processing), secondary (primary sector product processing), tertiary (service provision) and quaternary (research of science and technology). Currently, the needs and possibilities of WSNs in all the industry sectors and branches are widespread. The pace of industrial development pushes and encourages the development of IWSNs even more. In this chapter, we analyze and evaluate current applications of IWSNs and discuss possible future trends in this area.

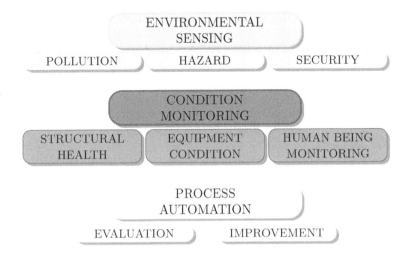

FIGURE 1.1
Taxonomy of industrial wireless sensor network applications.

TABLE 1.1
Main industry branches in 4 sectors.

Sector	Branch	Environmental			Condition		Process	
		Poll.	Haz.	Sec.	Struc.	Equip.	Eval.	Impr.
I	Mining, logging	•	•	•	•	•	•	•
I,II	Power/Energy		•	•	•	•	•	•
I,II	Agriculture	•	•				•	•
II	Chemical/Biotechnology		•	•			•	•
II	Civil engineering		•	•	•	•		
II	Electrical engineering		•		•	•	•	•
II	Mechanical engineering		•		•	•	•	•
II,III	Product processing		•	•		•	•	•
III	Transportation	•	•	•	•		•	
III	Military/Defense	•	•	•	•	•	•	•
III	Healthcare		•			•	•	
III	Communication		•			•	•	
III,IV	Security R&D		•			•		

This chapter is organized as follows. We address some of the biggest challenges in IWSN in Section 1.2. Then, we provide the taxonomy of the IWSN applications for environmental sensing in Section 1.3, condition monitoring in Section 1.4 and process automation in Section 1.5. In Section 1.6 we present the extensive list of IWSN solutions and service providers. Finally, we conclude the chapter by discussing some possible future IWSN trends in Section 1.7.

1.2   Technological Challenges

The deployment and the set up of Wireless Sensor Networks are extremely challenging tasks, which become even more challenging in industrial applications. The environment where IWSNs are deployed in order to monitor environmental or production processes is extremely dynamic, it can depend on the specific product, the phase of life of the product and the kind of service provision considered. In fact, each kind of product or phase of life has different requirements and imposes on the monitoring system different constraints. In this section we will try to give a general description that can be useful for IWSNs' designers.

One of the challenges to face is the impact of the **propagation environment**. When the IWSN is deployed inside a factory to assess the production process quality, the designer has to deal with the interference and the radio environment produced by the production machines. In this case, the IWSN has to be deployed and calibrated not only to guarantee the correct assessment of the production process, but above all not to interfere with the production process. The same logic holds for IWSNs used to monitor electricity, water, and gas consumption. Often nodes of the IWSN are immersed into the goods inside containers or any transportation means. For this kind of environment, the radio characteristics of the goods and the container have to be carefully investigated in order to determine the most efficient and effective way to make nodes communicate in spite of probable signal degradations.

In general, radio waves will not follow the same behavior according to the environments in which sensors are deployed. If the network is deployed outdoors, the radio propagation can be assimilated to a free-range perturbation, with an almost omnidirectional propagation. However, it will be impacted by the weather, more or less depending on the frequency used. For instance, frequencies around $2.4GHz$ may be stopped by a thick fog. When wireless sensor networks are deployed indoors, the data propagation is far from being omnidirectional. In this environment it is even harder to make the classical assumptions on the shape and extension of the communication range for sensors. In fact, sensors that are placed within the communication range might be invisible, whereas sensors that should be considered out of range are actually in the neighbors' set. This can be explained by the fact that waves can bounce on walls, machinery, etc. This amplifies the signal in some locations and cancels it in other locations. Indeed, metallic equipment may extend the propagation area for the radio signal or may prevent the signal to reach close areas beyond the equipment. Similarly, presence of metal and liquid greatly impacts the propagation. These challenges, strictly dependent from physical factors, are not easy to handle. The environment in which sensors will be deployed needs to be studied in order to determine the optimal locations for sensors.

Operation lifetime, as a result of the power management policy, is one of the key issues in all the WSN applications, including the IWSN. Several IWSN applications, especially in the field of environmental monitoring, require the autonomous power supply from alternative power sources, such as wind or solar power. Although it is possible to obtain a constant power supply in some industrial environments, sensors tend to be battery powered in order to keep the monitoring non-intrusive. However, in most cases, batteries are not expected to be reloaded or changed. Thus, energy should be preserved. There exist many ways to do so, both in software and hardware. From the hardware point of view, it is important to carefully choose the components. These latter should be low energy consuming while providing the needed capacity. In some particular applications, energy harvesting modules can be envisioned, like solar cells or kinematic sensors, etc., but their usage is still marginal.

From the software point of view, energy should be preserved by controlling the number of messages to be sent and the transmitting power. Indeed, radio activities, such as sending and receiving data are the activities that consume more energy in WSN compared with processing and sensing activities. Thus, it is important to monitor carefully the amount of data to send and the frequency at which it is sent, i.e., the number and size of messages, while preserving the quality of service expected by the application. Similarly, the further the messages are sent the more the energy needed and the more the interferences generated. Thus, it is important to monitor the transmission range based on the target to reach.

Another challenge in this field of application is the **heterogeneity** that the IWSN must deal with. In fact, heterogeneity is present in at least two different facets in industrial sensing applications: heterogeneity of *data collected* and heterogeneity of *objective network* to integrate with. The heterogeneity of data collected comes from the need of creating the dataset used to assess the quality of the production process/service provision by including a range spectrum of parameters. In turn, this requires to enable proper data fusion/data aggregation schemes for the acquisition/transmission of data, and powerful techniques of data analysis for the reception side. The heterogeneity of objective network is focused especially on logistics applications, where the IWSN used to monitor the transportation of some goods must be able to integrate with the IWSN deployed in both the production site and the delivery site, in order to acquire and exchange relevant information about the transported goods.

The third main technological challenge is that IWSNs for process and service monitoring have to **operate autonomously**. The operation of the IWSN should not represent an additional burden for the human involved in the sensor network operation, instead the IWSN should be able to autonomously configure and deploy when the deployment site is inaccessible for humans. In this context, **maintainability** is one of the important design challenges in the IWSN as well. The deployed IWSN and its components should be easily repairable or replaceable by maintenance personnel in the case of failures.

The requirements of the industrial process impose the need for reliable and secure IWSN. **Reliability** in this context refers to the monitoring system that has to provide accurate and real-time information regarding the monitored process even in harsh industrial environments under extreme vibration, noise, humidity, or temperature conditions. The information gathered during the monitoring process is vital for proper system operations, given that even the smallest errors can produce fatal consequences for both the system and the product. Therefore, system **security** concludes our list of main challenges in the IWSN design.

1.3   Environmental Sensing Applications

1.3.1   Concept and Objectives

Environmental sensing has been the basic WSN application, since WSN appeared in industrial processes. Nowadays, environmental sensing is widely spread in almost every field of industry. The common element that combines all the application scenarios together is the need for real-time information about the industrial environment, whether it is the production material, ambient or the process itself. Three different paradigms can be distinguished in this context: pollution, hazard, and security monitoring, that will be discussed further in this section.

General objective in the environmental monitoring is an efficient information gathering, used both for prevention (real-time or postponed) and analysis. The migration from the wired sensor networks to their wireless counterpart brings numerous advantages by facilitating the deployment and information gathering process. However, the compromise is always present - WSNs still have to cope with the problems of erroneous communication, robustness, lifetime, and cost constraints.

In this section we analyze some existing research works in the field of environmental sensing. Such environmental monitoring applications are the first ones that have been developed for wireless sensor networks. In fact, due to their small size, sensors can be easily and quickly deployed over large scales at low cost. Their wireless features that make them independent from any costly and fixed infrastructure also contribute to their success. Environment sensing is a very broad area that comes from monitoring for disaster prevention, like volcano monitoring, to healing operations when sensors indicates a critical area.

1.3.2 Existing Solutions

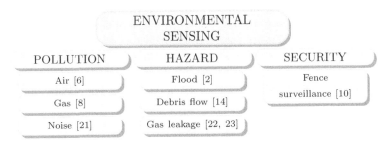

FIGURE 1.2
Environmental sensing applications.

1.3.2.1 Pollution

The importance of the pollution detection both in materials for the production as well as in the production ambients makes the pollution monitoring one of the most widely spread field of application in environmental sensing. Constant advance in the sensing technology supports this statement as well. In [8], authors propose a low-cost, fully automated, end-to-end in-sewer gas monitoring system based on floating-drifting embedded sensor platform. This system provides the user with accurate information regarding the gas readings and its localization. Data on air should be always available to citizens who want to know the level of pollution within their cities.

Some techniques to allow typical web users to access high-resolution pollution data gathered from a large number of vehicle-mounted mobile sensing devices coupled with highly-accurate static sensor data in an easy-to-use, intuitive interface are presented in [6].

Another source of pollution is the noise present in urban areas. Authors of [21] present a prototype of a platform for collection and logging of the outdoor noise pollution measurements. These measurements can be used for the analysis of pollution effect on manpower productivity and social behavior. There also exist many other kinds of applications for detecting and/or preventing pollution like also air or water pollution continuously or after an event like for instance a volcano eruption which may spread in the atmosphere some toxic gas.

1.3.2.2 Hazard

Industrial facilities are often localized in environments that are riskier than residential areas, especially in the case of oil, gas, and coal mining industries or agricultural companies. Therefore, proper early warnings or predictive disaster detection might be a valuable asset, resource, and life saver. In [2], authors implement the river flood detection. A sensor network is used for flood prediction based on the previous measurements. By using a small number of sensor nodes with self-monitoring for failure capabilities, they cover and secure large geographical area under the threat of disaster. In [14], authors propose the debris flow monitoring system that allows in-situ real-time debris flow tracking. A number of robust sensors with self-localization capabilities are thrown into the flow, thus providing the real time flow direction and volume information used for early warning issuing.

Autonomous wireless sensor systems can be used for early fire and gas leak detection [22]. In this work, a gas sensor module detects pyrolysis product in order to individuate a fire before inflammation, and a generic energy scavenging module, able to handle both alternating current and direct current based ambient energy sources, provides the power supply for the gas sensor module.

In [23], authors propose a differential gas measurement approach along with specific heating pulses for the sensor to secure substantial energy saving in hazardous gas leakage scenarios.

1.3.2.3   Security

The concept of security in industrial environments is indeed widely spread over almost all the industry branches in all sectors. The security itself can refer to the security of the information and the security of the people, products and equipment. In this chapter, we focus on the latter. In this context, the applications for security monitoring usually focus on the area, barrier, and point of interest monitoring. In [10], authors present a fence surveillance system that comprises the robot and camera sensor network and two types of nodes, ground, and fence nodes. The network reports the acquired data to the base-station with issues commands to mobile robots that extend the communication distance of the system.

1.4   Condition Monitoring Applications

1.4.1   Concept and Objectives

Why is there a need for WSN applications in condition monitoring? Every industrial system faces the problem of equipment amortization that introduces the maintenance cost into the equation. Moreover, there is a need for structural and equipment monitoring techniques that could provide a global picture on subject condition and accurately predict equipment failures and therefore improve component and equipment reliability and performance. Furthermore, structural monitoring system detects system damages before possible failures and minimizes the time that production line spends out of service, and thus increases the profit. Without automated monitoring system, it is necessary to schedule regular system checks and preventively replace production equipment, which can also include the risk of sending maintenance workers into hazardous environments.

WSN intended for condition monitoring avoids all the aforementioned problems and suits the application requirements in comparison with wired sensing systems, since:

- it is easily deployable and reconfigurable even in an inaccessible areas such as moving (rotating) machine parts,

- it is easily reconfigurable,

- it reduces the system installation and condition monitoring cost in general.

In this context, we distinguish two classes of condition monitoring applications: **structural health** monitoring (public, private and transportation infrastructure) and **equipment condition** monitoring (mechanical and manufacturing equipment). The third class that we introduce in this context is related to the condition monitoring of human beings. In this case WSN can be deployed directly on the body and be integrated with infrastructured networks.

1.4.2   Existing Solutions

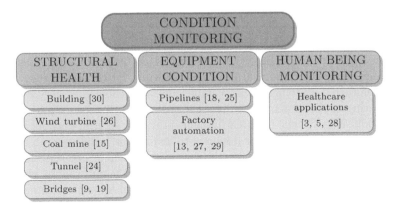

FIGURE 1.3
Condition monitoring applications.

1.4.2.1   Structural Health Monitoring

Wireless sensor networks are well suited for the structural health monitoring since they are easily deployable and configurable for this purpose. Here we cite some of the recent works that tackle the problem of structural monitoring (notably the vibration monitoring) on buildings [30], wind turbines [26], coal mines [15], tunnels [24], and bridges [9, 19].

In [30], authors deployed a wireless data acquisition system that is used for damage detection on the building. Their proposed system continuously collects structural response data from a multi-hop network of sensor nodes, displays and stores the data in the base station. They propose the system design and evaluate their approach by practical WSN implementation.

Prior to the monitoring system implementation, it is necessary to characterize the structure under consideration. In [26], authors deploy wireless sensor network on two wind turbines in order to gather the vibrational output data and to provide better models of wind turbine dynamic behavior and response to loading. Authors made the acquired data available for future structure health monitoring designs.

Authors in [15] discuss the design of a structure-aware self-adaptive approach that rapidly detects structure variations in coal mines caused by underground collapses. The goal of their system design is to detect the collapsed area and report it to the sink node, to maintain the system integrity when the structure is altered and to provide a robust mechanism for query handling over the network under unstable circumstances. They conducted field experiments and deployed a prototype system that proves the approach feasibility.

One of the hot topics in structural monitoring research are bridges, as a part of the public infrastructure. Numerous research papers are written on this topic and there are numerous attempts to propose solutions and practical implementations for bridge health monitoring. We present only most recent works in this field. In [9], authors describe requirement, challenges, their design, and implementation of WSN for health monitoring of Golden Gate Bridge in San Francisco Bay, USA. Authors identify requirements needed to obtain data of sufficient quality to have a real scientific value to civil engineering researchers for structural health monitoring. Furthermore, they propose the monitoring system that is scalable and applicable to different kinds of monitoring applications and address the typical problems encountered during the monitoring system practical implementations. In [19], authors present a low-cost wireless sensing unit designed to form dense wireless mesh networks. This device is developed for earthquake early warning projects, but it can be efficiently used as a tool for wireless structural monitoring as well. In order to determine the suitability of proposed system for structural monitoring applications, they conduct experiments on Fatih Sultan Mehmet Bridge in Istanbul, Turkey. Finally, in [24], authors tackle some problems and solutions for structural health monitoring (especially for tunnels and bridges), but also discuss the current state and limitations of the sensing and wireless technology that is currently used. Their work presents advice for future experimental practice and lessons learned from three deployment sites in United Kingdom: Ferriby Road Bridge, Humber Bridge, and London Underground tunnel.

1.4.2.2 Equipment Condition Monitoring

The production process speed and quality depends on the equipment condition and accuracy. In this section we present some of the recent works in the field of industrial equipment condition monitoring for pipelines [18, 25] and factory automation [13, 27, 29].

In [18], authors develop the wireless network system for a team of underwater collaborative autonomous agents that are capable of locating and repairing scale formations in tanks and pipes within inaccessible environments. Authors describe in detail ad-hoc network hardware used in their deployments, that comprises the pH, proximity, pressure sensors, and repair actuator. Furthermore, they describe the communication protocol and sensor/actuator feedback loop algorithms implemented on the nodes. Another work that focuses on WSN for pipeline monitoring is [25], where authors describe the WSN whose aim is to detect, localize, and quantify bursts, leaks, and other anomalies (blockages or malfunctioning control valves) in water transmission pipelines. Authors report the results and experiences from real deployment and provide algorithms for detecting and localizing the exact position of leaks that is tested in laboratory conditions. The system presented in this work is also used for monitoring water quality in transmission and distribution water systems and water level in sewer collectors. In this context, the work can be classified as the process evaluation group of works (Section 1.5.2.1) as well.

In [27], authors develop a WSN for factory automation condition-based maintenance and they present design requirements, limitations, and guidelines for this type of WSN applications. Furthermore, they implement their condition monitoring system in Heating & Air Conditioning Plant in Automation and Robotics Research Institute in University of Texas.

Authors of [29] propose the use of accelerometer based monitoring of machine vibrations and tackle the problem of predictive maintenance and condition-based monitoring of factory automation process in general. They demonstrate a linear relationship between surface finish, tool wear, and machine vibrations thus proving the usability of proposed system in equipment monitoring.

In [13], authors focus on preventive equipment maintenance in which vibration signatures are gathered to predict equipment failure. They analyze the application of vibration analysis for equipment health monitoring in a central utility support building at a semiconductor fabrication plant that houses machinery to produce pure water, handle gases, and process waste water for fabrication lines. Furthermore, authors deploy the same sensor network on an oil tanker in order to monitor the onboard machinery. In the end, they discuss design guidelines for an ideal platform and industrial applications, a study of the impact of the platform on the architecture, the comparison of two aforementioned deployments, and a demonstration of application return on investment.

1.4.2.3   Human Being Monitoring

WSN and wireless technology are a practical and convenient way to assess medical procedures and cares to patients and people in need. The key factors of innovative solutions that mix Wireless Body Area Networks (WBAN) and infrastructured networks are self-care, self-management, and cost effectiveness.

The objective of [5] is to provide a general architectural solution to provide wireless technology and self-learning solutions in planning and developing a wireless hospital or distributed hospital of which cognitive radios, smart components, and WBAN are the key components.

In [28], authors describe a policy-based architecture that utilizes wireless sensor, advanced network topologies, and software agents to enable remote monitoring of patients and elderly people; through the aforementioned technologies they achieve continuous monitoring of a patients condition and they can proceed when necessary with proper actions.

Body Sensor Networks (BSN), can remotely collect data and upload vital statistics to servers over the Internet [3]. In this work, authors propose to use BSNs to efficiently monitor and record data while minimizing the energy expenditure of nodes in the BSN. The collected data is transferred by connecting a BSN to a social network in order to create the unique ability to share health related data with other users through social interaction. The final objective of this work is to integrate social networks and BSNs to establish a community promoting well being and great social awareness.

1.5   Process and Service Monitoring Applications

1.5.1   Concept and Objectives

The last group of applications that we have considered are those concerned with monitoring processes and services. In this sense, the former is industry-oriented and it consists of all the activities performed by human workers or machines to produce goods; whereas the latter is user-oriented and it consists of the operations needed to provide end-users with specific services, such as electricity, water provision, or more integrated approaches involving building automation.

Process monitoring involves many industrial fields, since it focuses on tracking the quality of the entire life of a product, step by step from the materials provision used for its production till its disposal. Therefore, all the IWSN applications dealing with manpower and production materials tracking as well as logistics and transportation systems fall in this category. The role of IWSNs in these applications is driven by the need for evaluating and improving each and every step of goods production-distribution-consumption process as well as the cycle as a whole.

Service monitoring is mainly related to evaluating the quality of the provision of a specific service to end-users. The quality assessment can be performed in terms either of the efficiency of the provision line or the effectiveness of the service provision at the end-user's location. In the first case, service monitoring overlaps with the public/private infrastructure monitoring already investigated in Section 1.4, whereas the second is object of study in what follows. In service monitoring applications the IWSNs have the important role of offering both the provider and the consumer valuable information about the provision. From providers' perspective, remote monitoring/metering of electricity power, heat, water, or gas is a simple and effective solution, in fact it has gained great momentum in the last years. Several works integrate the evaluation of the quality detected by the WSNs with the capability to modify the environment by proper actuators. This case is referred as building automation. From consumer's perspective, IWSNs' applications represent a remarkable step in the direction of the green building deployment because they allow end-users to constantly estimate their energy expenditure as well as the quality of the environment where they live.

The applications for both the fields studied in this section would highly benefit from the integration of the actuator system in order to automatically improve the process/service provision and achieve the desired results. For this reason, we split the existing solutions subsection in two parts: the first describes research works aimed at evaluating and monitoring processes and services provision, whereas the second includes also the actuation, which allow the designer to close the feedback control loop and move towards completely automated systems.

1.5.2 Existing Solutions

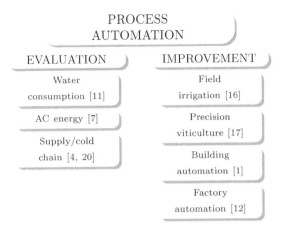

FIGURE 1.4
Process automation and service provision monitoring.

1.5.2.1   Process and Service Provision Evaluation

In this subsection we present some recent research works on developing IWSN applications aimed at gathering all the needed information from the product/service in order to assess the quality of the production/provision process. This kind of application is very useful from both the private citizen's or company's perspective and the public system's perspective. For example, in the United States, the Environmental Protection Agency claims that more than $18 billion a year could be saved by reducing the water consumption by 30% and that the American public water supply and treatment facilities consume about 56 billion kWh per year, which is enough electricity to power over 5 million homes [32].

In [11], authors present an easy-to-install self-calibrating system that provides users with information on when, where, and how much water they are using. The approach is non-intrusive, cost-effective, and easy to deploy. It is based on wireless vibration sensors attached to pipes, which are able to measure the water flow passing through the pipe and estimate consumption with a mean absolute error of 7%.

The United States Department of Energy in [31] estimated that 10% of the total energy used by commercial sector is wasted by parasitic energy use in commercial building HVAC systems. Green building deployment is the best response to this energy waste, but it will take some years before local and international communities and politics will accept and implement a different way to design and build public and private constructions. In the meantime, HVAC monitoring system are very useful to evaluate waste and consumption. For example, in [7], authors present a system for AC energy monitoring in large and diverse building environments. Their system provides real, reactive, and apparent power measurements and comprises the metering control interface, IP compatible network structure and software that provides various power-centric applications.

As we said, the monitoring activities can involve also the life-cycle of a product, from its production to its disposal. Specifically, we have selected two research works that fall in the subcategory of resource tracking and logistics, in order to show how IWSNs, because of their intrinsic capability to deal with dynamic process, are well suited to be used in highly constrained and resource demanding processes. In [4], authors present a way to apply the dynamic WSN in temperature controlled supply chain (cold chain) for fruit and pharmaceutical product storage and transport to avoid degradation and spoilage. The approach supports real-time monitoring and remote maintenance via wired and mobile wireless network access. In [20], author presents the architecture and implementation of a self-configuring WSN used in a cold chain management tool that contributes to quality improvement and waste reduction.

1.5.2.2 Process and Service Provision Improvement

WSN dedicated for industrial process improvement represent maybe the most important part of all the industrially applicable WSNs since they usually comprise the actuator components as well. In this way, these IWSNs could be observed as Automated Industrial Wireless Sensor and Actuator Systems, used for complete process automation that are, therefore, widespread and present in all the branches of industry. Backed up with the development of the sensing and communication component technology they represent the future of the Intelligent Process Automation Systems used in industry. A great number of works focuses on the agriculture production process amelioration. In [16], authors describe the design, development, and deployment of a WSN that improves the water efficiency in the field irrigation located in dry regions. In this way, with the use of temperature and humidity information it is possible to implement the automated control system that consume irrigation water in efficient manner.

Authors of [17] present the architecture, hardware, and the software of the platform used for precision viticulture. A major feature of this platform is its power-management subsystem, able to recharge batteries with energy harvested from the surrounding environment. It allows the system to sustain operation as a general-purpose wireless acquisition device for remote sensing in large coverage areas. The platform is currently being used as a simple and compact yet powerful building block for generic remote sensing applications, with characteristics that are well suited to precision viticulture.

The two mentioned works consider the automation of agriculture processes, the next will highlight the usage of IWSNs in building automation and the last in factory automation. In [1], authors tackle the problem of energy consumption control by presenting a novel control architecture that uses occupancy sensing to guide the operation of a building heating ventilation and air-conditioning systems. By interacting between sensing and actuating part of the HVAC control, they achieve significant results in energy saving.

Another example of WSN used for factory automation is given in [12]. In this case the emphasis is on highly dynamic processes in modern factory facilities. This work presents a conceptual study of a wireless real-time system dedicated for remote sensor/actuator control in production automation. It discusses the timing behavior and power consumptions and reviews the system design aspects such as network topology, multiple access schemes, and radio technologies suitable for these constrained application environments.

1.6   Commercial Solutions for IWSN

Commercial solutions for infrastructural and equipment monitoring include complete wireless system solution, comprising wireless data acquisition, vibration sensors, signal conditioning for vibration sensors, and signal processing software for equipment failure prediction and diagnostic. The sensors and network hardware used in these solutions are capable of coping with most of the aforementioned technological challenges. Table 1.2 provides a list of services and solution providers to which an interested reader should refer.

1.7   Conclusions

Following the above description of the available wireless sensor network solutions for industrial applications, we conclude this chapter by proposing a discussion about open issues and future trends for different types of industrial applications in environmental sensing, condition monitoring, and process automation.

Environmental sensing. Depending on the environment in which sensors will be deployed, the *hardware* should be carefully studied. Indeed, environment monitoring refers to outdoor applications. Thus, sensors should be water-proof, prone to shocks, etc. This has a cost which represents a real challenge since such applications require the deployment of a great number of sensors, which cannot be done if each single sensor is expensive. An important feature of wireless sensor networks deployed for environment monitoring is that the *environment is not controlled*, not controllable, and possibly not easily accessible, i.e., if a sensor is moved (by animals or wind for instance), the network must continue to work properly. Network basic mechanisms such as neighbor discovery or routing must adapt this mobility. This remains a great challenge.

Condition monitoring. Recent technological advances will permit the development of intelligent sensors and therefore the development of low-cost *intelligent monitoring systems* capable of reliable equipment failure prediction. Furthermore, technological advances allow the development of structures, production lines, and individual equipment components with built-in sensors, emergence of *self-monitoring equipment* that would behave as an ad-hoc network and that would easily communicate with other parts of condition monitoring system. In the context of automation and maintainability, the integration of complex software solutions onto sensing network and sensor firmware that would introduce the concept of *failure prediction* and real-time problem *diagnostics*. In the end, all the acquired condition monitoring information provide important guidelines and could serve the purpose of "to

TABLE 1.2
Commercial solution and service providers for IWSN

No.	Name	Environmental sensing			Condition monitoring		Process automation		website
		Pollution	Hazard	Security	Structural	Equipment	Evaluation	Improvement	
1	ABB						●	●	abb.com
2	Accutech Instruments						●	●	accutechinstruments.com
3	AES								aesolutions.com.au
4	Aginova					●	●		aginova.com
5	Apprion	●	●		●	●	●		apprion.com
6	Augusta	●	●	●		●	●	●	augusta-ag.de
7	AVIDwireless	●	●		●	●	●		avidwireless.com
8	Azbil				●	●	●		azbil.com
9	Banner		●			●	●		bannerengineering.com
10	Bridge Diagnostics Inc.				●	●			bridgetest.com
11	Coalesenses	●				●	●		coalesenses.com
12	Dust					●	●		dustnetworks.com
13	Electrochem	●	●		●	●	●		electrochemsolutions.com
14	Elpro	●	●			●	●	●	elpro.com
15	Emerson		●			●	●	●	emerson.com
16	Endress+Hauser	●		●		●	●	●	endress.com
17	Ferguson Beauregard	●	●			●	●	●	fergusonbeauregard.com
18	FreeWave			●		●	●		freewave.com
19	GridSense	●				●	●		gridsense.com
20	GST					●	●		globalsensortech.com
21	Hitachi	●	●		●	●	●	●	hitachi.com
22	Invensys				●	●	●	●	invensys.com
23	Invisible Systems					●	●		invisible-systems.com
24	Meastim				●	●	●		meastim.com
25	Memsic					●	●		memsic.com
26	National Instruments	●	●	●	●	●	●	●	ni.com
27	Nivis	●	●	●		●	●	●	nivis.com
28	On-Ramp Wireless			●		●	●		onrampwireless.com
29	Panasonic					●	●		panasonic.com
30	Pepperl+Fuchs	●	●			●	●	●	pepperl-fuchs.co.uk
31	RF Code	●				●	●	●	rfcode.com
32	Rockwell	●	●		●	●	●	●	rockwellautomation.com
33	Scanimetrics	●	●		●	●	●		scanimetrics.com
34	Sensormetrix	●	●			●	●		sensormetrix.co.uk
35	Siemens	●	●	●	●	●	●	●	automation.siemens.com
36	SKF				●	●	●	●	skf.com
37	Synapsense	●			●	●	●		synapsense.com
38	Timken				●	●	●		timken.com
39	Wessex Power	●	●			●	●	●	wessexpower.co.uk

do" list for future civil infrastructure and production equipment design and construction.

Process automation. Designers and developers of Wireless Sensor Networks have lost their initial view of WSN as a single system composed of simple and homogeneous devices. From the referenced works, we have seen that, especially in industrial applications of WSNs for process and service monitoring, the trend is to deploy pervasive systems of *heterogeneous capabilities*. These systems look at the nodes capabilities as the constituent blocks for designing new applications and, in turn, this calls for a standard communication paradigm. The process monitoring also offers the possibility to pave the way for the concept of *"plug-in network"*, which is a network of heterogeneous devices able to become part of a larger network, when some spatial proximity condition is satisfied. Therefore, we can imagine networks of devices deployed in a container to monitor the transportation conditions of some goods, which are able not only to communicate immediately with the "driver" if a problem occurs, but also to participate in successive phases of the consumption phase once they arrive on the delivery site. Specifically for service provision monitoring, as we have already mentioned the IWSN deployed to measure consumption and quality of the service provided represents an important step towards the *design of green building* where the consumption of energy is limited by new construction techniques, but IWSN can ensure a continuous and constant monitoring and fixing of consumptions, failures, wastage, etc. IWSN applications for process and service provision have gained a lot of market in the last years, because of the *complete control* they offer in the whole life-cycle of a product, it is very likely that the growth of this technology in the years to come will be even stronger, if designers and developers are able to create highly-specialized products for specific needs as well as cheap and customizable platforms for householders and small companies.

References

[1] Y. Agarwal, B. Balaji, S. Dutta, R.K. Gupta, and T. Weng. Duty-cycling buildings aggressively: The next frontier in hvac control. In *ACM/IEEE International Conference on Information Processing in Sensor Networks*, IPSN '11, pages 246–257, Chicago, IL, USA, 2011. ACM/IEEE.

[2] E.A. Basha, S. Ravela, and D. Rus. Model-based monitoring for early warning flood detection. In *Proceedings of the 6th ACM conference on Embedded network sensor systems*, SenSys '08, pages 295–308, New York, NY, USA, 2008. ACM.

[3] D. Bauschlicher, S. Bauschlicher, and H. ElAarag. Framework for the integration of body sensor networks and social networks to improve health

awareness. In *Proceedings of the 14th Communications and Networking Symposium*, CNS '11, pages 19–26, San Diego, CA, USA, 2011. Society for Computer Simulation International.

[4] D. Bijwaard, W. Kleunen, P. Havinga, L. Kleiboer, and M. Bijl. Industry: using dynamic WSNs in smart logistics for fruits and pharmacy. In *Proceedings of the 9th ACM conference on Embedded network sensor systems*, SenSys '11, pages 218–231, New York, NY, USA, 2011. ACM.

[5] M. Hämäläinen, J. Matti, and R. Kohno. Wireless communications in healthcare recent and future topics. In *Proceedings of the 4th International Symposium on Applied Sciences in Biomedical and Communication Technologies*, ISABEL '11, 3 pages, New York, NY, USA, 2011. ACM.

[6] W. Hedgecock, P. Volgyesi, A. Ledeczi, and X. Koutsoukos. Dissemination and presentation of high resolution air pollution data from mobile sensor nodes. In *Proceedings of the 48th Annual Southeast Regional Conference*, ACM SE '10, 6 pages, New York, NY, USA. ACM.

[7] X. Jiang, S. Dawson-Haggerty, P. Dutta, and D. Culler. Design and implementation of a high-fidelity ac metering network. In *Proceedings of the 2009 International Conference on Information Processing in Sensor Networks*, IPSN '09, pages 253–264, Washington, DC, USA, 2009. IEEE Computer Society.

[8] J. Kim, J.S. Lim, J. Friedman, U. Lee, L. Vieira, D. Rosso, M. Gerla, and M.B. Srivastava. Sewersnort: A drifting sensor for in-situ sewer gas monitoring. In *6th Annual IEEE Communications Society Conference on Sensor, Mesh and Ad Hoc Communications and Networks*, SECON '09, pages 1–9, June 2009. IEEE.

[9] S. Kim, S. Pakzad, D. E. Culler, J. Demmel, G. Fenves, S. Glaser, and M. Turon. Health monitoring of civil infrastructures using wireless sensor networks. In *ACM/IEEE International Conference on Information Processing in Sensor Networks*, IPSN '07, pages 254–263, 2007.

[10] Y. Kim, J. Kang, D. Kim, E. Kim, P. K. Chong, and S. Seo. Design of a fence surveillance system based on wireless sensor networks. In *Proceedings of the 2nd International Conference on Autonomic Computing and Communication Systems*, Autonomics '08, pages 4:1–4:7, ICST, Brussels, Belgium, 2008. ICST (Institute for Computer Sciences, Social-Informatics and Telecommunications Engineering).

[11] Y. Kim, T. Schmid, Z. M. Charbiwala, J. Friedman, and M. B. Srivastava. Nawms: nonintrusive autonomous water monitoring system. In *Proceedings of the 6th ACM conference on Embedded network sensor systems*, SenSys '08, pages 309–322, New York, NY, USA, 2008. ACM.

[12] H.-J. Korber, H. Wattar, and G. Scholl. Modular wireless real-time sensor/actuator network for factory automation applications. *Industrial Informatics, IEEE Transactions on*, 3(2):111 –119, May 2007.

[13] L. Krishnamurthy, R. Adler, P. Buonadonna, J. Chhabra, M. Flanigan, N. Kushalnagar, L. Nachman, and M. Yarvis. Design and deployment of industrial sensor networks: Experiences from a semiconductor plant and the north sea. In *SenSys '05: 3rd International Conference on Embedded Networked Sensor Systems*, pages 64–75. ACM, November 2005.

[14] H.-C. Lee, C.-J. Liu, J. Yang, J.-T. Huang, Y.-M. Fang, B.-J. Lee, and C.-T. King. Using mobile wireless sensors for in-situ tracking of debris flows. In *Proceedings of the 6th ACM conference on Embedded network sensor systems*, SenSys '08, pages 407–408, New York, NY, USA, 2008. ACM.

[15] M. Li and Y. Liu. Underground coal mine monitoring with wireless sensor networks. *ACM Trans. Sen. Netw.*, 5(2):10:1–10:29, April 2009.

[16] J. McCulloch, P. McCarthy, S. M. Guru, W. Peng, D. Hugo, and A. Terhorst. Wireless sensor network deployment for water use efficiency in irrigation. In *Proceedings of the workshop on Real-world wireless sensor networks*, REALWSN '08, pages 46–50, New York, NY, USA, 2008. ACM.

[17] R. Morais, M. A. Fernandes, S. G. Matos, C. Serdio, P.J.S.G. Ferreira, and M.J.C.S. Reis. A zigbee multi-powered wireless acquisition device for remote sensing applications in precision viticulture. *Computers and Electronics in Agriculture*, 62(2):94 – 106, 2008.

[18] F. Murphy, D. Laffey, B. O'Flynn, J. Buckley, and J. Barton. Development of a wireless sensor network for collaborative agents to treat scale formation in oil pipes. In *EWSN*, pages 179–194, 2007.

[19] M. Picozzi, C. Milkereit, C. Zulfikar, K. Fleming, R. Ditommaso, M. Erdik, J. Zschau, J. Fischer, E. Safak, O. Özel, and N. Apaydin. Wireless technologies for the monitoring of strategic civil infrastructures: an ambient vibration test on the fatih sultan mehmet suspension bridge in Istanbul, Turkey. *Bulletin of Earthquake Engineering*, 8:671–691, 2010. 10.1007/s10518-009-9132-7.

[20] R. Riem-Vis. Cold chain management using an ultra low power wireless sensor network. In *Proceedings of the 2nd international conference on Embedded networked sensor systems*, WAMES '04, 2004.

[21] S. Santini, B. Ostermaier, and A. Vitaletti. First experiences using wireless sensor networks for noise pollution monitoring. In *Proceedings of the workshop on Real-world wireless sensor networks*, REALWSN '08, pages 61–65, New York, NY, USA, 2008. ACM.

[22] A. Somov, D. Spirjakin, M. Ivanov, I. Khromushin, R. Passerone, A. Baranov, and A. Savkin. Combustible gases and early fire detection: an autonomous system for wireless sensor networks. In *Proceedings of the 1st International Conference on Energy-Efficient Computing and Networking*, e-Energy '10, pages 85–93, New York, NY, USA, 2010. ACM.

[23] A. Somov, A. Baranov, A. Savkin, M. Ivanov, L. Calliari, R. Passerone, E. Karpov, and A. Suchkov. Energy-Aware Gas Sensing Using Wireless Sensor Networks. In *9th European Conference on Wireless Sensor Networks*, EWSN '12, Trento, Italy, 2012.

[24] F. Stajano, N. Hoult, I. Wassell, P. Bennett, C. Middleton, and K. Soga. Smart bridges, smart tunnels: Transforming wireless sensor networks from research prototypes into robust engineering infrastructure. *Ad Hoc Networks*, 8(8):872 – 888, 2010. Elsevier.

[25] I. Stoianov, L. Nachman, S. Madden, and T. Tokmouline. Pipeneta wireless sensor network for pipeline monitoring. In *ACM/IEEE International Conference on Information Processing in Sensor Networks*, IPSN '07, pages 264–273, 2007. ACM/IEEE.

[26] R. A. Swartz, J. P. Lynch, B. Sweetman, R. Rolfes, and S. Zerbst. Structural monitoring of wind turbines using wireless sensor networks. *Environment*, 6(3):1–8, 2008.

[27] A. Tiwari, F.L. Lewis, and S.S. Ge. Wireless sensor network for machine condition based maintenance. In *8th Control, Automation, Robotics and Vision Conference*, ICARCV 2004, volume 1, pages 461–467, December 2004.

[28] D. Vassis, P. Belsis, C. Skourlas, and G. Pantziou. Providing advanced remote medical treatment services through pervasive environments. In *Personal Ubiquitous Computing*, volume 14, issue 6, pages 563–573, September 2010.

[29] P. Wright, D. Dornfeld, and N. Ota. Condition monitoring in end-milling using wireless sensor networks. *Transactions of NAMRI/SME*, 36, 2008.

[30] N. Xu, S. Rangwala, K. K. Chintalapudi, D. Ganesan, A. Broad, R. Govindan, and D. Estrin. A wireless sensor network for structural monitoring. In *Proceedings of the 2nd international conference on Embedded networked sensor systems*, SenSys '04, pages 13–24, New York, NY, USA, 2004. ACM.

[31] U.S. Department of Energy. *Energy Consumption Characteristic of Commercial Building HVAC System*. Office of Building Technology State and Community Programs, 2001.

[32] U.S. Environmental Protection Agency. Watershed assessment, tracking and environmental results, http://www.epa.gov/waters/index.html, 2012.

2

Machine Condition Monitoring with Industrial Wireless Sensor Networks

Neil W. Bergmann and Liqun Hou

School of Information Technology and Electrical Engineering, The University of Queensland

CONTENTS

2.1   Introduction

Modern industrial processes are built from a broad range and large number of machines and systems. To maintain the whole industrial process running in a reliable and efficient condition, suitable device monitoring, fault diagnosis, and process control techniques are necessary. These techniques are continually evolving to make best use of the latest technological advances, from the earliest watching, listening, feeling of skilled engineers and manual operating, to on-site analog instruments and analog controllers, and then to digital instruments and digital control, currently to large online wired monitored and controlled systems such as a Distributed Control System (DCS) or Field Bus Control System (FCS). Fault diagnosis techniques also developed from simple threshold to AI techniques, while control algorithms developed from classical PID control to many novel control algorithms.

Although online wired monitored and controlled systems are successfully employed to monitor and control the critical devices and process parameters, there are still many other non-critical devices that are not regularly monitored. Combining these into the wired monitoring system may significantly increase system cost because of the need for additional cabling. Sometimes the installation cost may be even higher than the cost of the sensor itself, in particular for remote monitoring. Adding monitors using a wired system is also inconvenient for doing temporary or specialized tests. Wireless sensor networks (WSNs) provide one potential solution to tackle these challenges and an opportunity for a new routine revolution of industrial processes monitoring and control techniques. Compared with a wired system, WSNs have many inherent advantages, such as lower cost, convenience of installation, and ease of re-location. These features make a low cost condition monitoring or control system for non-critical equipment possible. However common WSN sensor nodes are resource constrained, which is a problem for the high system requirements of industrial process monitoring and control.

2.2   System Requirements of Industrial Wireless Sensor Networks

2.2.1   Industrial Wireless Sensor Networks Application Cases

There have been relatively few reported instances of industrial wireless sensor networks (IWSNs) for industrial processes. Some typical application cases are given in this section.

A wireless sensor network for plastic machinery temperature monitoring is proposed in paper [14]. This application system with 4 temperature nodes at a cycle time of 128 ms has been successfully installed and tested on a plastic injection molding machine. The sampling and payload data transmission rates in this paper are relatively low (8 Hz) because the temperature is a slowly changing parameter.

The industrial monitoring system in [35] has both wired and wireless communication approaches to a supervisory system. A set of experiments were conducted to analyze the system's performance, in terms of packet losses, and the maximum achievable data acquisition and transmission rates considering a 60 Hz signal. The results show that the intelligent sensor modules can analyze up to the 5th harmonic of the fundamental 60 Hz signal.

A single hop wireless sensor network for machinery condition-based maintenance with commercial WSNs products is presented in [40]. This WSN is implemented in a heating and air-conditioning plant.

A scheme of applying WSNs in online and remote energy monitoring and fault diagnostics for industrial motor systems is proposed in [5]. A rotor eccentricity fault is tested to verify the feasibility of the proposed approach. A Fast Fourier Transform is applied to the stator current signals accumulated from multiple records received at the central supervisory station over the WSN.

Paper [25] proposes a ZigBee/IEEE 802.15.4 based wireless sensor network for health monitoring of an induction motor subjected to an imbalance fault. A three-axis accelerometer ADXL330 is employed to measure the motor vibration signature. Rotor imbalance faults with different levels were created on an induction motor to validate the proposed system. However, the sensor nodes only record and transmit the vibration data to the base station, all the signal analysis and processing functions have been implemented by MATLAB software on the base station.

As well as the research systems described above, a small number of commercial wireless condition monitoring systems are available, including WiMon from ABB, and Essential Insight.mesh from GE which are described below.

WiMon [8] is a wireless vibration monitoring system based on WirelessHART developed by ABB. WiMon comprises WiMon 100 sensor units, Gateway, WiMon Data Manger, OPC (Object Linking and Embedding for Process Control) server, and ABB Analyst. The WiMon 100 sensor unit consists of a vibration sensor, a temperature sensor, a long life battery, and a WirelessHART radio. The WirelessHART Gateway is especially developed by Pepperl+Fuchs GmbH to integrate with ABB's wireless systems. So the system supports the IEEE 802.15.4 radio standard and WirelessHART network standard. The battery lifetime is estimated to reach at least 5 years if vibration root mean square (RMS) and temperature values upload interval is once per hour and waveform upload interval is once per day.

Essential Insight.mesh is a wireless solution for condition monitoring from GE [11], which comprises wSIM (wireless mesh network node), wSIM Repeater, Manager Gateway, and Transducers. The system supports IEEE 802.15.4 compliant radio and wireless mesh communications, and the communication interfaces of Gateway is IEEE 802.15.4 (for future support of

ISA100.11a) [12]. One wSIM sensor node supports up to 4 temperature or vibration transducers. However for this product the minimum sample interval is 15 minutes for temperature or static vibration, and 24 hours for dynamic vibration. A battery pack provides up to 3 years of power with standard configurations, which is defined as static data from all 4 channels every 2 hours and dynamic data from each channel once per day [12]. Besides WiMon and Essential Insight.mesh there are a number of other commercial wireless monitoring systems, for example, TCLinK from Microstrain [38], which allows sample rates from 2 Hz to 1 sample every 17 minutes.

2.2.2 System Requirements of IWSNs

From looking at the example systems in previous section, some requirements for IWSN monitoring systems can be determined [17].

Processing heterogeneous sensor signals: An industrial monitoring system typically needs to measure both slowly changing scalar values such as temperature, and fast changing dynamic signals such as vibration. IWSNs need the capability of acquiring and processing heterogeneous sensor signals, and then transmitting the data.

Special sensors for IWSNs: Compared to the conventional wired monitoring systems, IWSNs monitoring systems can impose additional constraints on the sensors, such as small size, low energy consumption, and low cost, which are hard to fulfill with many conventional sensors. For example, a traditional proximity probe for vibration monitoring requires too much power for sustained battery-powered operation. Newly developed Micro-Electro Mechanical Systems (MEMS) sensors can provide an alternative to such expensive and power-hungry conventional sensors [2].

Higher sampling rate: The sampling rate of IWSNs typically has to be much higher than the sampling rate of WSNs for environmental or structural applications because of the need to perform accurate monitoring for high speed dynamic signals such as vibration.

Fast transmission rate: Higher sampling rates lead to a need for faster transmission rates if the sampled data is directly transmitted. Often the limited wireless bandwidth impedes high-speed data collection and transmission.

Energy efficiency: To achieve wireless operation, IWSN nodes are usually powered by batteries, and battery lifetime often determines the lifetime of the whole network. However, higher sampling rates and more transmitted data increase the energy consumption. So energy efficiency needs to be factored into the design of node hardware and communication protocols.

Higher reliability data transmission: Process-critical monitoring applications are more intolerant to loss of data and thus the communications have to be more reliable. However, the higher sampling rates make retransmission, which is often used for enhancing the reliability, more difficult.

Accurate time synchronization: Fault diagnosis and multi-sensor fusion algorithms require the joint study of the signals from several sensors. Time synchronization and clock drift become important for complex machine monitoring.

2.3 Resource Constraint versus Higher System Requirements

2.3.1 Resource Constrained Wireless Sensor Nodes

A wireless sensor node is characterized by its small size, its ability to sense environmental phenomena through a set of transducers and a radio transceiver with autonomous power supply [4]. Wireless sensor platforms have been developed and produced by different manufacturers, such as Crossbow [28] and CSIRO [15]. Several companies also provide some wireless modules including both the microcontroller and the RF components, and the relevant development kits (e.g., JN5148 from Jennic [22] and RCM4510W [30] from Rabbit Semiconductor), which enables users to implement WSNs with minimum time to market and relatively low development cost. There are also some wireless sensor nodes developed by individual researchers themselves using a commercial available microcontroller, RF transceiver, and other electronic components, for example the node presented in paper [44]. Table 2.1 compares some existing sensor node and module architectures.

Currently, commercial WSNs nodes and modules typically employ a RISC (Reduced Instruction Set Computer) microcontroller. Among the products in Table 2.1, four of them are 8-bit, while the JN5148 from Jennic is a 32-bit microcontroller with significantly higher processing capability. The program and data memory sizes of the sensor nodes or modules are also small and some products require additional external memory in addition to the memory on microcomputer - for example the prototype presented in paper [44] employs 128 KB external SRAM besides the 4 KB RAM on the microcomputer. Some types support the IEEE 802.15.4 standard operating at 2.4 GHz and having 250 Kbps data rate, while others operate with a data rate at 40 or 50 Kbps. The current draw on active status is from 8 mA to 150 mA, while the sleep state consumes current on the order of μA. The transmission range varies between about 90 and 4000 meters. Some manufacturers provide a broad variety of sensor boards to combine with the wireless modules. This approach has good expandability for measuring different phenomena. WSNs with transducers built directly on the sensor node are another choice which trades off expandability for robustness and compactness.

In summary, the low-end microcomputer, limited memory size and data rate, and limited battery energy constitute the resource constrained characteristic of common WSNs nodes.

TABLE 2.1
Comparison for various sensor node and module architectures.

	MICAz [28]	FLECK3B [15]	JN5148 [22]	RCM4510W [30]	PAPER [44]
Manufacturer	Crossbow	CSIRO	Jennic	Rabbit Semi-conductor	Stanford Univ Univ of Michigan
Microcontroller	ATmega128L	ATmega128L	JN5148	Rabbit 4000	ATmega128
CPU Clock	8 MHz	8 MHz	4 to 32 MHz	29.49 MHz	8 MHz
CPU bit	8	8	32	8	8
RAM (KB)	4	8	128	512	132
ROM (KB)	128	128	128	512	128
Radio	Chipcon CC2420 2.4 GHz 250Kbps	Nordic RF 905 433/915 MHz 50Kbps	Custom RF board 2.4 GHz 250Kbps	MaxStream XBee 2 2.4 GHz 250Kbps	MaxStream 9XCite 902-928 MHz 40Kbps
Wireless Protocol	IEEE 802.15.4	GFSK	IEEE 802.15.4	IEEE 802.15.4	FHSS
Max range	75-100 m	> 1000m	Up to 1000 or up to 4000 m	Up to 90 m	Up to 90 m
Current draw	8 mA active, <15 µA sleep	33 µA standby	17.5mA RX, 15mA TX, 1.25 µA sleep	150 mATX/RX, <20 µA sleep	77mA active, 100 µA standby
OS	TinyOS	TinyOS/FOS	N/A	N/A	N/A
Transducers	On sensor board	Temp sensor is on board, others on sensor board	N/A	N/A	On sensor board

2.3.2 Resource Constraints versus Higher System Requirements

This section investigates the tension between the higher system requirements of industrial monitoring and the resource constrained characteristics of IWSN nodes using the case of induction motor condition monitoring. These same principles can be more broadly applied to many other industrial machines.

In general, a motor fault leads to corresponding signal changes such as bearing vibration, stator current, and bearing and stator winding temperature. Supposing one stator current signal and four vibration signatures, viz. vibrations in x and y direction on both ends of the motor, are monitored. 3.1 kHz sampling rate and 12 bit ADC are used in our system to measure the parameters and obtain sufficient harmonic information for further fault analysis. The raw data rate for continuous monitoring is then:

$$5 \times 3100 \times 12 = 186 \; kbps$$

Although the data rate of a typical WSN radio channel, such as IEEE 802.15.4, can reach 250 kbps, transmission overheads reduce the actual achieved payload data rate to well below 250 kbps. In experiments reported in [37], the data rates observed by a sensor node with a streaming application using IEEE 802.15.4 non-beacon-enabled mode is about 42 kbps, and is about 4 kbps using beacon-enabled mode (Beacon Order: 8, Super frame Order: 5). So, even modest IWSN monitoring systems using raw data transmission can saturate the available payload bandwidth.

Large amounts of transmitted data also result in the wireless radio transceiver being in its active mode for a significant time. Because the radio transmitter is often the highest power component on an IWSN node, large transmission volumes limit node lifetime. If additional signals or higher resolution A/D converters are considered, the situation will become even worse.

So there is a tension between the higher system requirement of IWSNs such as higher sampling rate and faster data transmission rate, and the constrained resources of IWSNs nodes, such as the limited radio bandwidth and limited battery energy. The following sections will discuss several possible solutions for this tension, e.g., IWSNs protocol and standards, on-sensor data processing, and energy harvesting for sensor nodes.

2.4 Standards and Protocols of Industrial Wireless Sensor Networks

To apply WSNs successfully, it is necessary to select appropriate wireless protocols for various layers of the network protocol stack, in particular the physical (PHY), medium access control (MAC), and network (NWK) layers. For

PHY and MAC layers, the popular and commercially available wireless protocols for WSN include IEEE 802.11, IEEE 802.15.1, and IEEE 802.15.4 [23]. All of the three standards have been designed for use in different scenarios, thus each of them offers various advantages and disadvantages depending on their use [41].

IEEE 802.11 is the networking technology used for WiFi wireless local area networking. The design of IEEE 802.11 protocols, in general, is targeted towards high bitrates and the available transceivers require an order of magnitude more energy than is acceptable in low bitrate sensor network applications [23]. Another drawback of IEEE 802.11 is the relatively high cost, especially for simple WSNs.

The Bluetooth system (IEEE 802.15.1) is designed as a Wireless Personal Area Network (WPAN) with one major application, the connection of devices to a personal computer or mobile phone [23]. Two major drawbacks of Bluetooth are the need to constantly have a master node, which spends much energy on polling its slaves, and the rather limited number, up to seven, of active "slave" units (a new version of Bluetooth remedies this limitation) [23]. The data rate and the power consumption of Bluetooth are between that of IEEE 802.15.4 and IEEE 802.11. Bluetooth low energy (BLE) is a feature of Bluetooth 4.0 wireless radio technology for low-power and low-latency applications in areas such as healthcare, security, and home entertainment. The bit rate of BLE is less than 200 kbit/s. BLE was accepted as a part of Bluetooth 4.0 in December 2009 [6].

IEEE 802.15.4 is designed for WSNs requiring low cost, low power, low to medium data rate (up to some few hundred kbps), and scalability. Because of the low power consumption, longer battery life and cheaper cost of IEEE 802.15.4 compared to IEEE 802.11 and IEEE 802.15.1, it is more suitable for the industrial device monitoring system requiring long lifetime nodes powered by battery.

Recently, significant standardization efforts related to IWSNs have been conducted. Three industrial wireless network standards, ZigBee [46], WirelessHART [42], and ISA100.11a [18], have been ratified. They are all based on the IEEE 802.15.4 physical layer. There are also several non-standard protocols for industrial wireless sensor networks built on IEEE 802.15.4.

2.4.1   ZigBee

ZigBee is a specification for Low-Rate Wireless Personal Area Networks (LR-WPANs) [27]. ZigBee is widely used in many wireless monitoring and control application domains due to its low-cost, low-power, and implementation simplicity. ZigBee can work in a Non Beacon Enabled mode using Unslotted CSMA/CA (Carrier Sense Multiple Access/Contention Avoidance) or Beacon Enabled mode using Slotted CSMA/CA with or without Guaranteed Time Slots (GTSs). GTSs can be allocated by the network coordinator to devices which require specific bandwidth reservation [27]. Currently, there are many chip vendors, including Atmel [3] and Jennic [20], producing commercially available products adopting the ZigBee specification.

2.4.2 WirelessHART

WirelessHART is a wireless mesh network communication method designed to meet the needs for process automation applications [42]. The timeslots of WirelessHART are fixed length, 10 ms each, and organized by the superframe. Frequency hopping on a packet by packet basis and channel blacklisting techniques are used in WirelessHART to enhance the system robustness. Currently, WirelessHART is supported by many instrumentation suppliers, such as ABB [8] and Pepperl+Fuchs [32]. As an interoperable standard, WirelessHART provides an easy way to setup, operate, and maintain a wireless sensor network [39]. In addition, it is compliant with the existing HART devices and systems.

2.4.3 ISA100.11a

ISA100.11a is a wireless system standard for industrial automation, which is intended to provide reliable and secure wireless operation for non-critical monitoring or control applications in industrial applications [18]. ISA100.11a adopts a hybrid MAC layer, which combines TDMA (Time division multiple access) and CSMA/CA [33]. The timeslots of ISA100.11a are flexible with configurable length, which provides more flexibility for different system requirements. Frequency hopping and channel blacklisting techniques are also supported in ISA100.11a to ensure the robustness of wireless communications. ISA100.11a can be mixed with existing wired networks, including Modbus, PROFIBUS, Foundation fieldbus, and HART using tunnelling techniques [16]. Currently, there are some commercial ISA100.11a productions from different suppliers, for example, Yokogawa [45], and GE [12]. However, there are technical challenges to implement the full ISA100.11a stack architecture on low cost hardware [1].

2.4.4 Other Protocols for IWSNs

Besides the above-mentioned commercial industrial wireless sensor network standards, there are also some wireless protocols developed by individual researchers for industrial application.

By improving the ZigBee standard, paper [1] presents a complementary industrial specification called OCARI (Optimization of Communication for Ad hoc Reliable Industrial networks). This protocol was specially designed for industrial wireless application by supporting deterministic MAC layer, optimized energy-consumption routing, and HART application layer.

On a COTS hardware module (XBee module from Maxstream) a proprietary protocol stack has been developed in paper [14], in which a MAC layer with a hybrid medium access strategy employing TDMA and CSMA/CA based on an IEEE 802.15.4 compliant physical layer and star architecture are proposed. The beacon interval (Tcycle) is comprised of two parts, Join Period and Real Time Period. The former is used for network constitution and other acyclic transactions by utilizing CSMA/CA, while the latter is devoted to guarantee the cycle time deadline of sensory data transmission using a TDMA allocation scheme.

A hybrid MAC protocol named user-configured (UC) TDMA is developed in paper [40]. Scheduling characteristics of TDMA are used in this protocol to provide deterministic time slots for sensor nodes, while different length time slots can be employed for various signals. In one frame, the nodes may access the channel more than once. The TDMA slot assignment table is calculated and maintained by the base station to reduce the resource requirement such as the memory of the sensor node. A contention scheme liking the RTS (request to send) and CTS (clear to send) mechanism used in IEEE 802.11 is developed.

2.5 On-sensor Data Processing for IWSNs

This section introduces a novel IWSN for industrial device condition monitoring, in which on-sensor feature extraction and fault diagnosis are explored to address the tension between higher system requirements of IWSNs and resource constrained characteristic of sensor node. The prototype system is illustrated by an IWSN for induction motor condition monitoring and fault diagnosis. Experiment results show that on-sensor data processing is an effective approach to reduce data transmission, reduce energy consumption, and prolong node lifetime.

2.5.1 Experimental System Architecture

The architecture of the experimental induction motor monitoring system using IWSNs is illustrated in Figure 2.1. A star topology consisting of one coordinator and several end nodes is used to monitor motor parameters such as stator current and vibration. Many industrial processes will have more than one device which needs to be monitored. For monitoring multiple machines, a star network can be built with one coordinator for each machine, with all the coordinators constituting a mesh or tree network. Such a system will be a two-tier architecture; however, this section focuses on a star network for one monitored device. IEEE 802.15.4 and ZigBee protocols are used for the radio.

Signal conditioning module conditions the output of different sensors to the input voltage range requirement of the A/D converter. A low-pass filter is employed in this module for anti-aliasing and noise reduction. After being processed by the signal conditioning circuit, the sensor outputs are input to an

A/D converter integrated in the microcontroller of the IWSN node to achieve signal acquisition function and get a stream of digital samples.

Fault feature extraction module extracts the relevant fault features from the raw digital data stream for further fault diagnosis. Many fault feature extraction methods have been investigated by researchers in different domains. This system uses the frequency spectrum calculated by the Fast Fourier Transform (FFT) as the relevant feature. The FFT algorithm is suitable for implementation on the resource constrained sensor nodes. After feature extraction by FFT, Principal Component Analysis (PCA) can be employed to reduce the dimensionality of the fault features. The reduction of the dimensionality of a data set by PCA is achieved by transforming to a new set of variables, the principal components (PCs), which are uncorrelated, and which are ordered so that the first few retain most of the variation present in all of the original variables [19].

The purpose of fault diagnosis module is to give an estimation of the motor condition by analysis of the extracted fault features. This is basically a signal classification problem - to separate signals representing correct operation from those representing faulty operation. Many such analysis techniques exist, such as neural networks, fuzzy logic, and expert systems. For our specific application, training data belonging to different classes (healthy, loose foot fault, and mass imbalance fault) are well separated into clusters using simple clustering analysis. Prototype or template vectors are firstly chosen for each class, which lie at the center of each cluster of vectors. When the fault feature of new data is extracted, the distances between the new data point and each prototype vector will be calculated and compared. The new data sample is classified by its closest prototype vector. This method is quick and simple to implement. Other machine learning techniques, such as neural networks, can also be used, at the cost of slightly increased computational cost.

In many current condition monitoring and fault diagnosis systems, fault feature extraction and fault diagnosis algorithms are run on the centralized computer. However, considering the powerful calculation capability of the sensor node and the desirability of node energy saving, an alternative approach is to conduct fault diagnosis on the sensor node or coordinator, and only transmit diagnosis results to the centralized computer through the coordinator.

2.5.2 Experimental Validation

To verify the feasibility of the proposed IWSN machine condition monitoring and fault diagnosis system, a series of laboratory experiments were performed. In this section, the experimental setup is first introduced. The machine is run in three conditions - normal working condition and two typical motor faults, imbalance and loose feet. These conditions are then examined to investigate the monitoring and the fault diagnosis capability of the proposed system.

A. Experimental Setup

The experimental setup used in this section is shown as Figure 2.2(a). A

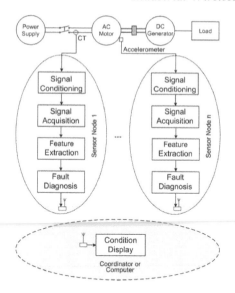

FIGURE 2.1
Schematic diagram of the proposed IWSN with on-sensor data processing.

single phase induction motor and a DC generator are installed on the same bench. A slide resistor is included as the variable load of the DC generator. A variable transformer provides the adjustable power for the motor. The key parameters of the AC motors are 0.56 kW, 110 V, and 1420 rpm. The parameters of DC generator are 0.56 kW, 2450 rpm maximum.

The Jennic JN5139 sensor board and controller board were selected as the hardware platform for the sensor nodes and the coordinator node, respectively. The JN5139 micro-controller supports IEEE 802.15.4 and ZigBee protocols, and it integrates a 32-bit RISC processor with a fully compliant 2.4 GHz IEEE 802.15.4 transceiver, 192 kB of ROM, and 96 kB of RAM [21]. The JN5139 represents a state-of-the-art small IWSN node suitable for on-sensor data processing.

A current transformer, DIGITECH OM-1565, is used to sense the stator current with a conversion ratio of 10 mV/A. The vibration signature is measured by ADXL335, a MEMS accelerometer, with a minimum full-scale range of 3 g and 300 mV/g typical sensitivity. After signal conditioning, the signatures are sampled by a 12-bit A/D converter, integrated in JN5139 with a sampling rate of 3.1 kHz.

To test the motor subjected to mass imbalance, a flywheel structure is mounted on the motor shaft. Mass blocks with different weights can be added on the flywheel through the drilled hole to simulate the various severity of fault. The details of the MEMS accelerometer, flywheel, and mass imbalance block are shown in Figure 2.2(b).

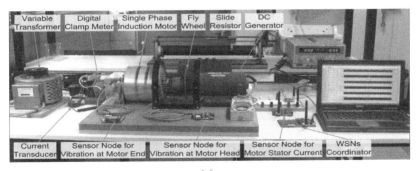

(a)

(b)

FIGURE 2.2
View of experimental setup. (a) Overall motor monitoring system using IWSN.
(b) Accelerometer, flywheel, and mass imbalance block.

B. Induction Motor Condition Monitoring

In this section three motor operating conditions, healthy motor, motor with imbalance, and loose feet, were created and monitored to evaluate the feasibility of the proposed IWSN motor monitoring and fault diagnosis system. In this experiment, raw data sampled by the sensor nodes were transmitted to the coordinator directly and then sent to laptop through a USB port, the signal processing was conducted on the laptop using LabVIEW to calculate the frequency spectrum of the signatures.

Figure 2.3(a) show the vibration signal and spectrum of a healthy induction motor. The main vibration concentrates at 100 Hz, twice the line frequency. The vibration signal and spectrum of an induction motor subjected to 12 g, 24 g, and 36 g mass imbalances are shown by Figure 2.3(b) - Figure 2.3(d). From the vibration waveforms, it can be seen that in the time domain the vibration amplitude increases with the weight of imbalance. In frequency domain, mass imbalance generally shows up as a vibration frequency exactly equal to the rotational speed [43], namely 16 Hz. Considering the 3.1 kHz sampling rate and 512-points FFT employed, the resolution of the frequency

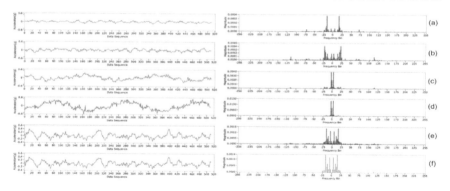

FIGURE 2.3

Vibration signal and spectrum of (a) healthy induction motor, (b) motor with 12 g mass imbalance, (c) motor with 24 g mass imbalance, (d) motor with 36 g imbalance, (e) motor with loose feet, (f) motor with loose feet by on-sensor fault feature extraction.

spectrum is about 6 Hz per bin. When 24 g and 36 g are added on the fly-wheel, Figure 2.3(c) and Figure 2.3(d) show that a single tall peak is present at 16 Hz, (bin 3), which explicitly indicates the existence of an imbalance fault. When the imbalance is decreased to 12 grams, in Figure 2.3(b) the fault feature at 3 bins can be seen, which is a bit larger than the vibration at twice of line frequency (100 Hz, at 16 bins) caused by magnetic force between stator and rotor, there also are vibration components at bin 14, 19, and 22.

The vibration and spectrum signal of an induction motor subjected to loose feet are shown in Figure 2.3(e). It can be seen that besides twice the line frequency, harmonics are detected at 2, 4, 5, 1, and 3 times rotational frequency, in order of decreasing amplitude.

The effectiveness of on-sensor fault feature extraction was also investigated. Figure 2.3(f) shows the fault features extracted by on-sensor data processing. After FFT on the sensor node, the top 10 peaks of frequency components and their location information were selected and transferred to the coordinator and centralized computer. Compared with the vibration spectrum obtained by data processing on the centralized computer in Figure 2.4(e), it can be seen that on-sensor fault feature extraction successfully extract the principal fault characteristics, which will be used for further fault diagnosis.

C. Induction Motor On-sensor Fault Diagnosis

The data sets of healthy motor and motor subjected to loose feet and 24 gram imbalance were used to verify the feasibility of on-sensor induction motor fault diagnosis. A total of 45 training data sets, 15 sets for each condition, were used to calculate principal component coefficients. Then 15 data sets, 5 for each condition, were used for testing and verification.

The motor fault features (10 key frequency components) are extracted,

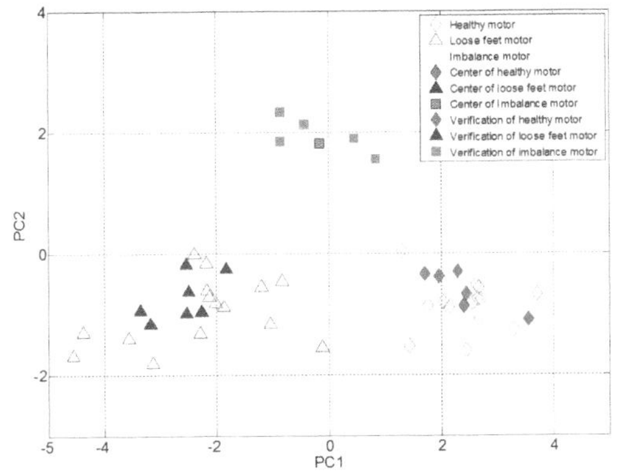

FIGURE 2.4
Two-dimensional plot of the first two principal components.

and then used as the input variables of PCA for reducing the dimensionality of the fault features. Finally, on-sensor fault diagnosis is carried out.

The first two PCs are plotted in Figure 2.4 to illustrate the classification capability by using principal components. It can be seen that the clusters for three different conditions are clearly separated.

In Figure 2.4, the blank diamond, upward-pointing triangle and square represent the training data set of healthy motor, motor with loose feet, and motor with imbalance respectively. The filled markers with black edge are the center of training data sets for different condition. The filled markers without black edge show the results of the test data set, 5 points for each condition. It is clear that the locations of the test data samples are close to the corresponding center of motor condition clusters.

Using the first two PCs, the distances from test data points to the prototypes from training data are calculated and given in Table 2.2. It is clear that the distance to the correct condition prototype is significantly shorter than the distance to the other conditions. By calculating and comparing the distances to the different working condition prototypes the working condition is correctly classified.

The diagnosis margins, defined as the distance to the second nearest center divided by the distance to the correct center, are also presented in Table 2.2. The minimum diagnosis margin is 2.7 for using the first two components, which displays the feasibility of the proposed on-sensor feature extraction and on-sensor fault diagnosis approach.

A set of tests to evaluate, compare and analyze the data transmission rate, energy consumption and node lifetime, for data transmitted after on-sensor data processing and direct raw data transmission, were also carried out. In this experiment 512 points data are used, the detailed results are given in Table 2.3.

1) Payload Transmission Data: For on-sensor feature extraction, after calculating the 512-point FFT, the 10 peaks of the frequency spectrum and corresponding location information are selected and transferred. The transmitted payload decreases from 1024 bytes to 40 bytes, a 96% reduction. For on-sensor fault diagnosis, only the device operating condition is transmitted to the coordinator and centralized computer, so the payload transmission data will decrease to 2 bytes.

2) Energy Utilization and Node Lifetime: Energy utilization was determined by measuring the predictable battery voltage droop over time, with the node waking at regular intervals to either raw data transmitting, or on-sensor data processing and transmitting summary results.

It is evident that node's energy consumption will reduce due to the large payload transmission data decrease. However, the data processing algorithm embedded on the sensor node will consume additional energy. The key question is whether the saving in energy from reduced radio transmission exceeds the increased energy for on-sensor computation. These consumptions are determined by the running time and complexity of the algorithm, and the current consumption for CPU processing and wireless radio transmission.

The typical current consumption for the JN5139 during CPU processing is 7.57mA when a 16 MHz system clock is used. Typical current consumption values for radio transmit and radio receive are 38mA and 37mA, respectively

TABLE 2.2

The distance from verifying sample to the center of training data sets of different working condition.

Test number	Motor operating condition	Distance to the centre of			Diagnosis margin	Diagnosis result
		Healthy motor	Loose feet motor	Imbalance motor		
1		0.10	4.47	3.39	33.9	
2		0.91	3.77	3.03	3.3	
3	Healthy	0.92	5.58	4.44	4.8	Healthy
4		0.64	4.26	3.07	4.8	
5		0.63	4.33	3.08	4.9	
6		4.71	0.73	2.94	4.0	
7	Loose Feet	4.08	0.94	2.55	2.7	Loose Feet
8		4.69	0.30	3.26	10.9	
9		5.05	0.64	3.35	5.2	
10		5.17	0.53	3.94	7.4	
11		2.77	3.79	1.00	2.8	
12		4.25	2.93	0.85	3.5	
13	Imbalance	4.14	3.47	0.44	8.0	Imbalance
14		4.34	3.46	0.65	5.3	
15		3.06	3.73	0.71	4.3	

[9]. The sensor nodes are powered by two general AAA alkaline batteries, the average voltage is about 2.7 V. Assuming T_{trans} is the time for data or results transmission, so the energy consumption for transmission, E_{trans}, can be calculated by:

$$E_{trans} = 2.7\,V \times 38\,mA \times T_{trans}$$

If T_{proc} indicates the running time for on-sensor data processing algorithm, the energy consumption for on-sensor data processing, E_{proc}, can be obtained by:

$$E_{proc} = 2.7\,V \times 7.57\,mA \times T_{proc}$$

The total energy consumption for this on-sensor feature extraction mode by 512 floating FFT is 22.4 mJ, reducing energy by 33.4% compared to 512 points raw data transmission. The total energy consumption for on-sensor fault diagnosis after on-sensor feature extraction by fixed FFT and PCA feature reduction is about 0.8 mJ, reducing by more than 97% energy compared to raw data transmission. The energy saving of on-sensor data processing is evident, particularly for the simplest data processing algorithm.

The node lifetimes under different operating modes were also measured. Figure 2.5 shows the node lifetime curves. When the sleep interval is 3 seconds compared to the 106 hours for raw data transmission, node lifetime reaches 120 hours and about 150 hours for on-sensor data processing by floating FFT and fixed FFT, respectively, extending lifetime by 13% and 43%.

There is a significant energy overhead in simply switching on the radio, even if just a few bytes are transmitted, so significant reductions in data

TABLE 2.3
Comparison of node lifetime and energy utilization of raw data transmission and on-sensor data processing.

Data processing and transmission mode	Reduced payload (Byte)	T_{trans} (ms)	E_{trans} (mJ)	T_{proc} (ms)	E_{proc} (mJ)	Total energy (mJ)	Energy saving (%)	Node lifetime (Hour)
Raw data transmission	1024	328	33.7	0	0	33.7	n/a	106
On-sensor floating FFT	40	6.3	0.6	1065	21.8	22.4	33.4	120
On-sensor fixed FFT	40	6.3	0.6	28.9	0.6	1.2	96.3	152
On-sensor fixed FFT & PCA fault diagnosis	2	0.3	0.032	8	0.764	0.796	97.6	153

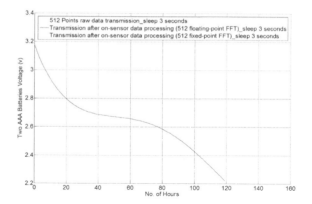

FIGURE 2.5
Node lifetime test on the implemented IWSN using 512 points data.

transmission result in only modest extensions to node lifetime. Another system operating mode is therefore explored, which only transmits the on-sensor fault diagnosis result to the coordinator when a motor fault happens or at a fixed interval, for example every one hour. Otherwise the sensor nodes do not operate their radios. The sensor node lifetime for no data transmission reaches 74 days for a 3-second sleep period. Together with the node lifetime of on-sensor fault diagnosis with results transmission every routine, namely 120 hours, the node lifetime using limited data transmission can be calculated. The results show that the node lifetime reaches 73 days if sensor nodes transmit diagnosis results once per hour, and reaches 70 days for one transmission every 15 minutes. This operating mode using the on-sensor fault diagnosis can significantly prolong the node lifetime, while still continuously monitoring the condition of the equipment.

2.6   Energy Harvesting for Wireless Sensor Nodes

The majority of wireless sensor nodes are powered by battery, because this method allows the sensor node to be compact. It does not need additional components for energy conversion, and it is convenient to install. However, sensor node lifetime will be impacted by the capacity of the battery. Periodic replacement of the battery also increases the cost of system operation.

Besides saving sensor node energy, another possible method to prolong node lifetime is energy harvesting, also known as energy scavenging. Recent review papers about energy harvesting for WSN provide good summaries of the field, such as papers [31], [29], [7]. The most commonly used sources of energy for scavenging include solar, wind or other fluid materials, vibration, and electromagnetic radiation.

For example, paper [26] proposed a wireless actuator node for irrigation systems using solar energy and super-capacitors. The node lifetime reached about 26 hours without solar energy support, if the synchronization period is 4 seconds.

Self-autonomous wireless sensor nodes with wind energy harvesting (WEH) for remote sensing of wildfire spread were proposed in paper [24]. The experimental results show that 7.7 mW energy can be harvested using the proposed WEH system with a wind speed at 3.62 m/s. Another self-powered wireless sensor for air temperature and velocity measurement using energy harvesting from the airflow in the duct is presented in paper [36]. The results indicate that the maximum power harvested reaches 45 mW at 9 m/s airflow and 500 Ω load, and the sensor node installed in the duct can operate without any battery.

Paper [34] presented a 1.9 GHz RF transmit beacon using environmentally harvested vibration and solar energy. The maximum power obtained from vibration (0.23 g at 60 Hz) by the piezoelectric generator is about 180 μW. A resonant packaged piezoelectric power harvester for machinery health monitoring is proposed in paper [10]. The experimental results show that the piezoelectric bimorph can provide 2.8 mW energy when it is strained to the maximum value, 700 $\mu\varepsilon$.

RF energy transmission for a low-power wireless impedance sensor node for structural health monitoring is reported in paper [13], in which 36 element rectennas are used to harvest the emitted energy from a parabolic grid antenna mounted on a car. The results show the voltage of the sensor node can be charged to 3.6 V in 27 s, when 1 W power was transmitted. The transmission range in field test is 1 to 2 meters.

In summary, energy harvesting is usually limited by the availability of the ambient sources. For example, solar and wind power are more suitable for outdoor applications, while energy scavenging based on vibration is suitable for the motioned or vibrated objects. Additional components for energy conversion will increase the system cost and impact the system reliability. In addition, the power obtained by energy harvesting is limited, from hundreds of μW to tens of mW, which is not sufficient for common sensor nodes.

2.7 Conclusions

In this chapter, the industrial wireless sensor networks application cases, including the general application cases and commercial systems, are reviewed. The system requirements for IWSN monitoring systems are then summarized. Several state-of-the-art wireless sensor platforms are introduced and compared. The low-end microcomputer, limited memory size and data rate, and limited battery energy constitute the resource constrained characteristic of common WSNs nodes. The tension between the higher system requirements of industrial monitoring and the resource constrained characteristics of IWSN nodes are demonstrated by the case of induction motor condition monitoring.

Three industrial wireless network standards, ZigBee, WirelessHART, ISA100.11a, and several protocols developed by individual researchers for IWSNs are reviewed. On-sensor data processing is also investigated to address this tension. A novel IWSN with on-sensor data processing for induction motor condition monitoring is presented for verification. Finally, energy harvesting for IWSNs is reviewed as another way for prolonging node lifetime.

References

[1] K. Al Agha, M. H. Bertin, T. Dang, A. Guitton, P. Minet, T. Val, and J. B. Viollet. Which wireless technology for industrial wireless sensor networks? The development of OCARI technology. *IEEE Transactions on Industrial Electronics*, 56:4266–4278, 2009.

[2] A. Albarbar, S. Mekid, A. Starr, and R. Pietruszkiewicz. Suitability of MEMS accelerometers for condition monitoring: An experimental study. *SENSORS*, 8:784–799, 2008.

[3] Atmel Corporation - MCU Wireless. http://www.atmel.com/products/ microcontrollers/wireless/default.aspx. [Accessed Oct 2012].

[4] P. Baronti, P. Pillai, V. W. C. Chook, S. Chessa, A. Gotta, and Y. F. Hu. Wireless sensor networks: A survey on the state of the art and the 802.15.4 and ZigBee standards. *Computer Communications*, 30:1655–1695, 2007.

[5] L. Bin and V. C. Gungor. Online and remote motor energy monitoring and fault diagnostics using wireless sensor networks. *IEEE Transactions on Industrial Electronics*, 56:4651–4659, 2009.

[6] Bluetooth low energy. http://en.wikipedia.org/wiki/Bluetooth_low_energy. [Accessed Sep 2011].

[7] P. C. P. Chao. Energy harvesting electronics for vibratory devices in self-powered sensors. *IEEE Sensors Journal*, 11:3106–3121, 2011.

[8] Condition Monitoring - Wireless vibration monitoring system. http://www05.abb.com/global/scot/scot267.nsf/veritydisplay/ 02cdc6632d9dfc40852578180053d492/$file/wireless%20vibration% 20monitoring%20system.pdf. [Accessed 26 Jul 2011].

[9] Data Sheet: JN5139-001 and JN5139-Z01. http://www.jennic.com/ files/support_files/jn-ds-jn5139-001-1v8.pdf. [Accessed 25 Aug 2010].

[10] Andries J. du Plessis, Marcel J. Huigsloot, and F. D. Discenzo. Resonant packaged piezoelectric power harvester for machinery health monitoring. In *Proc. SPIE 5762, 224 (2005); doi:10.1117/12.606009*, 2005.

[11] Essential Insight.mesh Bently Nevada Wireless Condition Monitoring System for Essential Assets. http://www.gepower.com/ prod_serv/products/oc/en/downloads/gea15019a_wirelesscond_r6.pdf. [Accessed 30 Aug 2010].

[12] Essential Insight.mesh Datasheet. http://www.ge-mcs.com/download/ monitoring/185301c.pdf. [Accessed 26 Jul 2010].

[13] K. M. Farinholt, P. Gyuhae, and C. R. Farrar. RF energy transmission for a low-power wireless impedance sensor node. *Sensors Journal, IEEE*, 9:793–800, 2009.

[14] A. Flammini, D. Marioli, E. Sisinni, and A. Taroni. Design and implementation of a wireless fieldbus for plastic machineries. *IEEE Transactions on Industrial Electronics*, 56:747–755, 2009.

[15] FLECK 3B. http://www.csiro.au/resources/smart-sensor-network-technology.html#8. [Accessed 2009].

[16] H. Hayashi, T. Hasegawa, and K. Demachi. Wireless technology for process automation. In *ICCAS-SICE international joint conference*, 2009.

[17] L. Hou and N. W. Bergmann. System requirements for industrial wireless sensor networks. In *15th IEEE International Conference on Emerging Technologies and Factory Automation (ETFA 2010), Bilbao, Spain*, pages 1–8, 2010.

[18] ISA. *ISA-100.11a-2009 Wireless systems for industrial automation: Process control and related applications*, 2009.

[19] I.T.Jolliffe. *Principal Component Analysis*. New York, U.S.A: Springer-Verlag, 1986.

[20] Jennic ZigBee Pro. http://www.jennic.com/support/zigbee_pro/. [Accessed Sep 2011].

[21] JN5139 Wireless Microcontroller. http://www.jennic.com/products/ wireless_microcontrollers/jn5139. [Accessed 23 Aug 2010].

[22] JN5148 Wireless Microcontroller Modules. http://www.jennic.com/ products/modules/. [Accessed 25 May 2012].

[23] Holger Karl and A. Willig. *Protocols and architectures for wireless sensor networks*. Chichester, England: John Wiley & Sons Ltd, 2007.

[24] T. Yen Kheng and S. K. Panda. Self-autonomous wireless sensor nodes with wind energy harvesting for remote sensing of wind-driven wildfire spread. *Instrumentation and Measurement, IEEE Transactions on,* 60:1367–1377, 2011.

[25] S. Korkua, H. Jain, W. Lee, and C. Kwan. Wireless health monitoring system for vibration detection of induction motors. In *2010 IEEE Industrial and Commercial Power Systems Technical Conference (I&CPS),* 2010.

[26] R. Lajara, J. Alberola, and J. Pelegri-Sebastia. A solar energy powered autonomous wireless actuator node for irrigation systems. *SENSORS,* 11:329–340, 2011.

[27] N. P. Mahalik. *Sensor networks and configuration: fundamentals, standards, platforms, and applications.* New York, U.S.A: Springer, 2007.

[28] MICAz. http://www.xbow.com/products/productdetails.aspx?sid=164. [Accessed 2009].

[29] P. D. Mitcheson, E. M. Yeatman, G. K. Rao, A. S. Holmes, and T. C. Green. Energy harvesting from human and machine motion for wireless electronic devices. *Proceedings of the IEEE,* 96:1457–1486, 2008.

[30] Models RCM4510W. http://www.rabbit.com/products/rcm4500w/. [Accessed 2009].

[31] M. T. Penella and M. Gasulla. A review of commercial energy harvesters for autonomous sensors. In *Instrumentation and Measurement Technology Conference Proceedings. IEEE IMTC,* 2007.

[32] Pepperl+Fuchs' WirelessHART products. http://www.pepperl-fuchs.com/india/hi/classid_2434.htm. [Accessed Sep 2011].

[33] D. Nguyen Quoc, K. Sung-Wook, and K. Dong-Sung. Performance evaluation of priority CSMA-CA mechanism on ISA100.11a wireless network. In *5th International Conference on Computer Sciences and Convergence Information Technology (ICCIT),* 2010.

[34] Shad Roundy, Brian P. Otis, Yuen-Hui Chee, Jan M. Rabaey, and P. Wright. A 1.9GHz RF transmit beacon using environmentally scavenged energy. In *International Symposium on Low Power Electronics and Design,* Seoul, Korea, 2003.

[35] F. Salvadori, M. de Campos, P. S. Sausen, R. F. de Camargo, C. Gehrke, C. Rech, M. A. Spohn, and A. C. Oliveira. Monitoring in industrial systems using wireless sensor network with dynamic power management. *IEEE Transactions on Instrumentation and Measurement,* 58:3104–3111, 2009.

[36] E. Sardini and M. Serpelloni. Self-powered wireless sensor for air temperature and velocity measurements with energy harvesting capability. *Instrumentation and Measurement, IEEE Transactions on*, 60:1838–1844, 2011.

[37] C. Suh, Z. H. Mir, and Y. B. Ko. Design and implementation of enhanced IEEE 802.15.4 for supporting multimedia service in Wireless Sensor Networks. *Computer Networks*, 52:2568–2581, 2008.

[38] TC-Link 6 Channel Wireless Thermocouple Node. http://www.microstrain.com/tc-link.aspx. [Accessed 23 Aug 2010].

[39] Technical White Paper - Wireless Technology WirelessHART. http://files.pepperl-fuchs.com/selector_files/navi/productInfo/doct/tdoct1841a_eng.pdf. [Accessed 24 Jul 2011].

[40] A. Tiwari, P. Ballal, and F. L. Lewis. Energy-efficient wireless sensor network design and implementation for condition-based maintenance. *Acm Transactions on Sensor Networks*, 3:1–23, 2007.

[41] A. Willig, K. Matheus, and A. Wolisz. Wireless technology in industrial networks. *Proceedings of the IEEE*, 93:1130–1151, 2005.

[42] WirelessHART Technical Data Sheet. http://www.hartcomm.org/protocol/training/resources/wiHART_resources/wirelesshart_datasheet.pdf. [Accessed 22 Jul 2011].

[43] V. Wowk. *Machinery vibration : measurement and analysis*. New York; Sydney: McGraw-Hill, 1991.

[44] Y. Wang, and J. P. Lynch, and K. H. Law. Design of a low-power wireless structural monitoring system for collaborative computational algorithms. In *Conference on Health Monitoring and Smart Nondestructive Evaluation of Structural and Biological Systems IV*, pages 106–117, San Diego, CA, 2005.

[45] Yokogawa's Wireless Instruments Based on ISA100.11a. http://www.yokogawa.com/rd/pdf/TR/rd-te-r05302-003.pdf. [Accessed Sep 2011].

[46] ZigBee Specification Overview. http://www.zigbee.org/specifications/zigbee/overview.aspx. [Accessed Sep 2011].

3

Wireless Sensor Networks for Intelligent Transportation Applications : A Survey

Kay-Soon Low and Marc Caesar R. Talampas

School of Electrical and Electronic Engineering,
Nanyang Technological University, Singapore

CONTENTS

3.1 Introduction

Wireless sensor networks are systems that are comprised of wirelessly connected heterogeneous sensor nodes that are spatially distributed across an area of interest. A multitude of different sensors can be used to measure and analyze a system's parameters. The sensor nodes are typically equipped with radio transceivers, microcontrollers, and batteries. They are designed to have small form factor, low in cost, and consume little power to operate for years with a small battery. They are integrated with various sensors and data fusion is usually performed to derive useful information. In recent years, WSN has been used in a wide variety of applications ranging from military, environment monitoring, healthcare system, agriculture, manufacturing, and automotive applications [1, 51].

For telerehabilitation applications, a measurement system based on a wearable wireless sensor network for tracking the human arm motion was developed in [46]. As compared with existing approaches, the developed system is portable and easy to use. It allows the patients to be monitored without restraint, and rehabilitation can be carried out in a home environment instead of a specialized laboratory in the hospital.

For inventory management, WSN has been used by British Petroleum (BP) to remotely monitor the liquid levels of factory tanks [65]. A WSN for cylinder inventory management in the packaged gas industry is presented in [53]. Apart from tracking the locations of gas cylinders, the WSN also uses pressure sensors to detect leaks and accelerometers to sense if the gas cylinders are stored in the proper position. General Motors (GM) also uses WSN for parts and vehicle inventory tracking at its plants and dealerships, respectively. WSN technology is also used by GM to monitor its manufacturing equipment and predict when the equipment might fail or require repair. This approach allows them to perform preemptive maintenance [34]. Predictive maintenance for factory machinery based on WSN is also the subject interest in [67]. In [67], the health information of the machinery is regularly monitored and any digressions from the tolerable behavior during operations are stored in RFID tags. The information can be retrieved by the maintenance personnel through a wireless querying of the tag. Similarly, a self-powered wireless sensor node was developed in [14] for detecting rotor asymmetries in induction motors which are indicative of damage. Intel and BP have also reported successful use of WSN on a crude oil tanker at Scotland to monitor machinery vibration. This study validated that the WSN can function well in a hostile shipboard environment [38].

In [3], a WSN-based system for monitoring the structural health of bridges was implemented using piezoelectric sensors to measure the ambient vibrations of the bridges. By identifying changes in a structure's modal parameters, damage can be detected and localized at an earlier time, and the remaining lifetime of the structure can be estimated.

WSN has also been developed for electric power distribution system [55, 39]. The WSN node can be attached to the conductor to be monitored and used in the Smart Grid. On the demand side of the Smart Grid, Intel Labs Wireless Energy Sensing Technology (WEST) is a device that plugs in to an electrical socket and wirelessly transmits electrical consumption data to a laptop, smartphone, or television. The consumption data can be used by the end user or electrical utilities for demand profiling. Intel Labs Eco-Sense Buildings project also aims to maximize the energy efficiency of buildings through sensor networks that detect ambient conditions such as temperature and room occupancy. This information is then used for intelligent control of a buildings lighting, heating, ventilation, and cooling.

For vehicular applications such as traffic monitoring or intra-vehicle sensors monitoring, the use of sensor networks for real-time data acquisition and control has been around for decades based on wired line solutions [26]. The key drawback to this approach is the high installation and maintenance costs. In wired networks, each sensor node requires a separate shielded-pair wire connection. This approach becomes expensive if the cabling across sensor nodes and the controllers are long. Moreover, configuration management would become difficult. The adoption of WSN in this application can potentially offer lower system and operating costs, improve the product reliability and the ease of upgrading. Furthermore, its self-configuring and self-organizing characteristics make the network more robust for harsh operating environment in the automotive applications.

The development of various transportation technologies has resulted in a more connected world. People are traveling frequently and are increasingly depending on vehicles. For developed cities, this leads to the traffic congestion and parking problems. To overcome these issues, various developments of technologies have emerged. Previously, these systems were connected through wired networks. With the advancement of RF technology, wireless monitoring and management of intelligent transportation applications have become attractive for the industry. Besides traffic and parking management, WSN has also been studied for improving the traffic safety and intra-car sensor network system. The importance of WSN in these areas can be witnessed by the increasing number of research papers in this field. To benefit the WSN research community, this chapter presents a survey paper covering the use of WSN for intelligent transportation systems.

The organization of this chapter is as follows. A survey of the traffic monitoring and control systems that are based on WSN is first presented, followed by a study of WSN applications for car park management systems. The use of WSN for intra-vehicle applications, traffic safety applications, and vehicle sensor networks are then discussed. A discussion of potential issues such as noise and interferences in the outdoor environments that are typical for transportation is then presented. The power consumption, sensor technologies, and comparison of the various techniques used by different papers are also discussed. Finally, we conclude the chapter.

3.2   Traffic Monitoring and Control System

Traffic congestion is a major concern in every big city around the world. The economic, social, and environmental costs of traffic congestion is quite

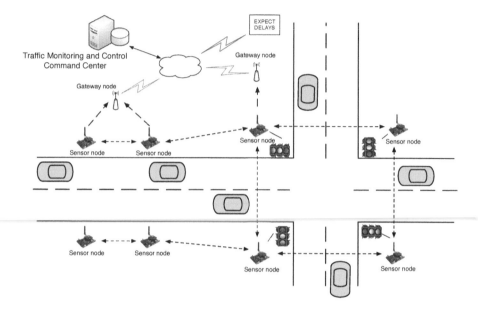

FIGURE 3.1
Diagram of a WSN-based traffic monitoring and control system.

substantial. In recent years, transport authorities have been placing increased emphasis on efficient operation of existing systems rather than increasing road capacity. Though it is essential to acquire accurate and reliable traffic information for driver guidance and traffic performance measurements, the cost of installing and maintaining the traditional sensor system has severely limited its widespread use [23]. Through the use of WSN, this problem could be overcome and traffic information can be collected and analyzed in a cost effective manner.

A typical configuration of a wireless sensor network for traffic monitoring and control is shown in Figure 3.1. Multiple sensors are placed along the highway to detect, count, and measure the velocity of passing vehicles. This information is then sent to a central server through a more powerful gateway node for storage and further analysis. The information is also used by the sensor network to determine lane congestion. If a lane is found to be congested, the traffic light signal timing is adjusted automatically. Road signs can also be updated automatically to inform commuters of traffic congestion. To minimize delays and transmission energy costs, in-network processing of traffic data is recommended.

In [71], a real-time vision system based on a network of autonomous tracking units was implemented for traffic monitoring. Each autonomous tracking unit (ATU) is equipped with standard CCTV cameras that are pre-calibrated and static (fixed field of view). Each ATU is composed of an embedded PC that performs image processing on the frames captured by the CCTV cameras attached to it. The image processing includes background extraction, foreground segmentation, and blob classification, among other things. The output of each ATU is then sent to a Sensor Data Fusion (SDF) server through wired or wireless transmission. For wireless transmissions, either WiFi or WiMax is used. The SDF uses the information from the ATUs to produce estimates of each moving target's position and velocity. The SDF is also responsible for tracking multiple targets. The performance of their system was evaluated in a tunnel and on an aircraft parking area. A major disadvantage of video-based traffic surveillance is the performance degradation experienced when there is poor visibility.

In [7], a system for real-time traffic data collection and traffic jam detection was developed. Unlike [71], an array of passive acoustic transducers is deployed along the roadside to sense the sound generated by passing vehicles. The vehicle angle of arrival relative to the transducer axis is computed by obtaining the time delay between the sound waves detected by two different transducers in the array through cross-correlation. The system was tested for 10 months of continuous operation along a highway, with its results showing good agreement with the results of a nearby magnetic loop detector.

The main objective of real time traffic controllers is optimization of traffic flow. An example of a real time traffic controller is the Best First controller, which prioritizes lanes with the maximum queue length, and thus relieves the most congested lanes. A wireless sensor-based traffic light controller was simulated in [88]. In this study, data from the wireless sensors are used to calculate the number of vehicles waiting at or approaching a traffic light. Simulation results show that the sensor-enabled traffic light controller performs on par with, and in some scenarios better than, a Best First controller. Only two sensors are needed in a single lane, and the controller performs better when the sensors are placed closer to the junction.

In [95], wireless magnetic sensors are used for a vehicle detection system that is capable of providing real time statistics of traffic flow, vehicle speed, and type. The eventual objective is to develop an algorithm that could categorize vehicles according to the Federal Highway Administrations 13 category classification scheme [89]. In [32], a commercially available vehicle detection system based on a wireless magnetic sensor network is described. The VDS240 [73] is envisioned to serve as a general-purpose sensor network for applications that require vehicle detection and counting. These applications include traffic monitoring, traffic signal control, and parking guidance. Estimating the travel time along arterial roads by vehicle identification was studied in [44]. By using the wireless magnetic sensor network described in [32], they perform a match of magnetic signatures at two different locations to estimate the travel time between those locations. The matching is based on a statistical model of magnetic signatures. In [69], a vehicle detection algorithm using information from roadside magnetic sensors was developed. Using both sensor response strength and velocity information, the system was able to distinguish between cars and bikes travelling along the road. However, no work was performed by the system to distinguish between different vehicle classes (i.e., between a car and a truck). Though the magnetic sensor is useful for vehicle detection, it is difficult to use for vehicle classification. To address this issue, a single magnetic sensor is used together with an improved support vector machine classifier in [45]. Their experimental results show a recognition rate of 90%, and that they can classify vehicles into three broad types: heavy tracked vehicles, light tracked vehicles, and light wheeled vehicles.

An alternative approach to using magnetic sensors for vehicle classification is the use of acoustic sensors based on sound spectrum analysis [97] and principal component analysis [98]. To support the research community, a compilation of data set based on real world measurements has been presented in [21] and made available for downloading. Four vehicle classes, namely assault amphibian vehicle, main battle tank, high mobility multipurpose wheeled vehicle, and dragon wagon have been used for the data collection. For the experiment, seventy-five WINS NG 2.0 nodes were used. For classification of urban ground vehicles such as buses, passenger cars, trucks etc., a distributed classification based on wireless audio sensor network has been reported [6]. The acoustic signals are recorded using Panasonic US395 microphones.

A method for vehicle passage and lane detection using Dedicated Short Range Communication (DSRC) transmissions between a vehicle and two anchor nodes is proposed in [2]. The two anchor nodes are placed five meters above the road and periodically broadcast packets containing their position, height, and packet transmission period to oncoming vehicles. By determining the relative acceleration difference between the vehicle and the two anchor nodes, the lane position of the vehicle and its time of passage between the anchor nodes can be derived. This information can then be transmitted to a control center for traffic monitoring purposes. A simulation of the proposed method was conducted and the vehicle detection rate was found to be constant at 95% for vehicle speeds ranging from 10 to 110kph. The lane detection rate is nearly 100% for speeds below 40kph. For speeds greater than 40kph, the lane detection rate starts to decrease to nearly 91% at 110kph. The advantage of this approach is that the lane and time of passage detection algorithms are performed by the vehicle, reducing the processing load on the anchor nodes and control center.

In [87], some of the issues concerning the use of WSN for highway and traffic applications have been reviewed. One of the issues is the optimal placement of sensors for best possible and most accurate measurements. The other issue is the activation policies of the sensor nodes for energy conservation. Four activation policies have been highlighted, namely the naive activation, randomized activation, selective activation based on prediction, and duty cycle activation. From the simulation study, it concludes that the best tracking strategy is the combined strategy based on duty cycle selective activation scheme.

WSN has been researched for traffic monitoring and control in China metropolises [59]. Its main research activity is in the distributed traffic information collection and processing to enable decision making for transportation planning and regulation. As the amount of collected data is huge, the centralized approach for data analysis is infeasible. Instead, distributed cooperative processing for raw local data is considered in their study.

In summary, acoustic, magnetic, and image sensors are the most commonly used devices in traffic monitoring systems. Among these sensors, the magnetic sensor has been used for vehicle detection and speed measurement. Various research has shown that it is reliable and has good accuracy. However, an additional acoustic sensor may be needed if the vehicle type is to be identified. A combination of the two sensors using data fusion algorithms could be researched in the future. Vehicular sensor networks for use in traffic monitoring applications will be discussed in a later section.

3.3 Intelligent Car Park Management System

Almost all modern cities today are facing the congestion problem. Furthermore, people who drive to the city always find it challenging to locate an unoccupied parking space in a large car park or along the street. As a result, the driver may end up spending considerable driving time searching for an unoccupied parking space. This is particularly common if some events are organized in the area. Finding an unoccupied parking space in the maze of a downtown area often works on a trial-and-error basis. The time needed and the driving distance could be significantly reduced if the drivers could be guided to an unoccupied parking space nearby. Besides saving time, it improves the environment by reducing the production of exhaust gas. In [77], a study was conducted at Westwood Village near the University of California, Los Angeles on how long it usually takes for drivers to cruise for a parking spot. It was found that on the average, drivers spend 3.3 minutes looking for a parking space. Given the number of drivers going through Westwood Village, the total amount of time spent looking for parking adds up to 95,000 hours per year. This results in the wastage of around 47,000 gallons of fuel and 730 tons of CO_2 emissions. Various research works have been conducted in recent years to explore new technical solutions that could provide information on individual parking space occupancy. The details are discussed in the following sections.

3.3.1 WSN-Based Car Park Management System

Most existing car parks employ simple vehicle detection and counting system. For such system, sensors are installed at the car park entrance to track the net number of vehicles entering the premises. Subsequently, this information is conveyed to the vehicle drivers via signboard mounted at the entrance or nearby streets. The key weakness to this approach is the lack of information on the actual location of the unoccupied lot. This poses a great challenge to the driver in a big car park. To overcome this limitation, car parks that track the occupancy of individual spaces have started to appear in recent years. For instance, the Siemens SiPark system [17] uses an ultrasound sensor mounted on the ceiling above a parking space for individual car-park lot detection. The sensor devices are connected using a bus-style network for information transmission and power supply.

A typical WSN-based car park management system is shown in Figure 3.2. In this setup, each car park lot is equipped with a wireless sensor node. Some sensors that can be used for parking lot occupancy detection include magnetic sensors, infrared sensors, ultrasonic sensors, and acoustic sensors. The sensor node is powered by a battery and is designed to operate with a life span of 3 to 5 years. Once the sensor node confirms the new status of a parking space, it transmits messages through its RF module. The node can be mounted on the berth surface or curb of the parking lot. The sensor data are transmitted to a cluster coordinator that is responsible for relaying the data to the management and control center via a gateway. Drivers can query the database through their mobile phones to inquire about the status and location of a parking berth. The sensor node also receives commands from the coordinator to carry out procedures such as time synchronization, debugging, working status reporting, and so on. For large car parks, multiple coordinators are needed. The network

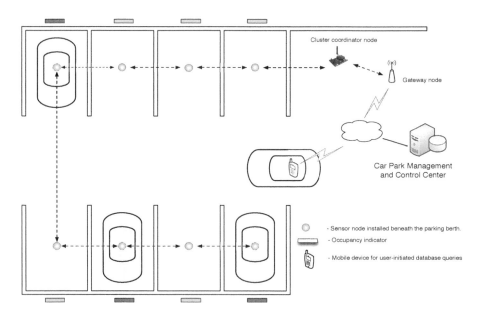

FIGURE 3.2
Diagram of a WSN-based car park management system.

topology varies across the different implementations, ranging from star, mesh, and multi-hop ad hoc networks [74, 11, 18, 100, 64, 82, 85, 48, 8, 13].

A WSN-based parking guidance system that guides a driver to an available parking lot has been described in [74]. The system consists of a vehicle detection sub-system that gathers information on the availability of individual parking lots, and a management sub-system that processes the available information for guiding the driver through a variable messaging system. Experimental trials have been conducted with 7 makes of cars for evaluation. A multi-hop ad-hoc routing algorithm was used for the network. From their measurements and calculations, it is estimated that the system can last for 5 years with traffic of 1 vehicle per minute using a 2400mAh AA sized lithium-ion battery.

In [11], the proposed WSN based car-park system comprises a large number of parking space monitoring nodes, a few parking guiding nodes, a sink node and a management station. Similar to [74], multi-hop routing is used. Experimental measurements have been conducted to study the packet loss rate, the throughput and propagation delay for up to 67 nodes placed in the building office. From the study, the use of link quality information for improvement of the routing protocol to reduce packet loss rate has been recommended.

For commercial developments, several companies offer WSN-based parking management systems. [64] and [82] offer city-wide parking solutions but do not provide much technical detail. Libelium's commercially available Waspmote platform for a smart parking application is discussed in [4]. A magnetic field sensor is used to detect the occupancy of a parking space. The electronics are placed inside a PVC cylinder and buried just under the parking space. Customers have the option of using either a 2.4GHz or 868/900MHz radio for communications.

3.3.2 Sensor Technology

The success of an intelligent parking application in detecting vehicles accurately and reliably is dependent on the types and locations of the sensors. In [100], an anisotropic magneto-resistive (AMR) sensor is used to detect the presence of a vehicle in the parking berth, and the time division multiple access (TDMA) scheme is adopted for the sensor nodes to communicate wirelessly to the gateway.

Owing to their ease of implementation, ultrasonic sensors are used in [16] to detect the occupancy of parking slots. A 2.4GHz radio was used for wireless communications, with the sensor nodes organized in a tree topology. While the ultrasonic sensors were able to reliably detect parking lot occupancy, a limitation on the radio transmission range was encountered. The maximum distance between a sensor node and a router node was found to be 8.04m. This makes the use of multiple router nodes placed at different locations in the parking lot necessary. In [48], a combination of magnetic and ultrasonic sensors is used together with detection algorithms to improve the detection accuracy. Their experiments also show that magnetic and ultrasonic sensors both perform better than infrared and other sensors at detecting vehicles. However, the application is only for counting vehicles entering and leaving the car park, as the sensors are placed only at the entrance and exit points of each floor. In [8], some practical issues in deploying WSNs for car park management systems have been reported. It was found that placement of the magnetic sensors along the East-West orientation yields stable results when detecting cars directly above the sensor. They conclude that a more sophisticated sensor system capable of self-calibration and correct operation regardless of orientation is highly desirable. Zigbee radio connectivity tests also showed that communication is reliable at a range of below 5 meters, unstable between 5 to 10 meters, and nearly impossible at greater than 10 meter range.

Instead of using magnetic, infrared, or ultrasonic sensors, an image-based approach is used in [13] to obtain car park occupancy information. Image processing techniques such as seeding, boundary search, object detection, and edge detection are integrated to form a complete structure of the system. The results show that accuracy can be as high as more than 93% under cloudy or heavy rain conditions. The main advantage of the image-based approach is that one video camera can potentially cover many car-park lots. However, the placement of the video cameras to avoid blockage and vandalism may be challenging. It is also difficult to develop a vision-based system to achieve near 100% accuracy, as its performance is highly dependent on lighting and visibility conditions. It should be noted that the implementation in [13] is based on a wired video camera, as using a video sensor in a WSN is impractical due to the high energy cost of transmitting and processing image data.

A combination of infrared sensors and a physical pressure belt is used to detect vehicles in [28]. This two-pronged approach to detecting a vehicle was used to increase the security of the system by making it harder for roadside attackers to unfairly reserve parking slots by posing as vehicles. In [28], a vehicle that intends to use the parking slot must be equipped with a short-range wireless transceiver and a microcontroller. As soon as the vehicle enters the parking area, it is assigned an available parking slot at the entrance booth. When the vehicle enters a parking space, its front wheels activate the pressure belt, initiating a handshake sequence with the belts wireless transceiver that validates if the vehicle is in its allocated parking space. The infrared sensor double-checks if the object pressing on the pressure belt is indeed a vehicle.

In [17], optical transmitter-receiver pairs are used to detect the presence of a vehicle in a parking space. One disadvantage to this approach is that the optical transmitter-receiver pairs will have to be perfectly aligned for the system to work reliably. Another factor that may affect this system is the buildup of dust or dirt on the optical heads, which may cause false positives to be reported. Finally, the setup described in [17] is prone to tampering, as the optical heads are readily accessible to passersby.

In conclusion, the authors recommend the use of AMR magnetic sensors for vehicle detection. For intelligent parking management systems, several studies have shown that it is more robust as compared to other sensors such as ultrasound, infrared, and imaging systems, especially in the outdoor environment. It is also easier to package and install just under the surface of the parking lot. Moreover, it can operate over a wider temperature range. In a car park, the transmission quality can be seriously hindered by various obstacles such as the parked vehicle, passing cars, pillars, and walls. This reduction in transmission quality may lead to routing problems, resulting in packet loss and disconnected networks. Hence, the network protocol design should use multi-hop ad-hoc routing. If necessary, additional routing nodes can be installed at strategic locations to forward packets and maintain routes.

3.4 Intra-Vehicle Applications

The number of sensors that are installed in a vehicle have increased significantly in recent years due to new applications as well as improved safety features. In the extreme case such as F1 vehicles, the sensors are numbered in the hundreds [93]. Typical measurements include:

- voltages and currents

- engine and gearbox oil temperature

- water levels

- strains on the suspension members

- positions and speeds of the servo valves

- wheel speeds

- engine and gearbox rpm

The addition of sensors and actuators to the modern vehicle results in an increase in the cost, complexity, and weight of the wiring harness. The wiring harness continues to be the most expensive, complex, and bulky electrical component in a vehicle. It can contribute up to 50kg to a vehicles mass, which can adversely affect its acceleration/deceleration and fuel efficiency [49].

Clearly, reducing the amount of wiring in a vehicle will result in wide ranging benefits in terms of cost and ease of system integration. These potential benefits have led to a growing interest over the past few years in replacing sections of automotive wiring with a wireless sensor network. An added benefit to using a wireless approach is that WSNs can provide an open sensor network architecture that is scalable with reprogrammable functionalities.

In [92], a network based on wireless sensing modules has been developed to measure strains, temperatures, and accelerations of a public transport vehicle. The objective is to log the data during its normal operation. This data provides the loading characteristics and is useful for designing the vehicle. The RF system of the sensing module is design to operate at 2.4GHz with a data rate of 1 Mbps. They communicate to the base station using a star network. The data is subsequently transmitted to the remote gateway via GPRS.

3.4.1   Intra-Vehicle Communication Link Quality Studies

Before any attempt to deploy WSNs for intra-vehicle applications can be made, it is necessary to investigate the communication link quality as the vehicle consists of many metallic parts and obstructions that may cause severe multipath fading effects. In [36], the use of ZigBee sensor nodes to perform packet transmission experiments in a car environment has been conducted. Various scenarios such as traveling on the road, waiting at the car-park lot or servicing at the maintenance garage have been studied. The results have demonstrated that the link quality with respect to locations of nodes in the car changes significantly. For instance, the engine noise can change the received sensitivity threshold by 24 dB. Moreover, Bluetooth interference can decrease the goodput performance by 340 percent. A similar result was obtained in [24] wherein the coexistence of Zigbee and Bluetooth networks within the same vehicle was investigated. In [36], detection algorithms were developed to adaptively adjust the transmit power of the sensor nodes depending on the current link quality, thus maintaining network performance while optimizing power consumption. Preliminary results show improved performance when these detection algorithms were used as opposed to keeping the transmit power of the sensor nodes at a fixed level. Although issues such as interference, sensor node placement, and engine noise could have adverse effects on the performance of an intra-vehicle sensor network, these problems do not seem to be unsolvable [36]. Thus, WSN is a plausible and promising technology for intra-vehicle sensor network.

In another study [37], statistical analysis based on the experimentally measured received power from 4 representative in-car wireless channels has been conducted. The four channels are (i) to/from sensor node on the hood (H); (ii) to/from sensor node in the trunk (TR); (iii) to/from sensor node inside the engine compartment (IE); and (iv) to/from sensor node under the engine (UE) compartment. In terms of average fade duration and probability of low received power, the TR and H channels are the best while the UE channel is the worst. Moreover, it was found that the TR and IE channels could support a packet reception rate of at least 98%. All four channels (UE, IE, TR, H) were also found to achieve the required maximum packet delay of less than 500ms. To further improve the performance of an all-wireless in-car network, the use of stronger forward error correction, higher transmit power, and automatic repeat request techniques are recommended.

3.4.2   UWB-Based WSN

From the studies in [36, 37], it is clear that the intra-vehicle environment features short-range and dense multipath. For the automotive industry, the key challenge is to explore ways to provide the same level of reliability, end-to-end latency and data rate as those offered by the wired system. One approach is to explore the use of new wireless technology such as the ultra-wideband (UWB) communication technique. UWB is a communication system that transmits short pulse width signal with a fractional bandwidth greater than 0.2 or occupying more than 500MHz of bandwidth. The FCC has allocated 7.5 GHz of free spectrum from 3.1GHz to 10.6GHz for commercial UWB devices provided they operate at low power with an EIRP less than -41.3 dBm/MHz. Several UWB based WSN systems have been studied [60, 81, 66, 30, 61, 62]. In [60], a communication testbed for UWB technology has been developed and its experimental data has been studied for its efficiency and reliability in transmitting automotive sensor data.

3.4.3   Other Issues

In [22], the IEEE 802.15.4 standard was evaluated with respect to its usability in an intra-vehicle wireless sensor network. Specifically, its performance in terms of meeting the latency requirements of an intra-vehicle application was analyzed. In a typical modern vehicle, the most demanding sensor latency requirement is approximately 1msec with a throughput of 12 kbps for up to 15 sensors. Based on a star network, the bounds on the slot duration of a TDMA scheme as a function of the sensor latency requirements among other parameters can be established. The study shows that the IEEE 802.15.4 could support up to 40 nodes with a guaranteed latency of not more than 100msec. However, latencies smaller than 15.9ms cannot be supported for a star network of any size, as the smallest superframe duration for IEEE 802.15.4 at a 250kbps bit rate is already 15.36ms.

Contention-free or scheduling-based medium access control (MAC) protocols were used in [36, 22, 19] to help ensure that the stringent latency requirements of the intra-vehicle application are met. The work in [19] presents a hybrid TDMA scheme with a flexible channel allocation mechanism that may be suitable for time- and safety-critical automotive applications. In [70], the performance benefits that may be obtained by using a frequency agile MAC for intra-vehicular sensor networks was assessed. Little research has been conducted on the MAC layer for intra-vehicle networks, even though its importance in the realization of intra-vehicle WSNs cannot be understated.

In summary, the development of intra-vehicle WSN is a very promising but challenging application. Due to the dense multipath environment and demanding latency requirement, traditional wireless communication approaches are inadequate for this application. UWB based WSN has many desirable characteristics such as robustness against multipath fading and high data rate capability [31]. Thus, it is one potential candidate technique for use in intra-vehicle communication applications and is a research area for further investigation.

3.5   Road Safety

Wireless sensor networks for road safety applications have the potential to prevent both human and economic losses arising from accidents on the road. Based on data released by the U.S. National Transportation Safety Board, there were 33,833 highway fatalities in 2009. Of these fatalities, 2175 were intersection-related and 4109 were pedestrians. The amount of non-fatal intersection crashes recorded for 2009 was 1,110,900 [91, 90]. Most accidents usually occur at intersections or wherever there is merging traffic.

Wireless sensor networks for road safety applications can be broadly classified into two types: road sensor networks and vehicular sensor networks (VSNs). In road sensor networks, the sensors are used to detect passing vehicles speed and direction. This information is then used to control roadside signals at intersections or traffic merges, warning pedestrians or other vehicles about approaching traffic. The main challenges for road sensor networks are accurate vehicle detection and timely transmission, reception, and processing of data. In vehicular sensor networks for road safety, each vehicle transmits information such as its speed, position, and heading to other neighboring vehicles or to a central server. This information is then processed to provide collision warnings and guidance to drivers.

3.5.1   Road Sensor Networks

A WSN based driving guidance system that provides speed, weather information, and road condition (e.g., icy, wet, etc.) has been described in [75]. The speed measurement subsystem measures the vehicle speed using two wireless magnetic sensor nodes. The accuracy is 99% at a speed of 80 kph as compared to a speed radar gun. For this study, three vehicle types namely a passenger car, Jeep, and SUV have been used. This speed information can be used to trigger a camera or a message signboard when the vehicle speed exceeds the speed limit. Drivers can be alerted to the presence of speeding vehicles on the road and can react appropriately.

The issue of providing early warning to the driver for two-way single carriageways based on WSN with magnetic sensors has been studied in [42]. The focus on such road is due to the fact that 80% of all road accidents in Ireland occur on this type of road. As the responsiveness and accuracy of the sensor node are highly important in such application, minimizing the duty cycle and communication of the sensor node for energy conservation could compromise the performance. Thus, the use of energy harvesting technique to supplement the battery may be desirable. For accurate speed estimations, the nodes should be localized accurately using GPS during the deployment.

In [43], a Kalman prediction algorithm that receives updated position and velocity estimates from deployed magnetometer sensors is reported. The system presents an early warning to the driver to mitigate collisions in highway intersections with non-stopping traffic. Particular attention was given to the latency of the system to validate that the data will be received and warnings will be sent in a timely manner. The maximum latency between initial vehicle detection to processing of warning was estimated to be 158ms. This is acceptable, as the vehicle would only move 4.39m at a speed of 100km/h. In [40], a prototype of safety control services using wireless magnetic sensor networks at a non-signalized intersection has been developed. To enhance energy efficiency, an event driven mechanism was adopted. The magnetic sensor nodes are normally in sleep mode until an external event (such as variations in the sensed magnetic fields) occurs, in which the sensor node will wake up and perform its operations. Moreover, a hybrid network topology using star and mesh was used to achieve reliable and efficient sensor data transmission. The magnetic sensor nodes are organized in a star topology with a by-pass node, while the by-pass nodes communicate with each other and the base station in a mesh topology. To help achieve the latency requirements of the application, a TDMA-based MAC was used.

Similar to [40], the work in [9] also implements a wireless magnetic sensor network for collision avoidance at non-signalized intersections. In this work, four sensor nodes spaced five meters apart are placed on each of the four roadsides leading to the intersection. A timestamp is generated by each sensor node as a vehicle passes by. These timestamps are then sent to the base station, which uses the timestamps to estimate each vehicle's entrance and exit times at the intersection. If the estimated entrance time of one vehicle is less than the estimated exit time of another vehicle, a collision is highly probable and a warning is given out. The latency of the collision detection algorithm was measured to have a maximum value of 100ms. However, the accuracy of the collision detection algorithm and its immunity to false detections was validated only with simulations.

3.5.2   Vehicular Sensor Networks (VSNs) for Road Safety

A popular approach to VSNs for road safety involves the collaboration of multiple vehicular sensor nodes to forecast and prevent collisions. Such VSNs are referred to by some as Cooperative Collision Warning (CCW) systems [83, 99, 72].

The accurate measurement of a vehicles speed, position, and heading is necessary for Cooperative Collision Warning systems. In [76], the measurement accuracy requirements for CCW systems were derived by using simple kinematic models. From their analysis, the lateral position error must be less than 50cm, speed errors must be less than 2m/s, and heading errors must be less than $5°$.

In [83], the feasibility of a future-trajectory-prediction-based cooperative collision warning system using a simple DGPS unit and basic motion sensors was explored. An Extended Kalman Filter was used to fuse the sensor readings from the DGPS unit, the speed sensor, and a MEMS yaw rate gyro. Vehicles were outfitted with sensors and data was gathered for various maneuvers and DGPS coverage conditions. From the gathered data, simulations were performed for 1720 different collision scenarios. The simulation results show that their future-trajectory prediction was effective in detecting all possible collisions at least 1.4s in advance.

A similar approach is presented in [68] for a real time position estimation scheme using an extended Kalman filter and integration of DGPS with vehicle speed sensors, steering angle encoder, and an optical yaw rate gyro. The objective is to provide potential collision warnings based on motions of neighboring vehicles. The work in [76] was used to set the requirements for the design of the position estimator. The filter performance was evaluated in 60km of driving experiments with both good and bad GPS coverage. The experiment results show that the requirements for CCW can be achieved by their system in most but not all scenarios.

In [63], increases in accuracy and reliability over GPS-based approaches are claimed with the use of radio-based ranging techniques for measuring inter-vehicle distances. A Kalman filter fuses information from RSSI-ranging and velocity readings to generate position estimates. Since vehicles are constrained to move along the confines of a road, road map data is used to constrain the estimates to road boundaries. For the purpose of simulation, the paper assumes that Gaussian random variable with a standard deviation of 6m describes the noise characteristics of the inter-vehicle distance measurements.

A cooperative driving system was proposed in [54], in which vehicles automatically detect road hazards, such as low road friction or reduced visibility, and relay this information wirelessly to other vehicles, particularly oncoming traffic. As an added feature, each vehicle verifies the positional relevance of the reported hazards and displays the information to the driver only when the hazard is in the driver's path.

The MAC and routing layer implementation requirements for a vehicular sensor network-based cooperative collision avoidance (CCA) system was presented in [12]. Their analysis shows that for vehicular safety applications, the existing MAC and routing protocols for traditional mobile ad-hoc networks are inapplicable. For such applications, the routing protocols should be broadcast oriented and use geographical or directional packet forwarding. Simulated intra-platoon vehicle crash experiments were performed to investigate the performance of a CCA system using different networking protocols. The sensitivity of the CCAs performance with respect to noisy channel conditions was also evaluated.

In a highway scenario where a string of vehicles are traveling at relatively similar speeds and inter-vehicle spacing, sudden braking of a vehicle may cause rear-end collisions especially in poor visibility conditions. In [56], emergency braking notifications through inter-vehicle WSN is proposed, taking into account the vehicles' speed, inter-vehicle spacing, driver reaction times, and vehicle deceleration rates.

The issue of network congestion due to frequent exchange of range information for cooperative positioning was investigated in [41]. Network congestion causes severe packet loss, which affects the reliability of cooperative positioning as used in safety applications. Improvements such as information piggybacking, data compression, and network coding were proposed as solutions to reduce the range information overhead.

To summarize, both road sensor networks and vehicular sensor networks for road safety applications share stringent latency requirements. Algorithms that process sensor data for determining collision probabilities must finish within a specific amount of time so that collision warning systems can give out timely warnings. For road sensor networks, accurate vehicle detection is necessary. For vehicular sensor networks, accurate lane level position detection and inter-vehicle distance measurements are required in order to realize a cooperative collision warning system. From the literature reviewed, no single technology will provide an optimal solution for position detection. GPS alone cannot meet the needed accuracies. Fusing data from GPS, in-vehicle sensors, and road maps greatly increases the positioning accuracy, integrity, availability, and continuity of service [80]. Due to the tight latency requirements, suitable MAC and routing protocols must be developed, as the protocols used for traditional mobile ad-hoc networks are inapplicable.

3.6   Implementation Issues

For the deployment of sensor nodes in a road sensor network, the precise locations of the sensor nodes are typically well studied and planned before the actual installation. The optimal placement of sensor nodes for hilly terrains and roads with sharp turns is particularly important, as the road's characteristics pose a challenge in maintaining radio connectivity between sensor nodes. In [25], road segments are broken down into cells which are categorized into six different types depending on their height, slope, and aspect or the direction in which the road faces. Using trigonometric and geometric methods, analytical expressions for the optimal number of sensor nodes per cell type were derived, taking into account the sensor nodes' known transmission ranges. These expressions were then used to develop algorithms for sensor placement.

In practice, a hand-held PDA with GPS capability can be used to determine the sensor node location and preferably transmit the information wirelessly to the node during the installation. Better localization can be achieved if the GPS data is fused with communication ranging information [29, 58]. Accurate localization is particularly important for large scale deployment.

Before deploying WSN, one should be aware that the coverage areas and reliability of data in the outdoor environment may suffer due to noise generated from equipment, co-channel interferences, multipath, and other interferers. Consequently, the signal can be severely affected [10, 57, 27]. Despite the presence of noises and interferences, the data integrity must be maintained in particular for the automotive applications. To ensure reliable delivery of messages from sensor to application host, optimization in the signal level and network connectivity need to be addressed [84, 47].

In an urban setting, the effect of buildings on the wireless channel must also be considered, as their presence may cause signal reflections and diffractions [52]. Buildings can also obstruct radio line-of-sight, especially in intersections. Systems that warn vehicles of the presence of oncoming vehicles at intersections must have the capability for non-line-of-sight reception (NLOS). While NLOS reception is achievable with DSRC, [52] contend that it is first necessary to qualitatively study if the NLOS condition is frequently encountered in urban settings.They analyzed building positions at intersections in the city of Munich and found that only 20% of the intersections provide radio line-of-sight. A data mining technique for classifying intersection types within city areas was also presented. These intersection types can aid in the generalization of DSRC simulations for city areas.

Other important issues for consideration are the power consumption and security. These will be addressed in the following sections.

3.6.1 Interference

The interference signal can be classified as narrowband and broadband. Narrowband are discrete and single frequency signals. Broadband interferences are signals with a constant energy spectrum over all frequencies and are usually unintentional radiating sources.

Both types of interferences have varying degradation effects on wireless link reliability. To reduce the interferences, spread spectrum radio modulation or UWB techniques are desirable as it offers anti-multipath fading capability, and anti-jamming capability [30, 61, 62, 31]. There are two main spread spectrum techniques, namely the DSSS (direct sequence spread spectrum) and FHSS (frequency hopping spread spectrum). Due to different physical mechanisms, DSSS and FHSS react differently in industrial settings. Both have their advantages depending on the applications and environments. DSSS distributes information combined with a higher data rate bit sequence (chipping code) over a wider bandwidth and lower power level. Chipping codes are usually 11 to 20 bits. In DSSS, data are transmitted over many carrier waves in parallel. The reliability is achieved through the redundancy where redundant chipping codes are added to the information signal. De-spreading process at the re-

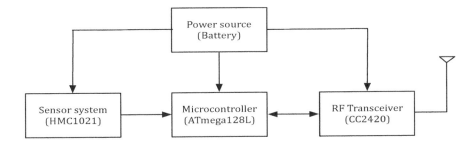

FIGURE 3.3
Diagram of a typical wireless sensor node for vehicle detection.

ceiver will also cause reduction in the power levels of interference. In contrast, FHSS sends the information through hopping different carrier frequencies at different times in pseudo-random fashion to avoid the interference.

3.6.2   Power Consumption

Batteries are the primary power sources for sensor nodes. In general, batteries are expected to last for years at a time to reduce the maintenance and replacement costs of the WSN installation. This is achieved through efficient power management and communication protocols. As sensor nodes are meant to be installed in outdoor and automotive environments, the impact of environmental factors such as temperature on the battery capacity cannot be neglected. As an alternative to batteries, energy harvesting techniques for self-powered wireless sensor nodes have also been studied [20, 94].

Figure 3.3 shows the block diagram of a typical wireless sensor node composed of four major blocks: the power source, the microcontroller, the sensor system, and the RF transceiver. Table 3.1 summarizes the various technologies used by the researchers surveyed in this paper. From Table 3.1, it can be seen that one of the most commonly used RF transceiver is the CC2420, and the microcontroller is the ATmega128L.

The overall energy consumption of a wireless sensor node is dependent on multiple factors. To optimize the energy consumption, factors such as active and sleep current, clock frequency, operating voltage, and duty cycle of operation must be considered.

The total energy E_{T_s} required for one sampling period can be computed using

$$E_{T_s} = V_{cc} \times (I_{active} \times T_{active} + I_{sleep} \times T_{sleep}) \tag{3.1}$$

where V_{cc} is the supply voltage, I_{active} and I_{sleep} are the active and sleep currents, and T_{active} and T_{sleep} are the active and sleep durations, respectively.

Table 3.2 shows the computed values for the overall energy consumption of a typical sensor node in mJ for one sampling period.

TABLE 3.1
Summary of technologies.

Reference	μP/μC	Radio	Comms./OS	Power source	Sensor
[95]	TI MSP430	CC2420	Point-to-point, TinyOS	No info	MicroMag2 magneto-inductive sensor
[32, 73]	VSN240	802.15.4	Sensys Networks Nanopower Protocol (SNP), TDMA	Li-SOCl2 3.6V 7.2Ah battery	3-axis magneto-resistive sensor
[69]	ATmega128L	CC1000	TinyOS	No info	HMC1021Z magnetic sensor
[98]	ATmega128L	No info	TinyOS	No info	Microphone
[87]	AtMega103L	RFMTR1000	TinyOS	No info	Mica Sensorboard
[75, 74]	ATmega128L	CC2420	Multihop	3.6V, 2.4Ah, AA Li-Ion	Magneto-resistive sensor
[11]	ATmega128L	CC1000	TinyOS	No info	Ultrasonic sensor
[18, 17]	8051	CC1010	Star topology	1.8Ah battery	Q45BB6DL Photoelectric sensor
[100]	MSP430-149	nRF905	TDMA/CSMA	No info	HMC1021 magneto-resistive sensor
[85]	ATmega128L	CC2420	XMesh, TinyOS	No info	MTS310 Sensor board
[48]	TI MSP430F161	CC2420	TinyOS	No info	HMC1052 magnetometer, SRF02 ultrasonic sensor
[8]	ATmega128L	CC2420	TinyOS	3V/9V battery pack	Speake FGM-3 magnetic field sensor
[92]	PIC16F88	nRF2401E	Star topology	4.8V battery pack	strain sensor, MMA7260 accelerometer, AD22103 temperature sensor
[36]	ATmega128L	CC2420	TinyOS	No info	N/A
[40]	ATmega128L	CC2420	Hybrid of star and mesh, multihop. TDMA	No info	HMC1021 and HMC1022 magnetic sensors
[16]	MSP430	nRF24L01	Tree topology	No info	Ultrasonic sensor
[50]	ATmega1281	XBee-802.15.4/Xbee-868/900	P2P/Tree/Mesh	13Ah Lithium	3-axis magnetic field sensor

TABLE 3.2
Energy consumption of a typical sensor node for vehicle detection.[a]

Component	Current, mA		Power, mW		Energy, mJ	
	active	sleep	active	sleep	active	sleep
ATmega128L [5]	2.30	0.008	6.90	0.024	0.69	0.24
CC2420 [86]	18.55[b]	0.020	55.65	0.060	5.57	0.59
HMC1021 [35]	3.75	0.00[c]	11.25	0.00	1.13	0.00
Total	24.60	0.028	73.80	0.084	**7.38**	**0.83**
			Total energy consumption for one sampling period, mJ			**8.21**

[a]For $V_{cc} = 3V$, $f_c = 1MHz$, $T_{active} = 100ms$, and $T_{sleep} = 9900ms$.
[b]This is the average of the current draw values for TX (17.4mA) and RX (19.7mA).
[c]It is assumed that the microcontroller disconnects the sensor from the voltage supply when in sleep mode.

The total energy consumed by the sensor node over a specified lifetime duration can be computed using

$$E_{total} = E_{T_s} \times \frac{T_{lifetime}}{T_s} \qquad (3.2)$$

where $T_{lifetime}$ is the lifetime duration and T_s is the sampling period in seconds.

From the total energy E_{total} in Equation 3.2, we can solve for the minimum required battery capacity in Ampere-hours by

$$C_{battery} = E_{total} \times \frac{1hr.}{3600s} \times \frac{1}{V_{bat}} \qquad (3.3)$$

where V_{bat} is the voltage of the battery during discharge.

As an example, for the typical sensor node describe in Table 3.2, and for a lifetime duration of three years together with a sampling period of 10s, the total energy consumed by the sensor node is

$$E_{total} = 8.21mJ \times \frac{94,608,000s}{10s}$$

$$E_{total} = 77,673.17J$$

For a battery voltage of 3V, the minimum required battery capacity for a lifetime duration of three years can now be computed as

$$C_{battery} = 77,673.17J \times \frac{1h}{3600s} \times \frac{1}{3V}$$

$$C_{battery} = 7.19A \cdot h$$

3.6.3 Security Issues

Wireless signals can be detected easily by data thieves or eavesdroppers. Thus, there are also security concerns over the deployment of wireless sensors [33]. Technology such as Zigbee uses 128-bit AES algorithm to provide strong authentication mechanisms that prevent unauthorized devices from joining and using the network key to control devices. However, the collected vehicle location data in traffic monitoring systems must be kept secure as well. This requires other schemes aside from strong authentication and encryption mechanisms. In [96], the location privacy of drivers is protected by anonymizing the IDs of cars. Their system uses short IDs which are derived from the vehicles full ID using periodically updated random patterns. The use of short IDs allow vehicles to be re-identified without revealing their full identities.

The addition of security and privacy mechanisms can introduce significant overhead to vehicular communications, especially when each vehicle periodically transmits beacons to other vehicles [15]. This may cause it to fail in meeting stringent latency requirements in critical applications such as collision avoidance. The effects of security on the overall vehicular communications performance must be studied, taking into account the communication technology, system computing resources, network configuration, environmental factors, security protocols, and supported applications.

3.7 Conclusions

In this chapter, a survey on the use of wireless sensor networks for intelligent transportation applications has been presented. The key application areas are the traffic and car park management, intra-vehicle monitoring systems, and road safety. The use of this technology promises significant advancements in the industry. The existence of a wide array of wireless technologies and system architecture require the users to evaluate and select the right technology for their applications. For successful deployment, several metrics in terms of power consumption, range, data rates, cost, scalability, security, robustness and software integration need to be adequately addressed.

For the near future in intelligent transportation system, it could be foreseen that more wireless sensor networks will be installed at the highway and traffic light junctions. Besides counting and classifying the vehicles, the signals can also be used to link the traffic signals at adjacent junctions to coordinate the start of their green times. This will allow the motorists to travel from one junction to another without having to stop often at the red light. With the wireless sensors to detect both the motorists and pedestrians at the traffic junction, the timings of the traffic light can be set adaptively according to the real time needs to balance the conflict demands between the motorists and pedestrians.

For a compact city such as Singapore, each vehicle is mandatorily installed with an in-vehicle-unit (IU) by the Land Transport Authority. The in-vehicle unit was initially meant for the purpose of electronic road pricing [79]. The driver is responsible for inserting a cash card into the IU before the vehicle passes through the charging gantries. Its application has lately been extended to car park payment system. By expanding the IU features with GPS, sensors and inter-vehicle communication capabilities, many possible emerging applications for large scale sensor network can be explored. The key concern here is to address the individual privacy adequately. For public vehicle such as bus or taxis, the privacy is less of an issue. These vehicles can be monitored to provide live traffic information such as the traveling speed or estimated arrival time through processing the GPS information from these vehicles.

References

[1] Akyildiz, I. F., W. Su, Y. Sankarasubramaniam, and E. Cayirci. "Wireless sensor networks: a survey." *Computer Networks* 38, 4: (2002) 393–422.

[2] Alam, N., A.T. Balaie, and A.G. Dempster. "A DSRC-Based Traffic Flow Monitoring and Lane Detection System." In *Vehicular Technology Conference (VTC Spring), 2011 IEEE 73rd*. 2011, 1–5.

[3] Araujo, A., J. Garcia-Palacios, J. Blesa, F. Tirado, E. Romero, A. Samartin, and O. Nieto-Taladriz. "Wireless Measurement System for Structural Health Monitoring With High Time-Synchronization Accuracy." *Instrumentation and Measurement, IEEE Transactions on* 61, 3: (2012) 801–810.

[4] Asín, A., and D. Gascón. "Smart Parking Sensor Platform enables city motorists save time and fuel." 2011. http://www.libelium.com/smart_parking/.

[5] Atmel Corporation. "8-bit Atmel Microcontroller with 128KBytes In-System Programmable Flash." ATmega128L Data Sheet, 2011.

[6] Baljeet, M., I. Nikolaidis, and J. Harms. "Distributed classification of acoustic targets in wireless audio-sensor networks." *Computer Networks* 52: (2008) 2582–2593.

[7] Barbagli, B., L. Bencini, I. Magrini, G. Manes, and A. Manes. "A real-time traffic monitoring based on wireless sensor network technolo-

gies." In *Wireless Communications and Mobile Computing Conference (IWCMC), 2011 7th International*. 2011, 820–825.

[8] Barton, J., J. Buckley, B. O'Flynn, S. C. O'Mathuna, J. P. Benson, T. O'Donovan, U. Roedig, and C. Sreenan. "The D-Systems Project - Wireless Sensor Networks for Car-Park Management." In *Vehicular Technology Conference, 2007. VTC2007-Spring. IEEE 65th*. 2007, 170–173.

[9] Basma, F., Y. Tachwali, and H.H. Refai. "Intersection collision avoidance system using infrastructure communication." In *Intelligent Transportation Systems (ITSC), 2011 14th International IEEE Conference on*. 2011, 422–427.

[10] Bello, L. L., O. Mirabella, and A. Raucea. "Design and Implementation of an Educational Testbed for Experiencing With Industrial Communication Networks." *Industrial Electronics, IEEE Transactions on* 54, 6: (2007) 3122–3133.

[11] Bi, Y. Z., L. M. Sun, H. S. Zhu, T. X. Yan, and Z. J. Luo. "Parking management system based on wireless sensor network." *Zidonghua Xuebao/Acta Automatica Sinica* 32, 6: (2006) 968–977.

[12] Biswas, S., R. Tatchikou, and F. Dion. "Vehicle-to-vehicle wireless communication protocols for enhancing highway traffic safety." *Communications Magazine, IEEE* 44, 1: (2006) 74–82.

[13] Bong, D. B. L., K. C. Ting, and K. C. Lai. "Integrated approach in the design of car park occupancy information system (COINS)." *IAENG International Journal of Computer Science* 35, 1: (2008) 7–14.

[14] Cabanas, M.F., F. Pedrayes, C.H. Rojas, M.G. Melero, J.G. Norniella, G.A. Orcajo, J.M. Cano, F. Nuno, and D.R. Fuentes. "A New Portable, Self-Powered, and Wireless Instrument for the Early Detection of Broken Rotor Bars in Induction Motors." *Industrial Electronics, IEEE Transactions on* 58, 10: (2011) 4917–4930.

[15] Calandriello, G., P. Papadimitratos, J.-P. Hubaux, and A. Lioy. "On the Performance of Secure Vehicular Communication Systems." *Dependable and Secure Computing, IEEE Transactions on* 8, 6: (2011) 898–912.

[16] Chang, T., M. Fu, Y. Gao, M. Chen, and C. Hu. "Wireless Sensor Network in Parking Guidance and Information System." *Energy Procedia* 13, 0: (2011) 4608–4614.

[17] Chinrungrueng, J., U. Sunantachaikul, and S. Triamlumlerd. "Smart Parking: An Application of Optical Wireless Sensor Network." In *Applications and the Internet Workshops, 2007. SAINT Workshops 2007. International Symposium on*. 2007, 66.

[18] Chinrungrueng, J., U. Sununtachaikul, and S. Triamlumlerd. "A Vehicular Monitoring System with Power-Efficient Wireless Sensor Networks." In *ITS Telecommunications Proceedings, 2006 6th International Conference on.* 2006, 951–954.

[19] Dat Tien, Nguyen, J. Singh, Le Hai Phuong, and B. Soh. "A Hybrid TDMA protocol based Ultra-Wide Band for in-car wireless communication." In *TENCON 2009 - 2009 IEEE Region 10 Conference.* 2009, 1–7.

[20] Dondi, D., A. Bertacchini, D. Brunelli, L. Larcher, and L. Benini. "Modeling and Optimization of a Solar Energy Harvester System for Self-Powered Wireless Sensor Networks." *Industrial Electronics, IEEE Transactions on* 55, 7: (2008) 2759–2766.

[21] Duarte, M. F., and Y. H. Hu. "Vehicle classification in distributed sensor networks." *Journal of Parallel and Distributed Computing* 64, 7: (2004) 826–838.

[22] ElBatt, T., C. Saraydar, M. Ames, and T. Talty. "Potential for Intra-Vehicle Wireless Automotive Sensor Networks." In *Sarnoff Symposium, 2006 IEEE.* 2006, 1–4.

[23] Fontaine, M. D., and B. L. Smith. "Investigation of the Performance of Wireless Location Technology-Based Traffic Monitoring Systems." *Journal of Transportation Engineering* 133, 3: (2007) 157–165.

[24] de Francisco, R., Huang Li, G. Dolmans, and H. de Groot. "Coexistence of ZigBee wireless sensor networks and Bluetooth inside a vehicle." In *Personal, Indoor and Mobile Radio Communications, 2009 IEEE 20th International Symposium on.* 2009, 2700–2704.

[25] Ghosh, S., S. Rao, and B. Venkiteswaran. "Sensor Network Design for Smart Highways." *Systems, Man and Cybernetics, Part A: Systems and Humans, IEEE Transactions on* PP, 99: (2012) 1–10.

[26] Gil-Castineira, F., F. J. Gonzalez-Castano, and L. Franck. "Extending Vehicular CAN Fieldbuses With Delay-Tolerant Networks." *Industrial Electronics, IEEE Transactions on* 55, 9: (2008) 3307–3314.

[27] Gnad, A., M. Kratzig, L. Rauchhaupt, and S. Trikaliotis. "Relevant influences in wireless automation." In *Factory Communication Systems, 2008. WFCS 2008. IEEE International Workshop on.* 2008, 341–348.

[28] Gongjun, Y., Y. Weiming, D. B. Rawat, and S. Olariu. "SmartParking: A Secure and Intelligent Parking System." *Intelligent Transportation Systems Magazine, IEEE* 3, 1: (2011) 18–30.

[29] Guo, H., K. S. Low, and H. A. Nguyen. "Optimizing the Localization of a Wireless Sensor Network in Real Time Based on a Low-Cost Microcontroller." *Industrial Electronics, IEEE Transactions on* 58, 3: (2011) 741–749.

[30] Guvenc, I., H. Arslan, S. Gezici, and H. Kobayashi. "Adaptation of two types of processing gains for UWB impulse radio wireless sensor networks." *Communications, IET* 1, 6: (2007) 1280–1288.

[31] Hancke, G. P., and B. Allen. "Ultrawideband as an Industrial Wireless Solution." *Pervasive Computing, IEEE* 5, 4: (2006) 78–85.

[32] Haoui, A., R. Kavaler, and P. Varaiya. "Wireless magnetic sensors for traffic surveillance." *Transportation Research Part C: Emerging Technologies* 16, 3: (2008) 294–306.

[33] Haowen, C., and A. Perrig. "Security and privacy in sensor networks." *Computer* 36, 10: (2003) 103–105.

[34] Hochmuth, P. "Case Study: GM cuts the cords to cut costs." *TechWorld*.

[35] Honeywell Corporation. "1- and 2-Axis Magnetic Sensors HMC1001/1002/1021/1022." HMC1001/1002/1021/1022 Data Sheet, 2008.

[36] Hsin-Mu, Tsai, O. K. Tonguz, C. Saraydar, T. Talty, M. Ames, and A. Macdonald. "Zigbee-based intra-car wireless sensor networks: a case study." *Wireless Communications, IEEE* 14, 6: (2007a) 67–77.

[37] Hsin-Mu, T., W. Viriyasitavat, O. K. Tonguz, C. Saraydar, T. Talty, and A. Macdonald. "Feasibility of In-car Wireless Sensor Networks: A Statistical Evaluation." In *Sensor, Mesh and Ad Hoc Communications and Networks, 2007. SECON '07. 4th Annual IEEE Communications Society Conference on*. 2007b, 101–111.

[38] Intel Corp. "Expanding Usage Models for Wireless Sensor Networks." *Technology@Intel Magazine* 4–5.

[39] Intel Labs. "Research@Intel 2011.", 2011. `http://download.intel.com/newsroom/kits/research/2011/pdfs/Research@Intel-2011_DemoFactSheet.pdf`.

[40] Jae Jun, Y., S. K. Bok, and J. J. Ah. "Intelligent Non-signalized Intersections Based On Magnetic Sensor Networks." In *Intelligent Sensors, Sensor Networks and Information, 2007. ISSNIP 2007. 3rd International Conference on*. 2007, 275–280.

[41] Jun, Y., A. T. Balaei, M. Hassan, N. Alam, and A. G. Dempster. "Improving Cooperative Positioning for Vehicular Networks." *Vehicular Technology, IEEE Transactions on* 60, 6: (2011) 2810–2823.

[42] Karpiński, M., A. Senart, and V. Cahill. "Sensor networks for smart roads." In *Pervasive Computing and Communications Workshops, 2006. PerCom Workshops 2006. Fourth Annual IEEE International Conference on*. 2006, 5 pp. 306–310.

[43] King, T. I., W. J. Barnes, H. H. Refai, and J. E. Fagan. "A Wireless Sensor Network Architecture for Highway Intersection Collision Prevention." In *Intelligent Transportation Systems Conference, 2007. ITSC 2007. IEEE*. 2007, 178–183.

[44] Kwong, K., R. Kavaler, R. Rajagopal, and P. Varaiya. "Arterial travel time estimation based on vehicle re-identification using wireless magnetic sensors." *Transportation Research Part C: Emerging Technologies* 17, 6: (2009) 586–606.

[45] Lan, J., Y. Xiang, L. Wang, and Y. Shi. "Vehicle detection and classification by measuring and processing magnetic signal." *Measurement* 44, 1: (2011) 174–180.

[46] Lee, Guo Xiong, Kay Soon Low, and T. Taher. "Unrestrained Measurement of Arm Motion Based on a Wearable Wireless Sensor Network." *Instrumentation and Measurement, IEEE Transactions on* 59, 5: (2010) 1309–1317.

[47] Lee, Jin Shyan. "A Petri Net Design of Command Filters for Semi-autonomous Mobile Sensor Networks." *Industrial Electronics, IEEE Transactions on* 55, 4: (2008) 1835–1841.

[48] Lee, S., D. Yoon, and A. Ghosh. "Intelligent parking lot application using wireless sensor networks." In *Collaborative Technologies and Systems, 2008. CTS 2008. International Symposium on*. 2008, 48–57.

[49] Leen, G., and D. Heffernan. "Vehicles without wires." *Computing & Control Engineering Journal* 12, 5: (2001) 205–211.

[50] Libelium Comunicaciones Distribuidas S.L. "Smart Parking Technical Guide." 2012. http://www.libelium.com/documentation/waspmote/smart-parking-sensor-board_eng.pdf.

[51] Low, K. S., W. N. N. Win, and M. J. Er. "Wireless Sensor Networks for Industrial Environments." In *Computational Intelligence for Modelling, Control and Automation, 2005*. 2005, volume 2, 271–276.

[52] Mangel, T., F. Schweizer, T. Kosch, and H. Hartenstein. "Vehicular safety communication at intersections: Buildings, Non-Line-Of-Sight and representative scenarios." In *Wireless On-Demand Network Systems and Services (WONS), 2011 Eighth International Conference on*. 2011, 35–41.

[53] Mason, A., A. Shaw, and A. I. Al-Shamma'a. "Inventory Management in the Packaged Gas Industry Using Wireless Sensor Networks." In *Advances in Wireless Sensors and Sensor Networks*, edited by Subhas Chandra Mukhopadhyay, and Henry Leung, Springer Berlin Heidelberg, 2010, volume 64 of *Lecture Notes in Electrical Engineering*, 75–100.

[54] Mitropoulos, G. K., I. S. Karanasiou, A. Hinsberger, F. Aguado-Agelet, H. Wieker, H. J. Hilt, S. Mammar, and G. Noecker. "Wireless Local Danger Warning: Cooperative Foresighted Driving Using Intervehicle Communication." *Intelligent Transportation Systems, IEEE Transactions on* 11, 3: (2010) 539–553.

[55] Moghe, R., F.C. Lambert, and D. Divan. "Smart 'Stick-on' Sensors for the Smart Grid." *Smart Grid, IEEE Transactions on* 3, 1: (2012) 241–252.

[56] Nekoui, M., and H. Pishro-Nik. "Analytical Design of Inter-Vehicular Communications for Collision Avoidance." In *Vehicular Technology Conference (VTC Fall), 2011 IEEE.* 2011, 1–5.

[57] Neumann, P. "Communication in industrial automation – What is going on?" *Control Engineering Practice* 15, 11: (2007) 1332–1347.

[58] Nguyen, H. A., H. Guo, and K. S. Low. "Real-Time Estimation of Sensor Node's Position Using Particle Swarm Optimization With Log-Barrier Constraint." *Instrumentation and Measurement, IEEE Transactions on* 60, 11: (2011) 3619–3628.

[59] Ni, L. M., L. Yunhao, and Z. Yanmin. "China's national research project on wireless sensor networks." *Wireless Communications, IEEE* 14, 6: (2007) 78–83.

[60] Niu, W., J. Li, S. Liu, and T. Talty. "Intra-Vehicle Ultra-Wideband Communication Testbed." In *Military Communications Conference, 2007. MILCOM 2007. IEEE.* 2007, 1–6.

[61] Norimatsu, T., R. Fujiwara, M. Kokubo, M. Miyazaki, A. Maeki, Y. Ogata, S. Kobayashi, N. Koshizuka, and K. Sakamura. "A UWB-IR Transmitter With Digitally Controlled Pulse Generator." *Solid-State Circuits, IEEE Journal of* 42, 6: (2007) 1300–1309.

[62] Oppermann, I., L. Stoica, A. Rabbachin, Z. Shelby, and J. Haapola. "UWB wireless sensor networks: UWEN - a practical example." *Communications Magazine, IEEE* 42, 12: (2004) S27–S32.

[63] Parker, R., and S. Valaee. "Vehicular Node Localization Using Received-Signal-Strength Indicator." *Vehicular Technology, IEEE Transactions on* 56, 6: (2007) 3371–3380.

[64] ParkingCarma Inc., `http://www.parkingcarma.com`.

[65] Premier 100 IT Leaders 2005 Best in Class. *ComputerWorld* 8–9.

[66] Qingchun, R., and L. Qilian. "Throughput and Energy-Efficiency-Aware Protocol for Ultrawideband Communication in Wireless Sensor Networks: A Cross-Layer Approach." *Mobile Computing, IEEE Transactions on* 7, 6: (2008) 805–816.

[67] Ramamurthy, H., B. S. Prabhu, R. Gadh, and A. M. Madni. "Wireless Industrial Monitoring and Control Using a Smart Sensor Platform." *Sensors Journal, IEEE* 7, 5: (2007) 611–618.

[68] Rezaei, S., and R. Sengupta. "Kalman Filter-Based Integration of DGPS and Vehicle Sensors for Localization." *Control Systems Technology, IEEE Transactions on* 15, 6: (2007) 1080–1088.

[69] Rui, W., Z. Lei, S. Rongli, G. Jibing, and C. Li. "EasiTia: A Pervasive Traffic Information Acquisition System Based on Wireless Sensor Networks." *Intelligent Transportation Systems, IEEE Transactions on* 12, 2: (2011) 615–621.

[70] Ruoshui, L., S. Herbert, L. T. Hong, and I. J. Wassell. "A Study on Frequency Diversity for Intra-Vehicular Wireless Sensor Networks (WSNs)." In *Vehicular Technology Conference (VTC Fall), 2011 IEEE*. 2011, 1–5.

[71] Semertzidis, T., K. Dimitropoulos, A. Koutsia, and N. Grammalidis. "Video sensor network for real-time traffic monitoring and surveillance." *Intelligent Transport Systems, IET* 4, 2: (2010) 103–112.

[72] Sengupta, R., S. Rezaei, S. E. Shladover, D. Cody, S. Dickey, and H. Krishnan. "Cooperative Collision Warning Systems: Concept Definition and Experimental Implementation." *Journal of Intelligent Transportation Systems* 11, 3: (2007) 143–155.

[73] Sensys Networks, Inc., `http://www.sensysnetworks.com`.

[74] Seong-eun, Y., C. P. Kit, K. Taehong, K. Jonggu, K. Daeyoung, S. Changsub, S. Kyungbok, and J. Byungtae. "PGS: Parking Guidance System based on wireless sensor network." In *Wireless Pervasive Computing, 2008. ISWPC 2008. 3rd International Symposium on*. 2008a, 218–222.

[75] Seong-eun, Y., C. P. Kit, P. Taisoo, K. Youngsoo, K. Daeyoung, S. Changsub, S. Kyungbok, and K. Hyunhak. "DGS: Driving Guidance System Based on Wireless Sensor Network." In *Advanced Information Networking and Applications - Workshops, 2008. AINAW 2008. 22nd International Conference on*. 2008b, 628–633.

[76] Shladover, S. E., and S.-K. Tan. "Analysis of Vehicle Positioning Accuracy Requirements for Communication-Based Cooperative Collision Warning." *Journal of Intelligent Transportation Systems* 10, 3: (2006) 131–140.

[77] Shoup, D. C. "Cruising for parking." *Transport Policy* 13: (2006) 479–486.

[78] Siemens AG. "SiPark SSD Car Park Guidance Systems.", `http://www.mobility.siemens.com/mobility/global/Documents/en/road-solutions/urban/solutions-around-parking/SIPARK-SSD-en.pdf`.

[79] Singapore Land Transport Authority. "Electronic Road Pricing.", `http://www.lta.gov.sg/content/lta/en/motoring/erp_.html`.

[80] Skog, I., and P. Handel. "In-Car Positioning and Navigation Technologies : A Survey." *Intelligent Transportation Systems, IEEE Transactions on* 10, 1: (2009) 4–21.

[81] Stoica, L., A. Rabbachin, H. O. Repo, T. S. Tiuraniemi, and I. Oppermann. "An ultrawideband system architecture for tag based wireless sensor networks." *Vehicular Technology, IEEE Transactions on* 54, 5: (2005) 1632–1645.

[82] Streetline Networks, Inc., `http://www.streetlinenetworks.com`.

[83] Tan, H. S., and J. Huang. "DGPS-based vehicle-to-vehicle cooperative collision warning: Engineering feasibility viewpoints." *Ieee Transactions on Intelligent Transportation Systems* 7, 4: (2006) 415–428.

[84] Tang, K. S., K. F. Man, and S. Kwong. "Wireless communication network design in IC factory." *Industrial Electronics, IEEE Transactions on* 48, 2: (2001) 452–459.

[85] Tang, V. W. S., Zheng Yuan, and Cao Jiannong. "An Intelligent Car Park Management System based on Wireless Sensor Networks." In *Pervasive Computing and Applications, 2006 1st International Symposium on*. 2006, 65–70.

[86] Texas Instruments Corporation. "2.4 GHz IEEE 802.15.4 / ZigBee-ready RF Transceiver." CC2420 Data Sheet, 2007.

[87] Tim Tau, H. "Using sensor networks for highway and traffic applications." *Potentials, IEEE* 23, 2: (2004) 13–16.

[88] Tubaishat, M., Qi Qi, Shang Yi, and Shi Hongchi. "Wireless Sensor-Based Traffic Light Control." In *Consumer Communications and Networking Conference, 2008. CCNC 2008. 5th IEEE*. 2008, 702–706.

[89] U.S. Federal Highway Administration. "Traffic Monitoring Guide, Section 4 Vehicle Classification Monitoring." 2001.

[90] U.S. National Safety Council. "Analysis of Intersection Fatal and Nonfatal Crashes from 2005 to 2009." 2011.

[91] U.S. National Transportation Safety Board. "2009-2010 U.S. Transportation Fatalities." 2011.

[92] Vilela, J. P. T., and J. C. M. Valenzuela. "Design and implementation of a wireless remote data acquisition system for mobile applications." In *Design of Reliable Communication Networks, 2005. (DRCN 2005). Proceedings.5th International Workshop on.* 2005, 8 pp.

[93] Waldo, J. "Embedded computing and Formula One racing." *Pervasive Computing, IEEE* 4, 3: (2005) 18–21.

[94] Wan, Z.G., Y.K. Tan, and C. Yuen. "Review on energy harvesting and energy management for sustainable wireless sensor networks." In *Communication Technology (ICCT), 2011 IEEE 13th International Conference on.* 2011, 362–367.

[95] Wilder, J. L., A. Milenkovic, and E. Jovanov. "Smart Wireless Vehicle Detection System." In *System Theory, 2008. SSST 2008. 40th Southeastern Symposium on.* 2008, 159–163.

[96] Xie, Hairuo, L. Kulik, and E. Tanin. "Privacy-Aware Traffic Monitoring." *Intelligent Transportation Systems, IEEE Transactions on* 11, 1: (2010) 61–70.

[97] Yang, S. S., Y. G. Kim, and C. Hongsik. "Vehicle identification using wireless sensor networks." In *SoutheastCon, 2007. Proceedings. IEEE.* 2007, 41–46.

[98] Yang, S., and J. N. Daigle. "A PCA-based vehicle classification system in wireless sensor networks." In *Wireless Communications and Networking Conference, 2006. WCNC 2006. IEEE.* 2006, volume 4, 2193–2198.

[99] Yang, X., L. Liu, N. H. Vaidya, and F. Zhao. "A vehicle-to-vehicle communication protocol for cooperative collision warning." In *Mobile and Ubiquitous Systems: Networking and Services, 2004. MOBIQUITOUS 2004. The First Annual International Conference on.* 2004, 114–123.

[100] Ying, W., Z. Guangrong, and L. Tong. "Design of a Wireless Sensor Network for Detecting Occupancy of Vehicle Berth in Car Park." In *Parallel and Distributed Computing, Applications and Technologies, 2006. PDCAT '06. Seventh International Conference on.* 2006, 115–118.

4

Design Challenges and Objectives in Industrial Wireless Sensor Networks

Johan Åkerberg, Mikael Gidlund, Tomas Lennvall, and Krister Landerns

ABB AB, Corporate Research

Mats Bjökman

Mälardalen University, School of Innovation, Design, and Engineering

CONTENTS

4.1   Introduction

In recent years the advances in wireless sensor networks have grown exponentially and WSNs have been deployed in diverse application areas such as agriculture, disaster management, intelligent transport systems, and industrial automation. In industrial automation, wireless sensor networks have so far mostly been considered within *building automation*, *factory automation*, and *process automation* in order to save cost in cable reduction and maintenance but also improved flexibility [14]. Recently, wireless sensor networks for smart grid applications have been discussed. Several market forecasts have recently predicted exponential growths in the sensor market over the next few years, resulting in a multi-billion dollar market in the near future. For instance, ABI research [18] predicts that in 2015 around 645 million IEEE 802.15.4 chipsets will be shipped and that the worldwide market for automation systems in process industries will grow to roughly $150 billion.

Building Automation Systems (BAS) can be used in schools, hospitals, factories, offices and homes, to enhance the quality of building services and reduce the operation and maintenance costs. Typical functionalities of BAS include the monitoring and controlling of heating, ventilation, and air conditioning (HVAC) systems, the management of building facilities and the automation of meter reading. KNX, LonWorks, and ZigBee are standards that are used and deployed in building automation. In building automation systems the requirements on latency and data throughput are more relaxed compared to process automation and factory automation. The main challenge for wireless sensor networks in BAS is to provide a low-cost solution which is robust against changing wireless channel conditions, coexist with other wireless technologies, and provide long battery lifetime.

Smart grids will provide more electricity to meet rising demand, increase reliability and quality of power supplies, increase energy efficiency, and be able to integrate low carbon energy sources into power networks [16]. Traditional electric-power-systems monitoring and diagnostic systems are mainly realized through wired communication but in the future it can realized with wireless sensor networks due to cost-efficiency [15]. It should be noted that most of the foreseen applications using wireless sensor networks in smart grids are pure monitoring with relaxed requirements. This implies that available commercial WSN chip sets are good enough for fulfilling the smart grid monitoring applications. For more demanding areas such as substation automation there is more challenging requirements with real-time communication and latency demands on some milliseconds which put high demands on reliable communication.

Wireless technologies for factory automation has been available since early 2000 when WISA was launched [13]. The requirements in factory automation are demanding with respect to determinism, reliability, and availability. With an update frequency of 100 ms to < 1 ms, the domain of factory automation is one of the most challenging to use wireless communication technology. With state of the art technologies, wireless solutions may serve applications with an update frequency of 10 ms to 20 ms. Because of the wide spreading of PLCs and decentralized peripherals, the ranges between devices may vary from a few meters up to some hundreds of meters. The network topologies for wireless solutions range from simple cable replacement point-to-point and point-to-multipoint connections up to cellular networks with roaming capabilities (production lines, automated guided vehicles). Some examples of domain and typical requirements are given in Table 4.1.

TABLE 4.1

Requirements for some typical automation domains

Application Domain	Update Frequency	Nodes / 10 m^2
Building Automation	seconds	$1 - 20$
Process Automation	$10 - 1000\ ms$	$1 - 20$
Factory Automation	$500\ \mu s - 100\ ms$	$20 - 100$
Substation Automation	$250\ \mu s - 50\ ms$	$1 - 10$
High Voltage DC control	$10 - 100\ \mu s$	$300 - 500$

The usage of industrial wireless sensor networks within Process Automation domain has gained a lot of interest in both academia and industry. The main concerns about reliability, security, and integration along with the lack of device interoperability have hampered the deployment rate although there exists a lot of research work regarding WSN based on IEEE 802.15.4 [11] and ZigBee for industrial automation applications with very relaxed requirements. However, for applications in the process automation domain these standards becomes obsolete [17] and in order to overcome the constraints in the Zig-Bee and IEEE 802.15.4 standards, WirelessHART, the first open and interoperable wireless communication standard designed specifically for real-world applications in process automation, was approved and released in 2007. WirelessHART builds on proven international standards - the HART Communication Protocol (IEC 61158), IEEE 802.15.4 radio and frequency/channel hopping, spread spectrum and mesh networking technologies to ensure reliable communication in harsh environments.

It is foreseen that wireless technology and especially wireless sensor networks has the potential to contribute significantly in areas such as cable replacement, mobility, flexibility, and scalability. It offers competitive advantages such as lower life-cycle cost and reducing connector failures which is one of the most common reliability problems. Below we summarize the major advantages of industrial wireless sensor networks.

- *Cost* - With capital at a premium, process manufactures are looking for quick investments that cost little and save even more. The major incitement for deploying WSN in process automation (all automation business) is that they are easier and less costly to install than traditional wired systems. Consider a green field installation today, it cost roughly $200 per meter to install wires in an ordinary process plant, and approximately $1000 per meter in offshore installations. As wires age they crack or fail. Furthermore, inspecting, testing, troubleshooting, and upgrading wires require time, labor, and materials.

- *Flexibility* - Many secondary process variables have long gone unmeasured, and expensive pieces of critical rotating equipment remain non-instrumented. With the advent of WSN, we are able to unlock stranded

FIGURE 4.1
Tumbling mills in mining automation creates severe fading dips.

FIGURE 4.2
Hot rolling mill.

information in instruments, gather information from where it previously has been economically unfeasible, such that the process can be enhanced with respect to quality and quantity, and reducing the possibility of mechanical failures. Furthermore, another advantage is that WSN enables temporary measurements of certain process values and quality indicators without installation of additional wires.

• *Emerging Applications* - With the advent of wireless infrastructure and WSN in process automation several new wireless applications are emerging such as empowering mobile workers, location of assets, safety mustering, integration on nontraditional signals such as video, bridging remote or isolated control systems, enabling wireless control applications, and allow connectivity for equipment that is "sealed-for-life."

• *Reliability* - Industrial applications require reliability and determinism in the automation systems to avoid serious consequences such as injury, explosions, and material losses. Industrial wireless sensor networks can offer built in redundancy and capabilities for anticipatory system maintenance and failure recovery. This could for example be achieved by designing meshed networks where there always exist available links to the control system.

Although there are several benefits with deploying wireless in industrial automation there are some challenges to consider. In most cases, the wireless sensor networks will be deployed in harsh environments (See Figures 4.1 and 4.2) which contain a lot of steel, metals, and rotating machinery which create multipath fading and deteriorate the performance of the wireless systems. These fading dips can be deep under a short time period and the received signal strength can drop as much as 30 - 40 dB and creates severe communication problems (could even lose the communication link temporarily). Typically in industrial environments the signal is static during a symbol period. In cellular communication the signals typically fluctuate within the symbol period and causes inter-symbol interference. Other problems that may occur in industrial plants is moving objects such as trucks parking in front of wireless sensor nodes and creating shadow fading which sometime can eliminate the communication completely and the only solution to that problem is using several communication links (i.e., using mesh network topology). The presence of electrical motors, cranes, and vehicles can also produce interference in communication systems. These interferences are a composition of random high energy spikes with randomly occurrence in terms of time and frequency which does not correspond to Gaussian noise and consequently affect the communication systems differently. Impulse interference typically occurs between 200 MHz - 1 GHz frequency band but can sometimes be present in higher frequency bands [8, 10]. The presence of multipath fading and impulse interference in industrial automation environments needs to be considered when designing industrial wireless sensor networks in the future.

4.2   Applications and Requirements for Industrial Automation

From the previous section we know that the general requirements, such as, reliability, low-powered, and enable to coexist with other wireless systems are the same for different industrial automation domains. However, some domains such as process automation and factory automation have much stricter requirements and in this section we highlight some important requirements that need to be met for a successful large scale deployment of WSNs in process automation and discrete manufacturing. Some typical examples of process automation industries are: pulp and paper (see Figure 4.3(a)), mining, steel, oil, and gas (see Figure 4.3(b)) to mention some. The main characteristic that groups them together is that the products are produced in a continuous manner, i.e., the oil is produced in a continuous flow. In discrete manufacturing, the products are produced in discrete steps, i.e., the products are assembled together using sub assemblies or single components. Typical examples of discrete manufacturing industries are automotive, medical, and the food industries. Discrete manufacturing heavily relies on robotics and belt conveyors for assembly, picking, welding, and palletizing. To generalize, discrete manufacturing normally have stricter requirements with respect to latency and real-time requirements compared to process automation. However, as always there are cases when this general assumption is not true. The main reason for this generalized assumption is that in order to pick, assemble, or palletize at high speed, the latency, refresh rates, and real-time requirements are stricter compared to a tank level control in process automation to achieve the re-

(a) Pulp and paper   (b) Oil and gas

FIGURE 4.3
Examples of process automation.

quired production quality. However, the focus in this chapter is on process automation, while keeping in mind that requirements might be even stricter for discrete manufacturing.

In the automation domain, many different communication protocols exist on various media such as fiber, copper cables, radio, or even power-line carrier communication. Since the automation equipment ranges from high-end server hardware from the IT domain, down to small tailored embedded systems with 8 bit processors and just a few kilobytes of memory, it is a challenging task to solve all needs with one single protocol. From an automation application point of view, communication is for example used for

- interconnection of automation equipment distributed over large geographical areas

- interconnection of dedicated real-time automation systems with operator work-places for control and supervision

- closed loop control, ranging from slow processes such as tank level control, to fast processes such as motion control

- interlocking and control, a major part of control applications in process control require discrete signaling. For example, a machine might have start, stop, and safety interlocks.

- monitoring and supervision, where large amount of data is transmitted and evaluated to predict and avoid interruption of production.

As illustrated in Figure 4.4 the automation network is divided into several different networks, with different demands and importance of various properties. Typically, the higher levels of the automation networks, i.e., server networks, have more relaxed constraints on for example latency and real-time properties, compared to the field networks. On the other hand, the field networks have in general relaxed constraints with respect to throughput, as real-time behavior, low latency, and low jitter are more important for process control.

4.2.1 Targeted Applications

Typical applications at the field network level that is targeted for industrial wireless sensor networks are monitoring and supervision, interlocking and control, and closed loop control. That means that the industrial WSNs should support all the requirements in order to fit the expected application scenarios. A design that for example only supports monitoring and supervision, but not interlocking and control will have a limited success probability in the domain of process automation.

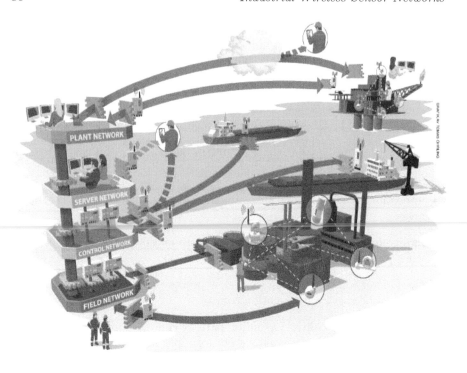

FIGURE 4.4
Future wireless infrastructure and its applications in the process automation domain.

Measuring the quality, production statistics, finding discrepancies and resolving them before unplanned shutdowns occurs are typical application scenarios for monitoring and supervision. As an example of preventive maintenance, a sensor is measuring the vibrations from an electrical machine in order to detect worn bearings such that maintenance can be scheduled at next planned production stop. Most of the measures are displayed for the plant operators such that they can monitor the production and tune the automation systems based on production rates and quality. Most measuring values are associated with alarm conditions, for example the tank is full or empty. Typically those alarms are collected in alarm lists that require the operator's attention in order to keep the production rate.

Another important use-case of physical measurements is for closed loop control where the automation system automatically stabilizes unstable processes without manual intervention. In some cases closed-loop control is convenient for the operators such that they do not need to manually open and close a valve to maintain a steady flow. But in most of the cases humans cannot control fast processes manually as for example in the case of controlling the speed and torque of electrical machines on a paper machine to keep the paper web stretched without dragging the paper or causing web breaks at high speeds (1800 m/min).

Furthermore, due to safety reasons, interlocking and control is another important area for process automation. If a control valve in an oil pipeline is closed the electrical machine controlling the pump needs to be stopped immediately in order to keep the oil pressure within the specifications. Thus, the pump is interlocked with the control valve. Another important scenario is that a control command needs to be issued to the oil pump for the gear box before starting the electrical machine that is connected to the gear box to avoid damages in the gear box. A more critical scenario is when sensors are located in the direction of a belt conveyor in a mining application. The sensors should indicate that the rocks will fit in the tunnel. If too large pieces of rocks are detected, the belt conveyor needs to be stopped before it jams the tunnel.

4.2.2 Requirements

Since fieldbuses are used to transmit data for interlocking, closed-loop control, and monitoring there exist several important properties. Some of the most important and sometimes even contradicting properties of industrial communication at various levels are given in Table 4.2.

In Table 4.3, the different types of applications are grouped and listed in three sub-categories:

- *Monitoring, supervision* - this sub-category collects different types of sensors that provide diagnostics and supervision that normally can be pre-processed and transmitted and updated on a period time from 1 second and more. This information is generally not sensitive for packet losses and jitter as it is used for supervision and condition monitoring. However, in some cases data consistency might be important.

- *Closed loop control* - sensors and actuators are connected to proportional-integral-derivative (PID) controllers that control the process with respect to the actual set-point. The purpose of the control loop is to continuously stabilize the (instable) process by controlling the actuators based on the actual sensor readings. Generally, closed loop control is sensitive to jitter and delays.

- *Interlocking and control* - normally the major part of control applications in process control require discrete signaling. For example, before a start command can be issued to a machine, several start conditions have to be fulfilled. A machine might have start, stop, and safety-interlocks. Therefore, interlocking and control signaling are sensitive to delays.

The reason for the separation of closed loop control and interlocking and control is that when closed control is used, the controller can compensate for

TABLE 4.2

Important properties of industrial communication

Property	Description
Safety	Failures in communication should not affect the health of persons, equipment, or environment
Security	The transmitted data should be secure against malicious manipulation to avoid loss of production or safety of person, equipment, and environment
High availability	Avoid a single-point-of-failure that can compromise the plant production. The production should be able to continue without major degradation in the case of a single failure
Deterministic	Data must always be delivered within given time constraints, for example to be able to stop a conveyor belt before any material transported on the conveyor will cause long time of production loss or affect safety
Low latency and jitter	Data must be transmitted with low latency and jitter for motion control in order to keep control performance and precision
High throughput	Plant operators will request production statistics from the control system that will be presented as a trend curve displaying different key performance indexes or to monitor the current production and react before any disturbances in production will occur
Efficient deployment and maintenance	The lifetime of automation equipment is expected to be longer than 10 years. In case of malfunctioning equipment the on-site maintenance staff should be able to replace faulty equipment and restore plant production without help from automation system experts. Another reason is that plants can be at inconvenient geographic locations and it would take a long time for an expert to reach the site, for example an offshore oil-rig or deep down in a distant mine
Flexible topology	The asset owners constantly seek new methods to improve the production speed and quality, and rearrangement or addition of automation equipment is not uncommon due to new discoveries in the production methodology.

delays and retransmissions without major degradation of control performance. On the other hand a process interlock, that should for instance stop a machine, is less tolerant for delays and retransmissions as the consequences of delays are most likely to introduce spurious problems in the production or

TABLE 4.3

Typical requirements for Industrial Wireless Sensor and Actuator Networks in the Process Automation Domain

Sensor Network Applications	Delay	Update Frequency
Monitoring and supervision		
Vibration sensor	s	sec - days
Pressure sensor	ms	$1\ s$
Temperature sensor	s	$5\ s$
Gas detection sensor	ms	$1\ s$
Closed loop control		
Control valve	ms	$10 - 500\ ms$
Pressure sensor	ms	$10 - 500\ ms$
Temperature sensor	ms	$500\ ms$
Flow sensor	ms	$10 - 500\ ms$
Torque sensor	ms	$10 - 500\ ms$
Variable speed drive	ms	$10 - 500\ ms$
Interlocking and Control		
Proximity sensor	ms	$10 - 250\ ms$
Motor	ms	$10 - 250\ ms$
Valve	ms	$10 - 250\ ms$
Protection relays	ms	$10 - 250\ ms$

even dangerous situations. Generally the update frequency, or period time, equals the deadline. This means that real-time communication in a meshed network should have a time-to-live (TTL) that equals the deadline in order to avoid network congestion. Furthermore, in order to be resilient to spurious transmission errors or deadline misses the control system typically accepts two out of three lost packets without any further action. If three consecutive packets are lost, the fail-safe state is activated in order to avoid uncontrolled processes and to signal that an error has occurred.

4.2.3 Design Objectives

As always, the key for success is in the details but more importantly the details needs to be hidden from the end-users as much as possible. One of the main design objectives is to design the system in a way such that the end-users have to interact with the communication system as little as possible. The motivation for this is that when a device or communication link fails, the production will be affected directly or indirectly, and in the worst case there will be a complete production stop. Thus the time to repair is a crucial parameter to consider. It should be possible to restore production with maintenance personnel who are typically not experts in WSN, but rather have a vast and broad experience of the multitude of different systems and components in order to restore production as fast as possible. There is not time to fly in an international recognized expert to restore production due to the cost of lengthy production losses in the case of a failure in a single device. Current systems are designed in such a way that the maintenance personnel identify and replace the faulty equipment and the system will download the previously used configuration in the new device in order to resume production without any further configuration or extensive testing. Furthermore, the real-time characteristics need to be preserved in case of maintenance and further motivates the central storage of configuration data that enables consistency in the system and eliminates the human factor. The major design objectives are to design for safety, security, test and maintenance, determinism, and reliability while keeping the total design as simple as possible. The main challenge is that some of the requirements are in contradiction with each other and many trade-offs are necessary in order to for example meet the real-time constraints.

The devices and systems needs to be designed to withstand harsh environmental environments such as high temperatures, designed for usage in areas with risk of explosion, small form factors, and that devices might be stacked in closed cabinets without forced ventilation. In other words, equipment needs to be reliable, rugged with a plug and play feeling, and cost sensitive due to relative small volumes and product life-cycles of up to 20 years. This typically implies that less is more, and the systems and devices are tailored for their specific needs without any additional nice to have features that you might find in the consumer market.

4.3   Research Challenges

In automation and control, sensors are just one part as the physical process measured is typically communicated to a control system that will react and actuate depending on the physical measurements. For example a robot needs to move based on the positioning information, or the flow of materials in pipes needs to be controlled in order to guarantee the expected mixture of chemicals in the process. In order to control a process the actuators are an important part of the total system. Some examples of actuators are control valves, discrete valves and electrical contactors, or variable speed drives. For economical reasons it is preferable to have the actuators as well integrated part of the WSNs, otherwise a parallel wired network needs to be deployed for the actuators.

Since the majority of available wireless sensor networks origins from the consumer market they are tailored for monitoring purposes. As mentioned earlier, there are several applications targeting closed loop control and interlocking which means that actuators are involved into the system. This implies that we need to be able to guarantee some kind of determinism and also that bandwidth is available when needed to close a valve etc. [4]. Unfortunately the real-time guarantees and quality of service aspects have been neglected in existing solutions which make it a real challenge to completely utilize the benefits that wireless brings into the process automation domain.

In the remainder of this section we will briefly outline some research areas that need attention in order to have real-time wireless capabilities that supports the requirements in process automation domain given in Section 4.2.

4.3.1 Safety

Safety of humans, environment, and property should always be the number one priority. In process automation some functions are safety-critical by definition, but most of them are not safety-critical. This does not mean that only the safety-critical functions should be designed, developed, and certified with care [1], as most process automation equipment depend on that the rest of the system operates within the boundaries of the specifications. Even if the functionality is not safety-critical by definition, there can be substantial production losses or damages to the property if the automation equipment is not designed to reduce the risk of uncontrolled or dangerous situations.

The prevention of uncontrolled processes is extremely important. As an example, if a set point from the control system to a control valve cannot be transmitted, the valve should fall back into a safe state (normally closed) after a time out. The time out depends on how long time the process can tolerate a malfunction of the actuator before a possibly dangerous situation occurs, ranging from milliseconds to seconds. In addition to this, the control system should detect this communication loss and indicate the failure. In this way, the rest of the equipment that depends on the correct operation of the control valve is signaled to avoid dangerous situations due to error propagation. Non-safety-critical automation equipment is designed such that if a problem can occur, it should be detected and force the process into a safe state. One of the worst scenarios that can occur is that the operator's view in the control room is not consistent with the actual state of the equipment on the factory floor. This implies that field devices cannot have extremely small duty cycles to preserve battery since both the operators and control system will neither get any life sign from the field devices, nor meet the required update frequency.

Some work with safety-critical communication using WirelessHART and PROFINET IO has been presented in [5] and the main result is that the WSNs need to have deterministic and synchronized communication in both the uplink and the downlink in order to avoid spurious fail-safe timeouts.

4.3.2 Security

Most information that is transmitted to and from field devices is usually normalized valued of the measured entity, i.e., 0-100% of the range of the measuring instrument. In some cases the actual measurement is transmitted along with the SI unit. The control system is collecting information of the state of the process, and the process is controlled based on the collected measurements and the control strategy by transmitting set points for actuators based on the actual state of the process. However, the main point is that it is not confidential information as such that is transmitted. As the confidential information resides inside the control system, i.e., recipes or control strategies, and those are not transmitted on the field networks. However, from a security perspective authentication, integrity, availability, and non-repudiation are important security objectives. With respect to this, the current situation is that confidentiality, authentication, and integrity are provided in WSNs. Here optimizations can be made with respect to security to improve energy consumption, latency, and security overhead. The most problematic situation besides deliberate manipulation of control data is denial-of-service attacks, which cannot be prevented by cryptography at all. In case of a denial-of-service attack, the automation system will transition into a safe state, if designed correctly.

Secondly, how to integrate the security mechanisms in the overall automation system in an efficient way with respect to key management and replacement of faulty field devices is an important issue. Users of control systems today are used to exchange faulty devices during operation without any additional configuration, i.e., "hot swap," to minimize the downtime.

4.3.3   Availability

In industrial large-scale production, availability is of significant importance. Even short and transient communication errors can cause significant production outages. This is mainly due to that the process has to be stopped in a controlled manner in case of a single communication problem, and it can take up to several hours to achieve full production rate again, with production losses in the range of hundreds of thousands dollars per hour. Self-healing mesh networks are appealing to use in industrial automation for several reasons, i.e., for redundancy, availability, etc. [19]. However, in literature it is commonly assumed that routing protocols in industrial settings should be able to deal with mesh networks containing thousands or tens of thousands nodes. Furthermore, it is assumed that all of the devices are battery operated, thus energy aware routing protocols are of significant importance in order to distribute the routing load in a fair manner amongst the nodes [23]. The first assumption is not correct, even though there are tens of thousands of instruments that need to communicate, they do not belong to the same network because of the availability concern. The nodes are distributed over a set of process controllers, divided in several process sections, in order to avoid a complete production stop in case of for example a "babbling idiot" failure in one node. Furthermore, energy optimized routing protocols might have a severe and negative impact on the latency and real-time performance of end-to-end communication using mesh networks in the case of link problems caused by fading. In industrial automation the mesh networks would rather benefit of rapid adoptions in routing, in the case of for example fading, while meeting the real-time constraints. Flooding might be one feasible alternative for usage in industrial automation [20].

4.3.4 Real-time Performance

Due to the nature of automation the data transmitted in the field networks is only valid for a short time. If the data is delivered too late it is of limited use, as in most real-time systems. Therefore new data should be propagated through the network instead of guaranteeing delivery of all transmissions. This is an important area of research, especially within the area of WSN where it can be mesh networks, multi-hop situations, and synchronized communication in both directions between the nodes. In addition, the automation systems will download configuration data to the field devices both at startup and during operation that should be end-to-end acknowledged and retransmitted to be able to guarantee delivery in case of communication losses.

Reliability and latency are the primary requirements in WSNs to guarantee deterministic real-time communication in industrial automation. A common method in wireless communication systems to improve the link reliability is to employ an automatic repeat request (ARQ) procedure (a retransmission procedure) but this gives a limited improvement in link reliability. Furthermore, an inherent drawback of ARQ is the increased latency of packet delivery, due to a number of packet retransmissions. This is in direct conflict with time-critical requirements of industrial applications. With respect to retransmissions in mesh networks and multi-hop situations, it has to be guaranteed that data is delivered in the correct order. It can easily occur that data is received in the wrong order if packets are not discarded in the mesh before the next periodic data is transmitted and delivered. The consequences can be substantial, as a single bit is used to start and stop electric motors ranging from kilowatts to megawatts. Another approach to ensure reliability is to apply forward error correction (FEC) strategies, reducing the bit error rate and consequently the number of retransmissions. In available IEEE 802.15.4 industrial wireless sensor network chips, FEC strategies have been omitted mainly due to power consumption of decoding operation. Nevertheless, by using FEC the overall energy consumption will become less since we will spend less energy on retransmission as shown in [22] and re-scheduling our network. The major trade off between the additional processing power and the associated coding gain need to be optimized in order to have a power, energy-efficient, and low-complexity FEC schemes in industrial wireless sensor networks [25]. In addition to that, we need to consider memory constraints in the embedded system and not jeopardizing the WSN requirements given in Table 4.3.

4.3.5   System Integration and Deployment

In order to make a smooth and efficient integration of industrial wireless sensor networks into existing automation infrastructures, the most critical point to consider is the gateways. Today there exist a small number of gateway vendors and they are proposing proprietary solutions, which prevents an open and efficient integration to existing infrastructure. Since all commercially available solutions today provide web-based configuration, everything is done manually, which slows down the engineering, commissioning, and maintenance of the system.

Today, most wired fieldbuses have services to download configuration and to simplify the integration work in general. What is currently missing is a standardized approach for WSN integration to the different fieldbuses in the IEC standards [2]. Some standardization efforts are currently ongoing [21, 19] as well as proposals for integration [3]. Therefore it is essential that the services provided by the industrial wireless sensor networks are possible to integrate efficiently in the automation system to have a seamless transition from wired to wireless communication, as well as simple deployment, commissioning, and maintenance.

Deployment of wireless sensor networks in industrial automation differs a lot compared to the consumer market. First of all, in process automation the sensors are placed on a fixed location to measure some entity and there is no option that this sensor could be placed ad hoc as in military and consumer market applications. This leaves only room for the gateway to be placed in an optimal place on the plant to cover as large areas as possible with good signal-to-noise ratio as possible. With this in mind, it becomes very important that the industrial wireless network can guarantee robust communication since some nodes are exposed to severe interference. It is also important that the network is scalable and can adapt to changing network size. Without this kind of support the network performance will degrade significantly as the network size increases. Although scalability is needed it is important to stress that network size for industrial automation will differ remarkably compared to consumer market. A large network in industrial automation will most likely contain 20-50 nodes while in consumer market it is about 200-300 nodes connected to one gateway. Network sizes of several hundred nodes will have severe problems with fulfilling the requirements given in Table 4.3 since the routing algorithms are not aimed for wireless real-time communication.

A way of building reliable networks is to consider mesh technologies such that there is always one available path from the sink to the gateway. However, building mesh networks with several hops and still maintaining requirements is a daunting challenge. A more appealing approach in industrial automation is to build star networks, but with available industrial wireless sensor networks we would jeopardize the communication reliability which is not an option.

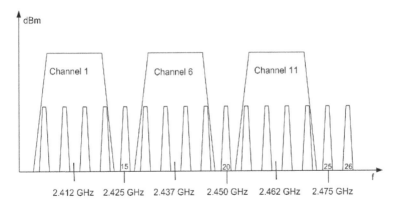

FIGURE 4.5
Distribution of IEEE 802.11 and IEEE 802.15.4 channels in the 2.4 GHz band.

4.3.6 Coexistence and Interference Avoidance

It is well known that performance of different wireless technologies can vary quite substantially within different environments and with several wireless systems deployed it is clear that the performance will degrade even more. A clear trend within industrial automation is the rapid deployment of many different wireless communications technologies such as WLAN, Bluetooth, and industrial wireless sensor networks that operate in the unlicensed 2.4 GHz ISM band. Thus, it is very important that a solution can coexist in a radio environment with a large amount of interferences from the harsh industrial environment as well as limit its own disturbance. Some sources of disturbances occurring in the harsh environment are multipath propagation [6], noise generated from the equipments or heavy machinery, or strong vibrations as discussed in the introduction of this chapter. Figure 4.5 show the distribution of IEEE 802.11 and IEEE 802.15.4 channels in the 2.4 GHz band and it is typically common in industrial deployments to configure WLAN access points to use non-overlapping channels 1, 6, and 11. It is obvious then to see that IEEE 802.15.4 can run relative interference-free operation in channels 15, 20, 25, and 26. In literature the effect of WLAN as interferer on IEEE 802.15.4 and vice versa have been studied in [24, 9, 7]. In general, the coexistence of industrial wireless sensor networks based on IEEE 802.15.4 with both WLAN and Bluetooth networks in the same area is possible with an acceptable performance, when nodes are not in a close proximity of each other and/or channels are adequately selected to prevent overlapping. In order to utilize the spectrum in a good way it is important to use simple means of coexistence management at the industrial plant. This is mainly comprised of following steps:

1. Registration of all applications using wireless communications in the plant

2. Assessment of coexistence situation, and if necessary,

3. Minimize the radio influences.

In order to handle coexistence and interference in an efficient way, many industrial wireless sensor network standards employ various spectrum-management techniques. A clear channel assessment (CCA) is performed before data transmission to ensure that the RF channel is free to use. For instance in WirelessHART, the CCA reports a busy medium if a signal compliant with IEEE 802.15.4 PHY modulation and spreading characteristics is detected. Other countermeasure for combating interference is to use different diversity (time, frequency, and space) schemes but also some more fancy innovative techniques such as interference cancellation and effective radio resource management.

4.3.7   Energy Consumption

Research on energy consumption and energy harvesting / scavenging is a hot topic today. However, due to refresh rates required in process automation it is difficult to save much energy by minimizing the duty cycle of a wireless field device. Secondly, as previously mentioned actuators are a mandatory part of process automation and they are sometimes pneumatic. Therefore, true wireless field devices are not foreseen in the near future in process automation. However, temporal installations of sensors used for validation of possible process optimization will greatly benefit since they can be true wireless, operated on battery power. Future research in energy scavenging will improve the situation and more kind of field devices can become true wireless. However, many devices are unlikely to become true wireless using today's technology, both sensors (i.e., acoustic or laser) and actuators (i.e., large valves or pneumatic control valves). Nevertheless, the advantages are still significant even if power supplies have to be wired to the sensors and actuators as the fieldbuses are not necessary to route between all sensors, actuators, and controllers in the specific process sections. Power is almost always found close to the field devices today and is likely to remain there, even in green field installations.

Assume for a while that all field devices could be battery operated today, a new significant problem would occur, namely scheduled battery maintenance of tens of thousands of nodes in a typical process industry. This would cause serious problems for the industries that are expected to operate flawlessly 24/7 with a few short schedule maintenance periods per year. In addition, the end-users need to keep batteries of various size and capacity on stock in order to quickly replace exhausted batteries. That means in the end that the life-cycle cost for a true wireless device would be more expensive compared to a device that only communicates wirelessly.

4.4 Conclusions

Industrial wireless sensor and actuator networks will become an important component in future industrial automation because it offers simplicity, flexibility, and less maintenance. Today most of available radio chips origins from consumer market with much more relaxed requirements than industrial automation requires. Furthermore, the available industrial wireless sensor network standards in process automation are tailored for condition monitoring applications with slow update frequency of measurement values. For future applications such as wireless control, it will become a major challenge to design industrial wireless sensor and actuator networks that are robust against interference, handle the tough real-time requirements that are given in process automation domain, and still be very cost effective.

References

[1] *IEC 61508 ed2,0. Functional safety of electrical/electronic/programmable electronic safety-related systems.* International Electrotechnical Commission, 2010.

[2] *IEC 61784-1 ed3,0. Industrial Communication Networks - Profiles - Part 1: Fieldbus profiles.* International Electrotechnical Commission, 2010.

[3] J. Åkerberg, M. Gidlund, T. Lennvall, J. Neander, and M. Björkman. Integration of wirelesshart networks in distributed control systems using profinet io. In *8th IEEE International Conference on Industrial Informatics (INDIN'10)*, pages 154–159, July 2010.

[4] J. Åkerberg, M. Gidlund, T. Lennvall, J. Neander, and M. Björkman. Efficient integration of secure and safety critical industrial wireless sensor networks. *EURASIP Journal on Wireless Communications and Networking*, 2011(1):100, 2011.

[5] J. Åkerberg, F. Reichenbach, and M. Björkman. Enabling safety-critical wireless communication using wirelesshart and profisafe. In *IEEE Conference on Emerging Technologies and Factory Automation (ETFA'10)*, pages 1–8, Sept. 2010.

[6] J. Åkerberg, F. Reichenbach, M. Gidlund, and M. Björkman. Measurements on an industrial wireless hart network supporting profisafe: A case study. In *Emerging Technologies Factory Automation (ETFA), 2011 IEEE 16th Conference on*, pages 1–8, Sept. 2011.

[7] L. Angrisani, M. Bertocco, D. Fortin, and A. Sona. Experimental study of coexistence issues between ieee 802.11b and ieee 802.15.4 wireless networks. *IEEE Trans. on Instrum. and Measurm.*, 53(8):1514–1523, 2008.

[8] P. Ängskog, C. Karlsson, J. Ferrer Coll, J. Chilo, and P. Stenumgaard. Sources of disturbances on wireless communication in industrial factory environments. In *Proc. Asia-Pacific International Symposium on Electromagnetic Compatibility*, 2010.

[9] M. Bertocco, G. Gamba, A. Sona, and S. Vitturi. Experimental characterization of wireless sensor networks for industrial applications. *IEEE Trans. on Instrum. and Measurm.*, 57(8):1537–1546, 2008.

[10] J. Ferrer Coll, J. Chilo, and S. Ben Slimane. Radio-frequency electromagnetic characterization in factory infrastructures. To appear in *IEEE Trans. On Electromagnetic Compatibility*.

[11] IEEE 802.15.4 Standard: Wireless Medium Access Control and Physical Layer Specifications for Low-Rate Wireless Personal Area Networks. http://www.ieee802.org/15/pub/TG4.html.

[12] F. De Pellegrini, D. Miorandi, S. Vitturi, and A. Zanella. On the use of wireless networks at low level of factory automation systems. *IEEE Transactions on Industrial Informatics*, 2(2):129–143, May 2006.

[13] D. Dzung, C. Apneseth, J. Endresen, and J.-E. Frey. Design and implementation of a real-time wireless sensor/actuator communication system. In *Emerging Technologies and Factory Automation, 2005. ETFA 2005. 10th IEEE Conference on*, volume 2, 433–442, Sept. 2005.

[14] V.C. Gungor and G.P. Hancke. Industrial wireless sensor networks: Challenges, design principles, and technical approaches. *Industrial Electronics, IEEE Transactions on*, 56(10):4258–4265, Oct. 2009.

[15] V.C. Gungor, B. Lu, and G.P. Hancke. Opportunities and challenges of wireless sensor networks in smart grid. *Industrial Electronics, IEEE Transactions on*, 57(10):3557–3564, Oct. 2010.

[16] V.C. Gungor, D. Sahin, T. Kocak, S. Ergut, C. Buccella, C. Cecati, and G.P. Hancke. Smart grid technologies: Communication technologies and standards. *Industrial Informatics, IEEE Transactions on*, 7(4):529–539, Nov. 2011.

[17] T. Lennvall, S. Svensson, and F. Hekland. A comparison of wirelesshart and zigbee for industrial applications. In *IEEE International Workshop on Factory Communication Systems*, pages 85–88, May 2008.

[18] ABI Wireless Sensor Networking Markets. http://www.abiresearch.com/research/1003936.

[19] TC2/WG12 Wireless Sensor/Actor Networks. `http://www.profibus.com/index.php?id=1314&tcwg_tc_uid=2&tcwg_wg_uid=14`.

[20] M. Soyturk and D.T. Altilar. Reliable real-time data acquisition for rapidly deployable mission-critical wireless sensor networks. In *IEEE INFOCOM Workshops*, pages 1–6, 2008.

[21] Wireless Cooperation Team. `http://www.hartcomm.org/hcf/news/pr2008/press_conf_interkama2008.pdf`.

[22] M.C. Vuran and I.F. Akyildiz. Error control in wireless sensor networks: A cross-layer analysis. *IEEE/ACM Transactions on Networking*, 17(4):1186–1199, Aug. 2009.

[23] A. Willig. Recent and emerging topics in wireless industrial communications: A selection. *IEEE Transactions on Industrial Informatics*, 4(2):102–124, May 2008.

[24] D. Yang, Y. Xu, and M. Gidlund. Wireless coexistence between ieee 802.11 - and ieee 802.15.4-based networks: A survey. *International Journal of Distributed Sensor Networks*, 2011.

[25] K. Yu, M. Gidlund, J. Åkerberg, and M. Björkman. Reliable and low latency transmission in industrial wireless sensor networks. In *First International Workshop on Wireless Networked Control Systems (WNCS)*, September 2011.

5

Resource Management and Scheduling in WSNs Powered by Ambient Energy Harvesting

Neyre Tekbiyik and Elif Uysal-Biyikoglu

Department of Electrical and Electronics Engineering, METU, Ankara, 06800, Turkey

CONTENTS

5.1 Introduction

The collaborative nature of industrial wireless sensor networks (WSNs) brings several advantages over traditional wired systems, including self-organization, rapid deployment, flexibility, and intelligent-processing capability [9]. Recently, WSNs have found their way into a wide variety of applications and systems with varying requirements and characteristics [1], [14]: ocean water

monitoring and bathymetry, avalanche rescue, object tracking, remote monitoring of oil and gas reservoirs, and, preventive and predictive maintenance
(PdM)[1], which is considered to be an important and relevant example of
the class of industrial sensor networking applications that provide measurable
value in real deployments.

Resource management is just as critical in the industrial environment as
in other deployment scenarios. This may seem counterintuitive since most
industrial plants have ample power supplies and distribution systems. However, as also discussed by Krishnamurthy et al. in [14], operating and safety
regulations typically call for each piece of equipment to have a dedicated
power circuit, requiring separate power connections for sensor nodes. Hence,
to reduce installation costs, the WSN must either be battery powered (and
employ aggressive resource management) or make use of energy harvesting.
Since industrial WSNs are expected to be deployed in harsh or inaccessible
environments for long periods of time, recently, employing energy harvesting
(via ambient energy sources such as solar [2], vibrational [13], [18], wind [30]
and thermal energy [29]) to replace/supplement batteries that power WSNs,
has earned much interest. Detailed information about different types of energy harvesting approaches can be found in [29], which is a useful source that
investigates the current energy harvesting WSN applications in several areas,
and, provides examples of leading development enterprises.

The most popular source of the ambient energy is the sun. Solar energy
is becoming widely used, due to its high power density (about 15 mW/cm^3)
compared to other sources of ambient energy [20]. Consequently, numerous
researchers have designed energy harvesting circuits to efficiently convert and
store solar energy [12], [23], [27], and, most of the studies mentioned in this
chapter focused on solar energy harvesting. As claimed by Yu and Yue [31],
solar energy harvesting is a comparatively full-fledged technology for WSNs
used for outdoor applications. However, for indoor applications, it is not suitable since the efficiency of photovoltaic cell is very low under low indoor light
luminous intensity. For indoor applications, one may prefer the micro-scale
indoor light energy harvesting system developed in [31], or, the energy optimized sensor node [24] designed to harvest the energy from indoor light, for
building climate control application.

Moser et al. argue in [19] that, depending solely on energy harvesting gives
rise to new challenges and will trigger the revision of conventional resource
management. If, e.g., the size of a solar cell limits the available power/energy of
an electronic device, decisions such as when to provide how much power, rate,
service, etc., have to be made in order to satisfy the needs of the user as well
as possible. Successful demonstration of perpetual operation with indoor EH
WSNs have already been made, that are drawing attention to the importance
of routing and scheduling mechanisms that are aware of the energy harvest

---

[1]Predictive maintenance is a general term applied to a family of technologies used to
monitor and assess the health status of a piece of equipment (e.g., a motor, chiller, or cooler)
that are in service [14].

process. For instance, in an indoor office environment, Hande et al. [10] used monocrystalline solar cells to scavenge energy from 34 W flourescent lightbulbs in order to supply (via supercapacitors) the routers of a WSN. The routers, operating in pairs, achieved virtually perpetual operation by resource-aware operation. The authors in [10] stressed that in scenarios with mobility, resource management mechanisms for other forms of energy scavenging (such as vibration-based or thermal) should be investigated in future work.

Resource (energy) management in WSNs equipped with energy harvesting capabilities is substantially and qualitatively different from resource management in traditional (battery-powered) WSNs. As stated by Mao et al. in [17], conservative energy expenditure in energy harvesting networks, may lead to (i) missed recharging opportunities because the battery buffer is full and, (ii) long delays because the energy is not being fully used to transmit at high enough data rates. On the other hand, aggressive usage of energy may result in reduced coverage or connectivity for certain time periods, not to mention complete battery discharges that could make the nodes temporarily incapable of transferring time-sensitive data. In industrial applications, this may lead to loss of production and may sometimes create hazardous situations [22]. Thus, new resource allocation and scheduling schemes need to be designed to balance these contradictory goals, in order to maximize the network performance.

5.2 Algorithms

In this section, we focus on resource allocation and scheduling schemes that could be used for sustainable operation of industrial WSNs.

5.2.1 SSEA and ASEA Schemes

In [21], the authors present both basic and advanced expectation models for solar energy harvesting. Based on these expectation models, they suggest energy allocation algorithms, SSEA (Simple Solar Energy Allocation), ASEA (Accurate Solar Energy Allocation), to achieve optimal use of harvested energy. Both algorithms operate based on time-slots. Assuming that the cycle of energy harvests has a period of T, and that T is divided into sub-periods (called slots) of equal length, the base energy harvest expectation increases during morning slots, decreases during afternoon slots, and stays nearly zero during the night. The basis of the expected harvest for each slot reflects relatively long-term tendencies such as seasonal or monthly trends. Nevertheless, short-term conditions, i.e., temporary environmental conditions are also important, especially in locations or seasons with frequent weather changes. Therefore, an "advanced" energy expectation that factors in faster weather dynamics is also computed.

Both algorithms focus on allocating energy fairly over time, leading to a more stable application performance, while at the same time maximizing utilization of the energy harvest. The SSEA algorithm is designed for a resource-constrained sensor. It uses a basic expectation model. Thus, it is simple, and has low overhead, but it sacrifices some degree of effectiveness in energy allocation. SSEA operates as follows: (1) Determine the amount of residual energy stored in the battery, and the expected amount of energy to be harvested during each slot. (2) Using this information, find an appropriate energy budget for every slot in a harvesting cycle. (3) Then, go to sleep until the start of the next harvesting cycle. ASEA algorithm, on the other hand, is based on an advanced expectation model, and is suitable for a node which needs a more precise energy allocation and has adequate resources to support additional computation, as it comes with a higher overhead than the SSEA scheme. Based on the expectation of harvested energy, ASEA solves a linear programming problem at every start of each slot [21], and calculates the energy to be allocated for the next slot.

Both algorithms are reported to dramatically reduce the number of occasions on which a node stays in sleep mode during an entire slot [21]. When compared to the ideal scheme (which assumes that the amount of energy that will be harvested during the harvesting period is known a priori), ASEA is shown to achieve results closest to those of the ideal scheme in all respects, and SSEA comes next.

5.2.2   A Practical Flow Control Scheme

Noh and Kang [20] develop a practical flow control (called PFC in this chapter) algorithm that aims to maximize the amount of data collected by both flow-centric[2] and storage centric[3] WSNs. The algorithm is distributed and operates in a time slotted system. It cooperates with an energy allocation algorithm (called Simple Solar Energy Allocation (SSEA) in [21]) so as to use the harvested energy optimally. Under the constraints of the energy allocated to each time slot, at the start of every time-slot, the algorithm determines an appropriate flow-rate of the outgoing links, while aiming to maximize its utilization of the energy budget for this slot. The algorithm tries to maintain an ABP (Adaptive Back Pressure) super-flow as long as possible, and thus, it can be seen as a modified version of the adaptive flow control algorithm proposed in [8].

Mainly, the algorithm operates as follows: In a time-slotted system where the unit block for energy allocation is slot, it is assumed that, each slot is divided into several sub-slots; the unit blocks used for determining the transfer rate. Under this setting, each node determines the transfer rate of each of its

---

[2]When there is a sink node in the network, WSN works as a flow-centric network and aims to maximize throughput to the sink node.
[3]When the network operates without a sink, it operates as a storage-centric network with the aim of minimizing the amount of data loss due to storage constraints.

outgoing links at the beginning of each sub-slot, and maintains this rate during the duration of a sub-slot. The transfer rate is computed by the transfer rate determination algorithm proposed in [20] which is designed to maintain the ABP superflow during a sub-slot. After the node determines the transfer rate for all outgoing links during a sub-slot, it checks whether it has enough energy to operate during that sub-slot. If it finds that there is not enough energy to sustain the node, the algorithm is terminated while making the node go into sleep mode and setting wake-up time to the start of the next slot. The flow-control scheme is shown to produce [20] the lowest amount of data loss (for the case of storage-centric WSN) as well as the highest throughput, proving that it can maximize the amount of collected data by the sink while balancing the data efficiently when the network operates in the flow-centric mode.

5.2.3 Fixed Power (FP), Minimum-Interference (MI), and Multi-Sink (MS) Power Allocation Schemes

FP, MI, and MS are simple, location-based power allocation algorithms [28] developed for structural monitoring applications with multiple sinks. Note that, all these schemes assume that energy harvesting nodes can only communicate with the sinks, not with each other. Moreover, as they do not consider energy harvesting statistics, all algorithms operate such that a data packet is sent to the sink(s) whenever sufficient energy is accumulated. FP is the simplest power allocation scheme since it assigns the same (fixed) transmit power (P) to all nodes. Tan et al. report in [28] that, for FP, a large P permits direct communication with more sink(s) (causing multi-sink redundancy), but, not only it results in a longer harvesting period, but also introduces the near-far effect[4]. Assigning low P, on the other hand, shortens the harvesting period and reduces the level of interference, at the expense of reducing the scope for exploiting multi-sink redundancy. MS and MI schemes are designed so that powers are assigned according to nodes' proximity from the sinks, i.e., P for each node depends on its communication range and distance from the sink. MS is a multi-sink scheme, where each node is assigned a power level just sufficient to communicate with its nearest j sinks. MI is the special case of MS, where j = 1, i.e., each node is able to communicate only with its nearest sink. This scheme minimizes the interference and near-far effect, while ensuring connectedness. According to the simulation results obtained for various node densities [28], FP poses a trade-off between throughput and fairness: throughput is maximized at lower powers at the expense of fairness and vice versa. The MS scheme does not perform as expected (close to MI) as the interference outweighs the potential benefits of multi-sink redundancy. Finally, by assigning the minimum P required for each node to communicate with its nearest sink, the MI scheme enables more nodes (from different locations) to

---

[4]Near-far problem: At higher node densities, contention becomes more severe resulting in nodes closer to the sink becoming more favored.

have successful simultaneous transmissions, causing its superior performance
in terms of throughput, data reliability, and fairness.

5.2.4   QuickFix/SnapIt Algorithms

QuickFix and SnapIt [16] were proposed as two different algorithms that work
in tandem, to maximize the network utility, i.e., the sum of the utility functions
of the nodes, with the aim of achieving proportional fairness in a slotted-time
system. The system is designed in such a way that the time during a day is
broken into multiple time intervals called epochs, where each epoch consists of
τ slots. QuickFix is an efficient dual decomposition and subgradient method
based algorithm that operates within each epoch, to reveal the feasible region
and the optimum solution differing in each epoch. It exploits the special structure of a DAG (Directed Acyclic Graph) to form an efficient control message
exchange scheme, which is motivated by the general solution structure of a
dynamic program. QuickFix offers a distributed solution that does not require
any knowledge of the future recharging rates. Moreover, it can efficiently track
instantaneous optimal sampling rates (for every slot) and routes in the presence of time-varying recharging rates. However, QuickFix's solution to the
proposed utility maximization problem depends on the average (long term)
energy replenishment rate of a node and not the state of the battery. Hence, if
fluctuations in recharging happen at a faster time-scale than the convergence
time of QuickFix, undesired battery outage and overflow scenarios may arise,
causing missed samples and lost energy harvesting opportunities respectively.
Therefore, Liu et al. introduce a localized scheme called SnapIt that uses the
current battery level to adapt the rate computed by QuickFix with the goal
of maintaining the battery at a target level, i.e., chosen as the half of the local
battery state in [16]. SnapIt chooses the rate, independently at each node i
based on the current state of the battery as follows: the rate found by Quick-
Fix is reduced by δ_i (different for each node) if the battery is less than half
full, and, is increased by the same amount when it is more than half full. We
refer the interested reader to [16], for the effect of δ on the performance of
QuickFix/SnapIt. In [16], QuickFix/SnapIt is compared to a modified version
of IFRC (Interference-aware Fair Rate Control) [25], a backpressure-based
protocol, which aims to achieve max-min fairness in WSNs. The results show
that the two algorithms, working in tandem, can increase the total data rate
at the sink by 42% on average when compared to IFRC, while significantly
improving the network utility.

5.2.5   DRABP and NRABP Schemes

Gatzianas et al. [5] model energy harvesting as a time-varying process and
consider jointly managing the data and battery buffers (queues). The authors
consider infinite data buffer and finite battery buffer sizes. They assume that
the energy harvesting process is memoryless (it is claimed that, for a more

general process, a slot analysis can be applied whose complexity will grow with the network size). Two policies (DRABP and NRABP) with decoupled admission control and power allocation are proposed with the goal of maximizing the total system utility (the long-term rate achieved per link) while satisfying energy and power constraints. They are carefully crafted modifications of the ABP-based policy of [7], which is known to achieve the optimal utility in the infinite battery scenario (non-rechargeable batteries). DRABP (Downlink Rechargeable Adaptive Backpressure Policy) is developed for downlink scenarios, whereas NRABP (Network Rechargeable Adaptive Backpressure Policy) is developed for multi-hop networks (ad hoc networks, sensor networks, etc.). DRABP is proven to be asymptotically optimal [5] when all nodes have sufficiently large battery capacities. Both schemes operate on virtual queues, which are constructed in such a way that any policy that stabilizes them also satisfies the appropriate long-term constraints. Since our main concern is WSNs, we focus on NRABP and refer the interested reader to [5] for details of DRABP.

NRABP operates as follows: (1) At the beginning of slot t, observe the virtual data queues and select appropriate packets for admission into the network layer, for every link, as the solution to the related problem described in [5]. (2) Observe the channel state, and, choose a power vector $P(t) = (P_1(t), ..., P_L(t))$ for each node (where $P_l(t)$ is the selected transmission power in link l during slot t), based on the constraints on the power-related virtual queues of each link, and the result of (1). (3) Update the states of all queues according to the number of bits that have arrived and departed in this stage. It is shown in [5] that, under NRABP, all queues are bounded and thus, NRABP stabilizes any multihop network. Performance bounds on NRABP can be found in [6].

5.2.6 Duty Cycling and Power Management Algorithm

Reddy et al. [26] develop a suboptimal duty cycling and power management algorithm (which we call DC-PM) for a single hop WSN, where K EHS (Energy Harvesting Sensor) nodes communicate with a powered destination over a wireless fading channel. The algorithm manages the power harvested at the individual nodes and duty cycle across them to avoid collisions in order to maximize the average sum data rate, subject to energy causality constraint, ECC (called energy neutrality constraint - ENC in [26]), at each node. The algorithm is build on two basic assumptions: (i) Time is slotted, with each constant channel (CC) slot of duration equal to the coherence time T_c of the channel. (ii) The harvested power at each node is assumed to remain constant for a constant power (CP) slot which contains a large number of CC slots. DC-PM consists of an inner stage (IS) of optimal duty cycling over the CC slots within each CP slot and an "outer stage" of power allocation across the constant-power slots while satisfying ECC at each of the nodes. Although suboptimal, the solutions to both stages are very simple in form and thus convenient for implementation.

The outer stage sets the short-term power constraints with the goal of maximizing the long-term expected sum data rate, subject to long-term energy causality at each node. It essentially solves the power management problem for a virtual sensor whose harvested power equals the sum harvested power across the nodes. The resulting power allocation scheme is to assign a clipped version of the sum harvested power across all the nodes, where the clipping thresholds are set to maximize the average sum throughput, subject to a sum power ECC. Hence, the average sum throughput depends only on the sum harvested power and its statistics. IS determines the duty cycles of the nodes that maximizes the average data rate (expected sum throughput) within a CP slot. It requires that the duty cycle allotted to each node be proportional to the power consumed by it in the CP slot, i.e., the duty cycle allocated to each node is the fractional allocated power of that node relative to the total allocated power. The transmission depends on the channel gain threshold at each node, which is noted to be the same [26] at all the nodes within a CP slot. DC-PM is shown to outperform other naive schemes mentioned in [26], such as equal duty cycling with scheduling, and optimal duty cycling without scheduling.

5.2.7   MAX-UTILITY and MAX-UTILITY-D Algorithms

MAX-UTILITY [32] is an epoch-based (harvested energy is modeled as an epoch-varying function), polynomial-time, and centralized rate allocation algorithm, designed to maximize total network utility, i.e., the aggregate utility of all nodes. It is applicable to arbitrary utility functions that are concave and non-decreasing. MAX-UTILITY exploits the concavity of the chosen utility function, and a special property of tree-based networks to allocate rates to nodes as evenly as possible for achieving the main goal of utility maximization, while maintaining the minimum sensing rate required by the application and energy neutral operation for every node. The algorithm is shown to be optimal [32] in terms of assigning rates to individual nodes to maximize overall utility, while ensuring energy-neutral operation. MAX-UTILITY runs in multiple iterations, assigning rates to a subset of nodes in each iteration. The algorithm uses one global variable and three per-node variables that are updated from iteration to iteration. The global variable is a set containing all the nodes in the network that have been assigned rates so far. The per-node variables are; the remaining capacity of node i, the set of unassigned nodes in node i's subtree, the maximum common rate for unassigned nodes in node i's subtree. In each iteration, MAX-UTILITY picks a critical node (the node with the least common rate among many unassigned nodes) of the current tree, and assigns its rate to the unassigned nodes in the critical node's subtree, then, it produces a pruned tree by removing any newly assigned nodes. MAX-UTILITY stops when rates are assigned to all N nodes. The distributed version of MAX-UTILITY is also available, MAX-UTILITY-D [32]. MAX-UTILITY-D is an entirely feasible alternative to MAX-UTILITY, and allows sensor nodes to collaboratively produce optimal rate assignments. It only requires a single coordinator node such as a routing tree root, which can be any node in the network. Each iteration of MAX-UTILITY-D consists of two stages: (1) determining the minimum common rate and the critical node in the tree, by requiring all the nodes to forward their maximum common rate for the unassigned nodes in their subtree, to the root. (2) Requiring the root to disseminate the minimum common rate discovered in (1), across the network, so that all unassigned nodes in the subtree of the critical node can receive and use this rate as their packet rate. MAX-UTILITY is a fast and efficient algorithm that can operate with various utility functions, and has a run time of $O(N^3)$, where N is the number of nodes. When compared to an alternative heuristic called Random Rate Augmentation (RRA), proposed by Zhang et al. [32], MAX-UTILITY is claimed to deliver superior utility improvement while ensuring energy neutral operation for all nodes.

5.2.8 NetOnline Algorithm

NetOnline is a distributed low-complexity algorithm heuristically developed for maximizing the throughput over a finite time horizon, in a sensor network with energy replenishment. The main motivation for this development [4] is the fact that, while the finite-horizon throughput optimization problem can be formulated as a convex optimization problem, its solution suffers from high complexity brought about by strong dependence of current decisions on future performance, *time coupling property*[5].

The NetOnline algorithm is comprised of two stages: (1) finding a throughput maximizing energy allocation through T slots, (2) routing. In Part 1, it is assumed that the energy replenishment (energy harvesting) profile can be estimated (predicted) for that period, ahead of time. Every node performs the following operations: Calculate the lower bound on the energy allocation from the lower-bound of the estimated replenishment profile, via the shortest-path solution (SPS)[6], i.e., SPS is shown to be optimal for a single node case [4], when the replenishment rate profile for the entire finite-horizon period is known in advance. Then, based on these estimations and current amount of recharging (harvesting), determine the energy to be allocated for each slot. In Part (2), the main concern is to determine the amount of data in the outgoing links of each node for the corresponding destination node in time slot t. The routing in each slot is determined by solving a simple linear programming (LP) problem. Since the defined problem is also a convex optimization problem, the authors use duality and the Lagrange multiplier method to get the optimal solution. The NetOnline algorithm is shown to

---

[5]In a time-slotted system, if energy is overused in a previous period, the total throughput attainable over the time horizon will decrease as a result. On the other hand, if energy is underused in a previous period, the total throughput will also decrease, even though there is no wasted energy.

[6]The shortest path is calculated using the linear time algorithm in [15], whose complexity is O(T).

be optimal under homogeneous replenishment profiles with perfect estimation for all nodes. Chen et al. reports in [4] that, in more general settings, the algorithm significantly outperforms a state-of-the-art infinite-horizon based scheme (NRABP proposed by Gatzianas et al. [5]), and it achieves empirical performance close to optimal.

5.2.9   The Joint Rate Control, Power Allocation, and Routing Algorithm

The joint rate control, power allocation and routing algorithm [17] (called RC-PA-R in this chapter) is a resource allocation algorithm developed for multihop networks operating in a time-slotted setting, under node-exclusive interference model. The algorithm jointly controls the data queue and battery (energy) buffer to maximize the long-term average sensing rate of an energy harvesting wireless sensor network under certain QoS constraints for the data and battery queues. The resource allocation part of the algorithm consists of two components: a rate control (RC) component and a power allocation (PA) component. Both components are index policies, i.e., the solutions depend on the instantaneous values of the system variables and thus, they are memoryless. The algorithm can either be implemented in a centralized or distributed manner depending on the algorithm used for the (RC) component. For the centralized version, the classical Maximal Weighted Matching (MWM) algorithm [11] is used whereas the distributed version employs the Maximal Matching (MM) based algorithm as in [3]. RC decides on the amount of data that will be sensed, by comparing all available data with a finite tunable approximation parameter that controls the efficiency of the algorithm. Thus, the rate controller makes sure that the data queue remains within a certain bound, making a positive effect on the battery (energy level) as well, since a certain portion of the data packets are not allowed into the transmitting node. PA solves a simple convex optimization problem in each time slot to determine the powers to be allocated so that no node transfers data of a flow to a relay node that is not the destination of that flow, unless the differential backlog for that flow is greater than a fixed value, which is chosen such that the resulting backlog of the receiving node is not larger than that of the transmitting node after the transmission. Thus, the data flow is pushed from the source node to the destination with a positive back pressure. It is shown through both analysis and simulation [17] that the performance of the proposed algorithm is close to that of the optimal solution. Specifically, as V increases, the average total sensing rates of the MWM and MM based algorithm are reported to keep increasing and get closer to the optimum and a value that is much larger than half optimum, respectively.

5.3   Comparison of the Algorithms

After having described several promising candidate algorithms above for possible application in energy harvesting industrial WSNs, we shall now comparatively address the drawbacks, advantages, and possible application areas of these algorithms. As a quick referral guide, a detailed comparison of all algorithms considered in this chapter is provided in Table 5.1.

TABLE 5.1
Comparison of algorithms considered in this chapter. (N.S. denotes "Not Specified")

Algorithms	Distributed / Centralized / Node level	Prediction	Type of Allocation	Battery Buffer (Finite, Infinite)	Harvesting Method
SSEA, ASEA	Node level	Yes	Energy	Finite	Solar
PFC	Distributed	Yes	Energy & Rate	Finite	Solar
FP, MI, MS	Centralized	No	Power	Infinite	N. S.
QuickFix/SnapIt	Distributed	No	Rate	Finite	Solar
DRABP, NRABP	Centralized	No	Power	Finite	N. S.
DC-PM	Centralized	No	Power	Infinite	N. S.
MAX-UTILITY	Centralized	Assumes	Rate	Finite	Solar
MAX-UTILITY-D	Distributed	Assumes	Rate	Finite	Solar
NetOnline	Distributed	Assumes	Energy	Finite	Solar
RC-PA-R with MWM	Centralized	No	Power	Finite	N. S.
RC-PA-R with MM	Distributed	No	Power	Finite	N. S.

Despite their simplistic design, FP, MI, and MS algorithms [28] operate only in a single-hop architecture, where a node can only be configured either as a source or a sink. Moreover, the algorithms require the location of each sensor node to be obtained from GPS or some other method, during deployment. The main drawback of these algorithms is that, when employed, nodes can transmit sensed data only when sufficient energy is harvested. This may cause long delays in terms of data delivery, i.e., when more energy is harvested, the packets will be sent, but, when low energy is harvested, packets will be kept waiting until required energy is accumulated. The best one of these three algorithms is known to be MI. Although not applicable for event-driven applications (e.g., detection of threats and oil spills) where data dissemination is only triggered upon the detection of abnormal phenomena, MI can be a good choice for predictive monitoring based WSN applications such as monitoring of road infrastructure, where sensed data is continuously being disseminated (e.g., periodically). The algorithm proposed by Reddy et al. [26], DC-PM, also operates on a single hop network where a bunch of energy harvesting sensor nodes communicate with a powered destination (sink), with the goal of maximizing the sum data rate. DC-PM is the only algorithm (among the ones mentioned in this chapter) that considers duty cycling as a part of the optimization process. Although centralized and suboptimal, DC-PM turns out to have a surprisingly simple form of power allocation and duty cycling. The algorithm is suitable for applications that require simple duty cycling and power management techniques. However, in order to implement solution, it requires the knowledge of the sum normalized power (sum harvested energy) for every slot (energy inter-arrival times form the slots) and its statistics.

In contrast to MI and DC-PM, QuickFix/SnapIt algorithms [16] can be used in well-structured networks with an underlying directed acyclic network graph (DAG). The algorithms working in tandem provide a distributed solution that does not require any knowledge of the future recharging rates. The combination (QuickFix/SnapIt) is suitable for WSN applications that demand proportional fairness and perpetual operation. Another advantage of QuickFix/SnapIt over MI is that, when solar energy harvesting is used, based on the application's minimum rate requirement, one can determine the minimum battery level that can support the minimum rate at night (when no, or, too little energy harvesting is available) and, trust on SnapIt algorithm to maintain the battery at that level to ensure the network remains active during the night time. However, although [16] target general multihop networks and offer an innovative solution, the proposed solution (QuickFix/SnapIt scheme) is not optimal, and can incur high control overhead and unpredictable running time, thus potentially limiting the practical implementation within resource-constrained WSNs. MAX-UTILITY algorithm, on the other hand, offers a time complexity of $O(N^3)$ for a system with N nodes. MAX-UTILITY-D, fully distributed version of MAX-UTILITY, allows resource-constrained sensor nodes to collaboratively produce optimal rate assignments. A common limitation of the algorithms is that they apply only to tree-based WSNs. As they require energy prediction, MAX-UTILITY and MAX-UTILITY-D algorithms are not suitable for WSNs powered by unpredictable energy sources (such as vibration). [21] and [20] also depend on energy prediction. The proposed algorithms, SSEA, ASEA [21], and PFC [20], are developed for WSNs that use solar energy harvesting. SSEA and ASEA are suitable for WSNs that require minimizing variations in the energy allocation. Note that, there is no energy to be harvested when the sun is down. However, in some industrial applications, data needs to be collected at the same rate at all times. SSEA and ASEA allow sensor nodes to reserve an adequate amount of energy to operate at a constant level at all times. Hence, the target application of SSEA and ASEA is time-driven WSNs, not the event-driven WSNs.

PFC algorithm cooperates with the SSEA energy allocation scheme to maximize long-term performance, especially the amount of data collected at the system level. It is suitable both for flow-centric and storage centric industrial WSNs. In storage centric networks, the acquired data has to be stored in the network temporarily (may be couple of days) until the sink node is connected to the network in order to gather it. Note that, this algorithm is the only one (among the ones mentioned in this chapter) considering storage centric networks. Hence, if the IWSN has the ability of solar energy harvesting, and the sink node is not usually, but only periodically, connected to the network, PFC algorithm seems to be the best choice in terms of minimizing the amount of data loss due to storage constraints. The algorithm is a good alternative for flow-centric networks (the aim is maximizing the throughput) as well, since it can operate in a distributed manner. However, the algorithm can only operate in solar-based networks and when a reasonable amount of solar data is available for prediction process (SSEA).

When prediction is not possible (or available), approaches that dynamically adapt to instantaneous energy and data buffer states are recommendable. For example, Gatzianas et al. [5] model energy harvesting as a time-varying process and consider jointly managing the data and battery buffers. The authors consider infinite data buffer and finite battery buffer sizes. They assume that the harvesting process is i.i.d, and, show that under the proposed policy, DRABP, the probability of battery state being less than the peak power or close to the full battery state vanishes as the battery size grows. NRABP, the multi-hop version of DRABP, is also shown to stabilize [5] any multi-hop network. Note that, if the data buffer size is infinite, the concern is the stability of the data queue, while for finite data buffer, excessive data losses should be avoided. A common drawback of the proposed schemes is that, although Gatzianas et al. claim that for non i.i.d processes, a slot analysis can be applied, this operation has high complexity and depends on the network size.

In [17], Mao et al. consider all combinations of finite and infinite data and battery buffer sizes by defining minimum number of virtual queues in a general format. In addition to the constraint on the stability of the data queue (constraint on the data loss ratio when the data buffer size is finite), they also impose a constraint on the frequency of battery discharge. Rather than assuming an i.i.d energy harvesting process as in [5], they allow for a general harvesting process without assuming ergodicity, and, consider jointly managing the data and battery buffers to deal with the coupling between them. The developed algorithm is more advantageous over previously mentioned algorithms, as it has a built-in routing algorithm, and can be used in industrial WSNs requiring high long-term average sensing rate. However, Chen et al. argue in [4] that the infinite-horizon based solutions, such as those proposed in [5] and [17], may be highly inefficient, especially in the context of networks with energy harvesting. The stated reason is that the harvesting profiles are time varying and may not even be stationary and ergodic. Note that, the finite-horizon problem is important and challenging as well because it necessitates optimizing performance metrics that are exhibited in the short term rather than metrics that are averaged over a long period of time. One difference between the finite horizon problem in [4] and the infinite-horizon problem in [5] is, in the finite horizon problem inefficiencies cannot be made to vanish to infinitely small values. This implies that new techniques, such as

NetOnline [4], need to be developed to mitigate these inefficiencies. Although no comparison of NetOnline to RC-PA-R [17] exists, NetOnline is shown to outperform the NRABP algorithm proposed in [5].

5.4 Conclusions

This chapter investigated the state-of-the-art resource management and scheduling algorithms that can be used in energy harvesting industrial WSNs. Detailed operation of the algorithms, along with their drawbacks, advantages, and possible application areas are discussed. It should be noted that, depending on the requirements of the chosen WSN application (whether long-term or short term metrics are more appropriate), battery capacity, and the type of application, the relative performances of the various proposals surveyed in this chapter will be perceived differently. Depending on the type of the industrial setting, the network size and the performance criteria, we believe that one of these solutions, when appropriately tuned, will provide an efficient resource management and scheduling solution.

References

[1] Akhondi, M. R., Talevski, A., Carlsen, S., Petersen, S.: Applications of Wireless Sensor Networks in the Oil, Gas and Resources Industries. In: *Proceedings of the 24th IEEE International Conference on Advanced Information Networking and Applications (AINA)*, pp. 941–948. (2010)

[2] Bogue R.: Solar-powered sensors: a review of products and applications. Sensor Review. vol. 32, no. 2, 95–100 (2012)

[3] Chen, L., Low, S. H., Chiang, M., Doyle, J. C.: Cross-layer congestion control, routing and scheduling design in ad hoc wireless networks. In: *IEEE INFOCOM*, pp. 1–13. (2006)

[4] Chen, S., Sinha, P., Shroff, N. B., Joo, C.: Finite-Horizon Energy Allocation and Routing Scheme in Rechargeable Sensor Networks. In: *IEEE INFOCOM*, pp. 2273–2281. (2011)

[5] Gatzianas, M., Georgiadis, L., Tassiulas L.: Control of Wireless Networks with Rechargeable Batteries. *IEEE Transactions on Wireless Communications*. vol. 9, no. 2, 581–593 (2010)

[6] Gatzianas, M., Georgiadis, L., Tassiulas L.: Asymptotically optimal control of wireless networks with rechargeable batteries. Technical report, Available: http://users.auth.gr/~leonid/public/TechReports/tecreport_rabp.pdf

[7] Georgiadis, L., Neely, M. J., Tassiulas L.: Resource allocation and cross-layer control in wireless networks. Foundations and Trends in Networking. vol. 1, no. 1, (2006)

[8] Georgiadis, L., Tassiulas, L.: Optimal overload response in sensor networks. IEEE Transactions on Information Theory. vol. 52, no. 6, 2684–2696 (2006)

[9] Gungor, V.C., Hancke, G.P.: Industrial Wireless Sensor Networks: Challenges, Design Principles, and Technical Approaches. *IEEE Transactions on Industrial Electronics.* vol. 56, no. 10, 4258–4265 (2009)

[10] Hande, A., Polk, T., Walker, W., Bhatia, D.: Indoor solar energy harvesting for sensor network router nodes, Microprocess. Microsyst. vol. 31, no. 6, 420–432 (2007), doi: 10.1016/j.micropro.2007.02.006.

[11] Hoepman, J.-H.: Simple distributed weighted matchings. In: CoRR (2004)

[12] Jiang, X., Polastre, J., Culler, D. E.: Perpetual environmentally powered sensor networks. In: *Proceedings of the Fourth International Symposium on Information Processing in Sensor Networks (IPSN)*, pp. 463–468. (2005)

[13] Khaligh A., Zeng P., Zheng C.: Kinetic Energy Harvesting Using Piezoelectric and Electromagnetic Technologies State of the Art. *IEEE Transactions on Industrial Electronics.* vol. 57, no. 3, 850–860 (2010)

[14] Krishnamurthy, L., Adler, R., Buonadonna, P., Chhabra, J., Flanigan, M., Kushalnagar, A., Nachman, L., Yarvis, M.: Design and deployment of industrial sensor networks: experiences from a semiconductor plant and the north sea. In: *Proceedings of the 3rd International Conference on Embedded Networked Sensor Systems*, pp. 64–75. (2005)

[15] Lee, D. T., Preparata, F. P.: Euclidean shortest paths in the presence of rectilinear barriers. Networks. vol. 14, no. 3, 393–410 (1984)

[16] Liu, R.-S., Sinha, P., Koksal, C. E.: Joint energy management and resource allocation in rechargeable sensor networks. In: *Proceedings of the 29th Conference on Information Communications (INFOCOM)*, pp. 902–910. (2010)

[17] Mao, Z., Koksal, C. E., Shroff, N. B.: Near optimal power and rate control of multi-hop sensor networks with energy replenishment: Basic limitations with finite energy and data storage. *IEEE Transactions on Automatic Control.* vol. 57, no. 4, 815–829 (2012)

[18] Mohamed, M. I., Wu, W.Y., Moniri, M.: Power harvesting for smart sensor networks in monitoring water distribution system. In: *IEEE International Conference on Networking, Sensing and Control (ICNSC)*, pp. 393-398. (2011)

[19] Moser, C., Chen, J.-J., Thiele, L.: Optimal service level allocation in environmentally powered embedded systems. In: *Proceedings of the ACM Symposium on Applied Computing*, pp. 1650-1657. (2009)

[20] Noh, D. K., Kang, K.: A Practical Flow Control Scheme Considering Optimal Energy Allocation in Solar-Powered WSNs. In: *Proceedings of 18th Internatonal Conference on Computer Communications and Networks* (ICCCN), pp. 1–6. (2009)

[21] Noh, D. K., Kang, K.: Balanced energy allocation scheme for a solar-powered sensor system and its effects on network-wide performance. Journal of Computer and System Sciences. vol. 77, no. 5, 917–932 (2010)

[22] Paavola, M., Leiviska, K.: Wireless Sensor Networks in Industrial Automation. In: Chapter 10, Factory Automation. InTech (2010)

[23] Park, C., Chou, P.: Ambimax: Autonomous energy harvesting platform for multi-supply wireless sensor nodes. In: *Proceedings of the Sensor and Ad Hoc Communications and Networks (SECON)*, pp. 168–177. (2006)

[24] Raisigel, H., Chabanis, G., Ressejac, I., Trouillon, M.: Autonomous Wireless Sensor Node for Building Climate Conditioning Application. In: *4th International Conference on Sensor Technologies and Applications (SENSORCOMM)*, pp. 68–73. (2010)

[25] Rangwala, S., Gummadi, R., Govindan, R., Psounis, K.: Interference-aware fair rate control in wireless sensor networks. In: *SIGCOMM Proceedings of the Conference on Applications, Technologies, Architectures, and Protocols for Computer Communications.* vol. 36, no. 4, 63–74 (2006)

[26] Reddy, S., Murthy, C.R.: Duty cycling and power management with a network of energy harvesting sensors. In: *4th IEEE International Workshop on Computational Advances in Multi-Sensor Adaptive Processing (CAMSAP)*, pp. 205–208. (2011)

[27] Simjee, F., Chou, P. H.: Everlast: long-life, supercapacitor-operated wireless sensor node. In: *International Symposium on Low Power Electronics and Design (ISLPED)*, pp. 197–202. (2006)

[28] Tan, H.-P., Valera, A., Koh, W.: Transmission power control in 2-D wireless sensor networks powered by ambient energy harvesting. In: *IEEE 21st International Symposium on Personal Indoor and Mobile Radio Communications (PIMRC)*, pp. 1671-1676. (2010)

[29] Vullers, R. J. M., Schaijk, R. V., Visser, H. J., Penders, J., Hoof, C. V.: Energy Harvesting for Autonomous Wireless Sensor Networks. *IEEE Solid-State Circuit Magazine.* vol. 2., no. 2, 29–38 (2010)

[30] Yen K. T., Panda S. K.: Energy Harvesting for Autonomous Wireless Sensor Networks. *IEEE Transactions on Power Electronics.* vol. 26., no. 1, 38–50 (2011)

[31] Yu, H., Yue Q.: Indoor Light Energy Harvesting System for Energy-aware Wireless Sensor Node. In: *International Conference on Future Energy, Environment, and Materials*, pp. 1027–1032. (2012)

[32] Zhang, B., Simon R., Aydin, H.: Maximal utility rate allocation for energy harvesting wireless sensor networks. In: *Proceedings of the 14th ACM International Conference on Modeling, Analysis and Simulation of Wireless and Mobile Systems (MSWIM)*, (Best Paper Award), pp. 902–910. (2011)

6

Energy Harvesting Techniques for Industrial Wireless Sensor Networks

Gurkan Tuna and Vehbi Cagri Gungor

Bahcesehir University

Kayhan Gulez

Yildiz Technical University

CONTENTS

6.1 Introduction

Advancements in technology have made it possible to realize low-cost wireless sensor network (WSN)-based industrial automation systems [16]. Wireless sensor networks can be used for various industrial applications including factory automation, process control, real-time monitoring of machineries, monitoring of contaminated areas, detection of liquid or gas leakage, and real-time inventory management. In these systems, sensor nodes monitor and gather the parameters critical to automation processes and transmit these data to a control center or an operator. Since sensor nodes are generally battery-powered devices, their operational lifetimes are limited. Energy harvesting techniques have a good potential to solve this constraint. This approach is already used in passive Radio Frequency Identification (RFID) tags [30],[50].

Energy harvesting is a technique which harvests or scavenges unused ambient energy and converts the harvested energy into electrical energy [47]. In WSNs, energy harvesting from any sources near to sensor nodes such as electric fields, magnetic fields, solar, thermal, air flow, vibrations, etc., can be utilized to charge the batteries or to operate without using the batteries. While energy harvesting techniques offer the ability of extracting energy from the environment, a proper and efficient integration is necessary. In this way, harvested energy can be translated into increased application performance and system lifetime. In addition to energy harvesting, energy efficient design techniques have been studied for WSNs. The use of energy management and conservation schemes with proper node and network level adaptations can considerably increase the lifetime of sensor nodes [4].

6.2   Wireless Sensor Networks for Industrial Applications

Industrial applications of WSNs can be mainly categorized into two broad classes, wireless monitoring applications and wireless control applications. Though these two application classes exhibit different characteristics, common challenges for wireless sensor networks for industrial applications are as follows [2],[3].

1. Limited bandwidth
2. Latency
3. Coverage of the network
4. Power
5. Environmental conditions
6. Robustness
7. Cost

In general, WSN-based monitoring applications call for low-power sensor platforms with low-power transceivers. In the applications of this category, low-to-medium data rates are acceptable and medium-to-high latencies are tolerable

[2]. Main challenge of these applications is the use of low-power sensor platforms. Hence, efficient energy harvesting techniques with energy management schemes can be utilized effectively to increase the operational time of these applications. On the other hand, WSN-based control applications call for sensor platforms with high-throughput low-latency radios [3]. In this type of applications, medium-to-high data rates are acceptable and low latencies are tolerable. Since, different from the WSN-based monitoring applications, main challenge resides in the trade-off between robustness and latency, aggressive energy conservation schemes in addition to efficient energy harvesting techniques can be utilized.

6.2.1   Challenges

Implementation of WSNs for industrial applications is constrained by technical challenges of sensor platforms. The major technical challenges can be outlined as follows.

1. Environmental conditions: In industrial environments, sensors may be subject to high temperatures, humidity, liquids, dirt, dust, oil, corrosive agents, or other conditions which challenge performance. As a result of these harsh environmental conditions, a number of sensor nodes may malfunction. Another common challenging problem in industrial environments is radio frequency (RF) interference.

2. Quality-of-service (QoS) requirements: Depending on goals of an industrial application, different QoS requirements and specifications may be required. QoS requirements play an important role in improving the quality of supplied services by industrial applications.

3. Deployment: For most industrial applications, a large number of sensor nodes need to be deployed. Deployed sensor nodes form a wireless mesh network whereby if source and destination nodes are not in direct communication range, then intermediate nodes relay the traffic. Hence, WSNs deployed in industrial environments establish connections and maintain network connectivity autonomously.

4. Resource constraints of sensor platforms: Sensor platforms are small-sized low-cost electronic devices with limited memories and computational capabilities [16]. Due to their limited physical size, they have limited energy supplies provided by internal/external batteries.

5. Topology changes, packet errors, and link capacity: In industrial environments, due to link and sensor node failures, the topology and connectivity of the network may vary [16]. Due to obstructions and noisy industrial environments, capacity of wireless links varies continuously. Another common problem with the deployment of WSNs

in industrial environments is that high bit error rates are observed in communication [16].

6. Integration issues and security: Integration with existing networks is of fundamental importance for industrial applications to allow data exchange between a WSN and an existing company network. Currently gateways are used for the integration but researchers are going on providing IP connectivity to sensor platforms [31]. The integration of WSNs to existing networks or the Internet brings security concerns. Therefore, security is a critical feature in the design of WSNs for industrial applications.

6.2.2 Design Goals

To deal with the challenges and the requirements of existing and potential WSN-based industrial applications, the following design goals need to be followed.

1. Energy efficiency: Sensor nodes are powered through batteries which often cannot be replaced. To deal with this problem, various energy harvesting techniques depending on the conditions of the industrial environment can be used to recharge batteries. In addition to the energy harvesting techniques, minimizing the power usage of sensor platforms by energy conservation schemes such as adaptive sampling is a promising technique. Energy saving can also be accomplished with energy-efficient protocols [16].

2. Reliability: Due to the harsh conditions of industrial environments, sensor nodes may not function properly. To deal with node failures, node redundancy is a basic solution. Transmission errors due to fading, interferences caused by devices or machineries, and obstacles may be observed. Collisions and congestions are other frequently encountered problems in WSNs in industrial environments. To deal with transmission related problems, mechanisms such as forward error correction (FEC), packet retransmission, routing through redundant paths, etc., can be utilized.

3. Timeliness: Since industrial applications are usually time-sensitive, data gathered by sensor nodes need to be processed timely. Due to communication or processing delays resulting from lack of infrastructure and dynamic topologies, accumulated data may not be up-to-date and may lead to wrong decisions [16]. As a result, these latencies may delay taking preventative maintenance steps. To meet the deadlines of industrial applications, in addition to protocol level improvements, new time synchronization strategies should be designed. In-network processing, processing of data locally by sensor nodes can also help reducing communication overhead significantly

to transmit only the processed data and thereby meet the deadlines of industrial applications [16].

4. Adaptive scalability: An adaptive scalable architecture can support the requirements of heterogeneous industrial applications [16]. In this way, the robustness and flexibility of industrial applications can be enhanced. In addition to scalability, integration with existing communication infrastructures is required.

5. Self-configuration and self-organization: To deal with dynamic topologies and power-downs, and to minimize operational efforts in large scale deployments, sensor nodes can turn on via auto-configuration and can form a WSN. These capabilities also allow new sensor nodes to participate in an existing WSN.

6. Meeting application specific requirements: Various industrial automation and industrial monitoring applications have different specific requirements and constraints. To find a solution which meets the requirements of all these applications is not possible. Instead, alternative designs should be developed [16].

7. Security: The integration with existing legacy networks and the Internet brings several security concerns. Hence, in the design of WSNs for industrial applications, security primitives at different levels should be addressed.

8. Cost: Large scale deployments of WSNs in industrial environments necessitate using low-cost sensor platforms. In addition to pre-deployment costs, there are other costs such as implementation costs, training costs, and maintenance costs which need to be considered.

6.3   Energy Harvesting Techniques for Industrial Wireless Sensor Networks

WSN-based industrial applications are designed with energy related constraints to meet desired design goals in addition to being optimized for various parameters depending on the application-specific requirements. For example, WSNs optimized for increased lifetime, such as monitoring applications, may operate at low duty cycles. On the other hand, WSNs optimized for coverage and robustness may need to operate with high-capacity batteries, or periodic maintenance may be required to change depleted batteries.

Power consumption of wireless sensor nodes depend on several factors listed below [23].

1. Battery related factors (battery dimensions, electrode material, diffusion rate in the electrolyte, discharge rate, supply voltages, relaxation effect)

2. Sensor related factors (signal sampling, signal conditioning, signal conversion)

3. Analog-to-digital converter (ADC) related factors (aliasing, dither, sampling rate)

4. Radio related factors (duty cycle, transmission power, data rate, modulation scheme)

5. Microprocessor/Microcontroller related factors (operating frequency, power, ambient temperature, peripheral utilization, code)

In [30], a realistic scenario which evaluates the battery lifetime of a sensor node is given. In this scenario, it is assumed that the transmit and receive mode consumptions of the sensor node with a 9V 1200mAh battery are 35mA and the sleep mode consumption is $75\mu A$. Then, if the node transmits or receives data once every 15 minutes, its battery life will be less than a year.

The goal of energy harvesting techniques is to generate power to sensor nodes. Typical forms of ambient energies which can be harvested are mechanical energy, sunlight, wind energy, thermal energy, and RF energy. Energy sources have different characteristics in terms of magnitude, predictability, and controllability [22], [18]. Controllable energy sources can provide harvestable energy whenever required. With this kind of sources, it is not required to predict energy availability before harvesting. Non-controllable energy sources may not provide harvestable energy all the time. Therefore with these kinds of sources, it is required to harvest energy whenever available by using a prediction model to forecast the availability of the source [42]. Though it depends on the harvesting technique, average power which can be supplied is between tens of microwatts and several hundred milliwatts. A comparison of the efficiency of energy harvesting solutions is given in Table 6.1. Since it is not possible to harvest energy all the time and the voltage of the harvested energy is not stable, a rechargeable battery is required to store the harvested energy and to provide a stable voltage to the sensor node. While throughput and latency are usually traded off to extend network lifetime in WSNs consisting of battery-operated wireless sensor nodes [40], longer lifetime can be achieved using energy harvesting solutions without compromising performance. Predicting future energy availability with a prediction module helps determining sleep and wake-up schedules of energy-harvesting sensor nodes.

Energy harvesting solutions provide several benefits to the industry where WSNs will be deployed. Major benefits are as follows.

1. Energy harvesting solutions reduce the dependency on battery power.

2. Since energy harvesting offers supplementing the internal batteries,

TABLE 6.1

The comparison of energy harvesting solutions

Solution	Power density in outdoor implementations	Power density in indoor implementations	Commonly available in the market	Safe
Solar energy harvesting	15mW/cm^2 [37]	100μW/cm^2 (at 10W/cm^2 light density) [47]	Yes	Yes
Thermal energy harvesting	100μW/cm^2 at 5°C gradient, 3.5mW/cm^2 at 30°C gradient [47]	100μW/cm^2 at 5°C gradient, 3.5mW/cm^2 at 30°C gradient [47]	Yes	Yes
Vibration based energy harvesting	500μW/cm^2 (piezoelectric method), 4μW/cm^2 (electromagnetic method) 3.8μW/cm^2 (electrostatic method) [8]	500μW/cm^2 (piezoelectric method), 4μW/cm^2 (electromagnetic method) 3.8μW/cm^2 (electrostatic method) [8]	Yes	Yes
RF energy harvesting	15mW (with a transmitted power of 2–3W at a frequency of 906 MHz at a distance of 30 cm) [46]	15mW (with a transmitted power of 2–3W at a frequency of 906 MHz at a distance of 30 cm) [46]	Yes	Questionable
Air flow energy harvesting	3.5mW/cm^2 (wind speed of 8.4m/s) [44]	3.5μW/cm^2 (air flow speed is less than 1m/s) [44].	No	Yes
Electromagnetic wave energy harvesting	0.26μW/cm^2 (from an electric field of 1V/m) [30]	0.26μW/cm^2 (from an electric field of 1V/m) [30]	No	Questionable
Acoustic energy harvesting	960nW/cm^3 (acoustic noise of 100dB) [37]	960nW/cm^3 (acoustic noise of 100dB) [37]	No	Yes
Biochemical energy harvesting	0.1-1mW/cm^2 [27]	0.1-1mW/cm^2 [27]	No	Yes

the batteries will last longer so the maintenance cost to change the depleted batteries is reduced.

3. Energy harvesting reduces environmental impact by eliminating the need of numerous batteries [45].

4. Long-term solutions can be provided by energy harvesting solutions.

5. Continuous monitoring and automation services can be provided in hazardous industrial environments.

6.3.1 Solar Energy Harvesting

Harvesting solar energy is ideal for outdoor WSN deployments due to the omnipresent nature of sun light. Solar energy is an uncontrollable but predictable type of energy source. Solar energy harvesting is achieved by solar panels which capture the sun's energy using photovoltaic cells. Solar panels are made up of solar cells connected in series/parallel, and are designed based on the voltage and current level required. Conversion efficiency and quantum efficiency are the measures which characterize solar cell performance [53], [1]. Two parameters, the open circuit voltage (V_{oc}) and the short circuit current (I_{sc}), characterize solar panels. V_{oc} remains almost constant all the time, though I_{sc} depends on the amount of incident solar radiation.

Though a continuous supply of sunlight is needed, solar cells can generate some electricity on cloudy days [54], [56]. Solar energy can provide 15mW/cm^2 [9] out of the estimated available 100mW/cm^2. Both for the outdoor and for the indoor solar harvesting, power which can be harvested depend on the light intensity. Indoor solar panels can harvest 100μW/cm^2 at 10W/cm^2 light density [47].

The conversion efficiency of solar cells is around 15% [9], [37]. Besides the low conversion efficiency of solar cells, solar panels exhibit current source-like behavior. Hence, to provide a stable voltage to sensor nodes, rechargeable batteries or ultracapacitors are used to store the harvested energy. The advantage of this technique is that there are various commercially available solar cells including silicone, thin film, plastic based solar cells in the market. Also, there are specific solar panels which are small enough to fit the form factor of wireless sensor nodes [42]. Among many types of solar cells, in spite of the low efficiency in the range of 8 – 13%, thin film cadmium telluride cells are commonly preferred for the reason that they give satisfactory performance under various light conditions [30].

6.3.2 Thermal Energy Harvesting

Thermal energy harvesting offers harvesting energy through heat transfer. Thermoelectric generators produce electrical energy by following the principle of thermoelectricity which is the phenomena of generating electric potential with a temperature difference [9]. Basically, two dissimilar metals joined at two junctions kept at different temperatures generate an electrical voltage [30]. The maximum efficiency of this technique is determined by the Carnot cycle [30]. Thermoelectric generators are relatively small, light weight devices without any vibration and noise. They can work for long hours under harsh industrial environments with little or no maintenance [47]. Commercially available products require a temperature difference of 10-200°C. The power density of harvested energy using thermoelectric generators is $100\mu W/cm^2$ at the temperature difference of 5°C and $3.5mW/cm^2$ at the temperature difference of 30°C [47].

6.3.3 Vibration-Based Energy Harvesting

Mechanical vibrations can be harnessed to produce electrical energy. Vibration based energy harvesting techniques can be classified broadly into three types [30]:

1. Piezoelectric technique: This technique uses the property of a piezoelectric material, and produces electric potential under mechanical stress. The power density of harvested energy using this technique is $500\mu W/cm^2$ [8].

2. Electromagnetic technique: This technique uses Faraday's law of electromagnetic induction. Vibration of the magnet fixed to a spring inside a coil causes an induced voltage. The power density of harvested energy using this technique is $4\mu W/cm^2$ [8].

3. Electrostatic technique: Electrostatic technique is based on changing the capacitance of a variable capacitor which is vibration dependent. The power density of harvested energy using this technique is $3.8\mu W/cm^2$ [8].

The efficiency of vibration-based energy harvesting systems mainly depends on the resonant frequency of vibrations. There are several commercially available products utilizing piezoelectric, electrostatic, or electromagnetic harvesting techniques in the market.

6.3.4 Air Flow Energy Harvesting

Air flow can be used to produce electric energy. Some ambient sources of air flow such as wind are uncontrollable but predictable energy sources. Wind energy can provide 1200mWh/day [42]. Since the strength of wind changes, the same amount of energy cannot be generated all the time [47]. To harvest energy using air flow techniques, there are various methods including micro wind turbines, oscillating wings, and flapping wings. In [44], a very small-scale windmill prototype is given. Though the efficiency of the air flow harvesting techniques is questionable in practice, technological advancements can make it possible to manufacture small-sized effective air flow power converters. Wind turbine generators can provide $3.5\text{mW}/\text{cm}^2$ at the wind speed of 8.4m/s. Indoor turbine generators can generate up to $3.5\mu\text{W}/\text{cm}^2$ when the air flow speed is less than 1m/s [44].

6.3.5 Acoustic Energy Harvesting

If sound waves encounter a barrier while travelling through air, then this sound energy can be converted into electrical energy. When acoustic noise comes into an acoustic energy converter which is a membrane-type receiver, the electrical power will be produced. It was shown that noises equal to 160dB can generate power up to 100kW [47]. With an acoustic noise of 100dB, the average power density which can be harvested is $960\text{nW}/\text{cm}^3$ [37]. If it is possible to manufacture small scale acoustic energy converters, this solution seems promising.

6.3.6 Magnetic Field Energy Harvesting

Since there is magnetic energy everywhere on the earth, it is a green energy source [47]. The magnetic field existing near power lines can be harvested. There are commercially available products based on the transformer action in the market to harvest energy from magnetic fields [30]. There is a clamp around the conductor in these products.

6.3.7 Electromagnetic Wave Energy Harvesting

Theoretically $0.26\mu\text{W}/\text{cm}^2$ energy can be harvested from an electric field of 1V/m [30]. However, electric fields of this order may only be encountered very close to powerful transmitters. Also, this conflicts with the fact that sensor nodes need to be located away from high voltage conductors to function properly. Instead of harvesting energy from electric fields, RF energy can be used to power wireless nodes [30].

6.3.8 Radio Frequency Energy Harvesting

Since intense RF signals can block other types of radio transmissions, governments do not allow the transfer of them [33]. On the other hand, radio signals and TV broadcast signals are usually transmitted by using intense RF signals. RF energy harvesting is a promising solution if the harvested energy is sufficient for powering wireless sensor nodes. In [32], it is shown that that Mica2 sensor platforms can be operated using RF energy harvesting when their duty cycle is set based on the incident RF power. Ambient RF energy available through GSM, WLAN, and TV broadcasts [46],[33] can be harvested. For distances between 25–100 m from a GSM base station, power density levels of $0.1–1.0\text{mW}/\text{m}^2$ can be harvested for single frequencies [6]. In an experimental study, for 4 km to a nearby TV station, $0.1\mu W/cm^2$ power density was achieved [38]. Instead of ambient RF energy sources, a dedicated AC/DC powered RF source positioned close to wireless sensor nodes can be used. Using this approach, higher power levels can be harvested from RF sources with low transmission power [55]. Using a commercial RF energy harvesting system [55] with a transmitted power of 2–3W at a frequency of 906 MHz, 15mW of power can be received at a distance of 30 cm in ideal conditions [46].

A similar technique is used by Passive RFIDs, which harvest electrical energy from the received RF signal for data transmission. However, this technique differs from RF energy harvesting in several aspects.
In passive RFID systems, a reader generates an intense radio emission, and the tag replies to the reader using the backscattered radio signal [33]. In these systems, the tag maintains silence until the reader scans it [33]. On the other hand, wireless sensor nodes operate continuously for active sensing. Another proposed solution aiming to reduce energy consumption due to data transmission is modulated backscattering (MB) [16], [5]. MB allows battery-operated sensor nodes to send data by switching the impedance of their antennas and by reflecting the incident signal coming from an AC/DC powered RF source [5].

6.3.9 Envisaged Energy Harvesting Solutions

Using current transformer sources [14] and optical sources [43] are envisaged solutions for high voltage conditions. Several other techniques including motion-based [16], biochemical energy harvesting [17], and biomechanical energy harvesting [17], [11] are currently being developed. In [17], an enzymatic biofuel cell (BFC) is used to convert the chemical energy of glucose and oxygen into electrical energy. The highest theoretical voltage which can be obtained from a glucose oxidase/laccase-based BFC depends on thermodynamics, and is 1V [12].The maximum power density of a BFC is $0.1\text{-}1mW/cm^2$[27]. Microbial fuel cells [MFCs] can provide power density from less than $1mW/m^2$ to $6.9W/m^2$ [48].

6.4 Open Research Issues

Energy harvesting techniques offer extending the lifetime of sensor platforms by acquiring energy from the environment. In addition to harvesting energy from the deployment field, energy management and conservation schemes can be used to minimize the energy usages of sensor platforms, which, in turn, further extend the lifetime of each sensor node. Basically, energy management schemes switch of node components which are not temporarily needed [4].

These schemes may play a key role since a great deal of energy is consumed by node components. Interestingly, even if the node components are idle, they consume some energy [4].

Node and network-level adaptations provide mechanisms to reduce energy consumption. Both node-level and network-level adaptations such as hardware design [29], tiered system architectures [15], redundant node placement [49], [24], power-aware medium access control (MAC) protocols [36], [52], [7], topology control and management [39], [51], data gathering and aggregation strategies [21], energy efficient routing [41], [19], duty cycling strategies [13], [12], and adaptive sensing [26] have been proposed. Setting node-level parameters such as sampling rate, duty cycle, transmission power, sensing reliability, data processing, etc., and applying other node-level adaptations including hierarchical sensing, prediction of measurements, reducing the resolution of measurement samples, reducing the sampling rate, sampling interesting regions of space and intervals of time in accordance with application requirements by using a prediction module offer extending sensor lifetime while increasing performance. Network-level adaptations offer various mechanisms such as energy-efficient routing protocols, cluster-based routing, energy-efficient data collection techniques, energy-aware MAC protocols, etc., in order to maximize throughput and minimize delays during network activities while reducing energy consumption.

Although energy harvesting techniques supplement batteries, it does not eliminate the requirement of replacing the batteries when they are depleted. But advances in energy harvesting techniques, power electronics, and storage devices make it possible to deploy WSNs which only rely on harvested energy. For instance, [28] introduces a smart power unit which can utilize multiple energy sources to power a sensor node. For a successful integration of energy harvesting technologies into WSNs, there are three major factors:

1. The cost of an energy harvesting device has to be optimized relative to the cost of sensor nodes.

2. Power levels delivered by an energy harvesting device should be enough to power sensor nodes.

3. By means of optimization techniques and technological advancements, power consumption of sensor nodes should be minimized [46].

Currently, the power levels available from small-sized energy harvesting devices are in the order of tens to thousands of microwatts or several milliwatts [40]. But these power levels are only from 1% to 20% of the operating power of wireless sensor nodes [8], [40]. Hence, it is not possible to power the sensor nodes continuously by using direct energy-harvesting.

6.5   Conclusions

The contribution of WSNs in improving the efficiency of monitoring and control applications in industrial automation systems has been recognized by the industry. To provide uninterrupted service to customers, preventative maintenance needs to be done to avoid failures and to improve the quality of supplied services. Recently, supplementing battery supplies with energy harvesting techniques have emerged as viable options to extend the lifetime of WSNs. Considering the effectiveness and commercial availability of energy harvesting products, vibration based energy harvesting and solar energy harvesting seem as feasible methods. They are also safe for humans and environment. On the other hand, harvesting electromagnetic wave energy is questionable when health related concerns are taken into consideration.

In this chapter, the potential uses of WSNs for industrial applications have been introduced along with the related technical challenges. Specifically, to improve the operational time of WSNs, various energy harvesting techniques together with different energy management and conversation schemes are investigated. Compliance with green energy requirements is an important design criterion for envisaged harvesting solutions.

References

[1] A. Ahnood and A. Nathan. Flat-Panel Compatible Photovoltaic Energy Harvesting System. *Journal of Display Technology*, 8(4):204–211, 2012.

[2] I. F. Akyildiz, T. Melodia, and K. Chowdhury. A survey on wireless multimedia sensor networks. *Computer Networks*, 51(4):921–960, 2007.

[3] G. Anastasi, M. Conti, and M. D. Francesco. Extending the Lifetime of Wireless Sensor Networks through Adaptive Sleep. *IEEE Transactions on Industrial Informatics*, 5(3):351-365, 2009.

[4] G. Anastasi, M. Conti, M. D. Francesco, and A. Passarella. Energy conservation in wireless sensor networks: A survey. *Ad Hoc Networks*, 7(3): 537–568, 2009.

[5] A. Bereketli and O. B. Akan. Communication Coverage in Wireless Passive Sensor Networks. *IEEE Communications Letters*, 13(2): 133-135, 2009.

[6] U. Bergqvist, G. Friedrich, Y. Hamnerius, L. Martens, G. Neubauer, G. Thuroczy, E. Vogel, and J. Wiart. Mobile telecommunication base sta-

tions - Exposure to electromagnetic fields, report of a short term mission within COST-244bis. Technical Report, 2000.

[7] M. Buettner, G. V. Yee, E. Anderson, and R. Han. X-MAC: A Short Preamble MAC Protocol for Duty-cycled Wireless Sensor Networks. In *Proceedings of the 4th International Conference on Embedded Networked Sensor Systems*, pages 307–320, 2006.

[8] B. H. Calhoun, D. C. Daly, N. Verma, D. F. Finchelstein, D. D. Wentzloff, A. Wang, S. H. Cho, and A. P. Chandrakasan. Design Considerations for Ultra-Low Energy Wireless Microsensor Nodes. *IEEE Transactions on Computers*, 54(6):727–740, 2005.

[9] S. Chalasani and J. Conrad. A Survey of Energy Harvesting Sources for Embedded Systems. In *Proceedings of IEEE Southeastcon*, pages 442–447, 2008.

[10] A. Dolgov, R. Zane, and Z. Popovic. Power Management System for On-line Low-Power RF Energy Harvesting Optimization. *IEEE Transactions on Circuits and Systems*, 57(7):1802–1811, 2010.

[11] J. M. Donelan, Q. Li, V. Naing, J. A. Hoffer, D. J. Weber, and A. D. Kuo. Biomechanical Energy Harvesting: Generating Electricity During Walking with Minimal User Effort. *Science*, 319(5864):807-810, 2008.

[12] P. Dutta, M. Grimmer, A. Arora, S. Bibyk, and D. Culler. Design of a Wireless Sensor Network Platform for Detecting Rare, Random, and Ephemeral Events. In *Proceedings of the 4th International Conference on Information Processing in Sensor Networks*, pages 497–502, 2005.

[13] S. Ganeriwal, D. Ganesan, H. Shim, V. Tsiatsis, and M. B. Srivastava. Estimating Clock Uncertainty for Efficient Duty-cycling in Sensor Networks. In *Proceedings of the 3rd ACM Conference on Sensor Networking Systems*, pages 130–141, 2005.

[14] Z. Gang, L. Shaohui, Z. Zhipeng, and C. Wei. A novel electro-optic hybrid current measurement instrument for high-voltage power lines. *IEEE Transactions on Instrumentation and Measurement*, 50(1):59–62, 2001.

[15] O. Gnawali, K.-Y. Jang, J. Paek, M. Vieira, R. Govindan, B. Greenstein, A. Joki, D. Estrin, and E. Kohler. The Tenet Architecture for Tiered Sensor Networks. In *Proceedings of the 4th International Conference on Embedded Networked Sensor Systems*, pages 153–166, 2006.

[16] V.C. Gungor and G. Hancke. Industrial wireless sensor networks: challenges, design principles, and technical approaches. *IEEE Transactions on Industrial Electronics*, 56(10):4258–4265, 2009.

[17] B. J. Hansen, Y. Liu, R. Yang, and Z. L. Wang. Hybrid Nanogenerator for Concurrently Harvesting Biomechanical and Biochemical Energy. *ACS Nano*, 4(7):3647–3652, 2010.

[18] A. Harb. Energy harvesting: State-of-the-art. *Renewable Energy*, 36(2011):2641–2654, 2011.

[19] W. Heinzelman, A. Chandrakasan, and H. Balakrishnan. Energy-Efficient Communication Protocol for Wireless MicroSensor Networks. In *Proceedings of the 33rd Annual Hawaii International Conference on System Sciences*, volume 8, pages 1–10, 2000.

[20] A. Heller. Miniature Biofuel Cells. *Physical Chemistry Chemical Physics*, 6(2):209–216, 2004.

[21] K. Kalpakis, K. Dasgupta, and P. Namjoshi. Maximum lifetime data gathering and aggregation in wireless sensor Networks. In *Proceedings of the IEEE International Conference on Networking*, pages 685–696, 2002.

[22] A. Kansal, J. Hsu, S. Zahedi, and M. B. Srivastava. Power Management in Energy Harvesting Sensor Networks. *ACM Transactions on Embedded Computing Systems*, 6(4):1–35, 2007.

[23] C. Kompis and S. Aliwell. Energy Harvesting Technologies to Enable Remote and Wireless Sensing. Technical Report, 2008.

[24] S. Kumar, T. H. Lai, and J. Balogh. On k-coverage in a Mostly Sleeping Sensor Network. *Wireless Networks*, 14(3):277–294, 2008.

[25] T. Le, K. Mayaram, and T. Fiez. Efficient Far-Field Radio Frequency Energy Harvesting for Passively Powered Sensor Network. *IEEE Journal of Solid-State Circuits*, 43(5):1287–1302, 2008.

[26] H. Liu, A. Chandra, and J. Srivastava. eSENSE: Energy Efficient Stochastic Sensing Framework for Wireless Sensor Platforms. In *Proceedings of the 5th International Conference on Information Processing in Sensor Networks*, pages 235–242, 2006.

[27] B. E. Logan, B. Hamelers, R. Rozendal, U. Schrder, J. Keller, S. Freguia, P. Aelterman, W. Verstraete, and K. Rabaey. Microbial fuel cells: methodology and technology. *Environ. Sci. Technol.*,40:5181–5192, 2006.

[28] M. Magno, S. Marinkovic, D. Brunelli, E. Popovici, B. O'Flynn, and L. Benini. Smart Power Unit with Ultra Low Power Radio Trigger Capabilities for Wireless Sensor Networks. In *Proceedings of DATE 2012*, pages 75-80, 2012.

[29] R. Min, M. Bhardwaj, S.-H. Cho, A. Sinha, E. Shih, A. Wang, and A. Chandrakasan. An architecture for a power-aware distributed microsensor node. In *Proceedings of the IEEE Workshop on Signal Processing Systems*, pages 581–590, 2000.

[30] R. Moghe, Y. Yang, F. Lambert, and D. Divan. A scoping study of electric and magnetic field energy harvesting for wireless sensor networks in power system applications. In *Proceedings of the IEEE ECCE*, pages 3550-3557, 2009.

[31] G. Montenegro, N. Kushalnagar, J. Hui, and D. Culler. Transmission of IPv6 packets over IEEE 802.15.4 networks. Internet Engineering Task Force RFC-4944, 2007.

[32] P. Nintanavongsa, U. Muncuk, D. R. Lewis, and K. R. Chowdhury. Design Optimization and Implementation for RF Energy Harvesting Circuits. *IEEE Journal on Emerging and Selected Topics in Circuits and Systems*, 2(1):24–33, 2012.

[33] H. Nishimoto, Y. Kawahara, and T. Asami. Prototype implementation of ambient RF energy harvesting wireless sensor networks. In *Proceedings of the IEEE Sensors 2010 Conference*, pages 1282–1287, 2010.

[34] T. Paing, J. Shin, R. Zane, and Z. Popovic. Resistor Emulation Approach to Low-Power RF Energy Harvesting. *IEEE Transactions on Power Electronics*, 23(3):1494–1501, 2008.

[35] J. A. Paradiso and M. Feldmeier. A compact, wireless, self-powered push-button controller. In *Proceedings of the ACM International Conference on Ubiquitous Computing*, pages 299–304, 2001.

[36] J. Polastre, J. Hill, and D. Culler. Versatile Low Power Media Access for Wireless Sensor Networks. In *Proceedings of the 2nd International Conference on Embedded Networked Sensor Systems*, pages 95–107, 2004.

[37] V. Raghunathan, A. Kansal, J. Hsu, J. Friedman, and M. Srivastava. Design Considerations for Solar Energy Harvesting Wireless Embedded Systems. In *Proceedings of the Fourth International Symposium on Information Processing in Sensor Networks*, pages 457–462, 2005.

[38] A. Sample and J. R. Smith. Experimental results with two wireless power transfer systems. In *Proceedings of the 2009 IEEE Radio and Wireless Symposium*, pages 16–18, 2009.

[39] C. Schurgers, V. Tsiatsis, S. Ganeriwal, and M. Srivastava. Optimizing sensor networks in the energy-density-latency design space. *IEEE Transactions on Mobile Computing*, 1(1): 70–80, 2002.

[40] W. K. G. Seah, Z. A. Eu, and H.-P. Tan. Wireless Sensor Networks Powered by Ambient Energy Harvesting (WSN-HEAP) – Survey and Challenges. In *Proceedings of the Wireless VITAE'09*, pages 1–5, 2009.

[41] R. C. Shah and J. M. Rabaey. Energy aware routing for low energy ad hoc sensor Networks. In *Proceedings of the IEEE Wireless Comm. and Networking Conference*, pages 350–355, 2002.

[42] S. Sudevalayam and P. Kulkarni. Energy Harvesting Sensor Nodes: Survey and Implications. *IEEE Communications Surveys & Tutorials*, 13(3):443-461, 2011.

[43] C. Svelto, M. Ottoboni, and A. M. Ferrero. Optically-supplied voltage transducer for distorted signals in high-voltage systems. *IEEE Transactions on Instrumentation and Measurement*, 49(3):550–554, 2000.

[44] Y. K. Tan and S. K. Panda. Self-Autonomous Wireless Sensor Nodes with Wind Energy Harvesting for Remote Sensing of Wind-Driven Wildfire Spread. *IEEE Transactions on Instrumentation and Measurement*, 60(4):1367–1377, 2011.

[45] Y. K. Tan and S. K. Panda. *Review of Energy Harvesting Technologies for Sustainable WSN, Sustainable Wireless Sensor Networks*. W. Seah and Y. K. Tan (Ed.), ISBN: 978-953-307-297-5, InTech, 2010. Available from: http://www.intechopen.com/articles/show/title/review-of-energy-harvesting-technologies-for-sustainable-wsn.

[46] R. J. M. Vullers, R. V. Schaijk, H. J. Visser, J. Penders, and C. V. Hoof. Energy Harvesting for Autonomous Wireless Sensor Networks. *IEEE Solid-State Circuits Magazine*, 2(2):29–38, 2010.

[47] Z. G. Wan, Y. K. Tan, and C. Yuen. Review on Energy Harvesting and Energy Management for Sustainable Wireless Sensor Networks. In *Proceedings of the IEEE International Conference on Communication Technology (ICCT'11)*, pages 362–317, 2011.

[48] H. Wang, J.-D. Park, and Z. Ren. Active Energy Harvesting from Microbial Fuel Cells at the Maximum Power Point without Using Resistors. *Environ. Sci. Technol.*, 46(9):5247-5252, 2012.

[49] X. Wang, G. Xing, Y. Zhang, C. Lu, R. Pless, and C. Gill. Integrated coverage and connectivity configuration in wireless sensor networks. In *Proceedings of the 1st International Conference on Embedded Networked Sensor Systems*, pages 28–39, 2003.

[50] R. Want. An Introduction to RFID Technology. *IEEE Pervasive Computing*, 5(1):25–33, 2006.

[51] Y. Xu, J. Heidemann, and D. Estrin. Geography-informed energy conservation for ad hoc routing. In *Proceedings of the ACM International Conference on Mobile Computing and Networking*, pages 70–84, 2001.

[52] W. Ye, J. Heidemann, and D. Estrin. An Energy-efficient MAC Protocol for Wireless Sensor Networks. In *Proceedings of the 21st Annual Joint Conference of the IEEE Computer and Communications Societies*, volume 3, pages 1567–1576, 2002.

[53] http://www.eere.energy.gov/basics/renewable_energy/pv_cell_conversion_efficiency.html

[54] http://www.energysavingtrust.org.uk/Generate-your-own-energy/Solar-panels-PV

[55] http://www.powercastco.com/products/powerharvester-receivers/

[56] http://www.renewableenergyworld.com/rea/tech/solarpv

7

Fault Tolerant Industrial Wireless Sensor Networks

Ataul Bari and Jin Jiang

The University of Western Ontario, London, Ontario, Canada

Arunita Jaekel

The University of Windsor, Windsor, Ontario, Canada

CONTENTS

7.1 Introduction

Although sensors and/or networks of sensors are widely used in many industries, most of these still use wires for signal transmission, which incur high installation and maintenance costs [21]. Recently, wireless sensors have started being deployed in industrial and manufacturing applications because of their many advantages such as reduced infrastructure and operating costs, easier deployment and upgrade, and greater mobility and placement freedom [28], [53]. Sensors, installed on industrial equipment, can monitor critical parameters such as temperature, pressure, level and vibration, and can transmit these information to a sink node through wireless links. This allows the plant operators to learn about any potential problems associated with the equipment, and can be used as an advanced warning system [16]. A wireless sensor network (WSN) can also be deployed in hazardous or not-so-easily-accessible environments where the wired alternatives may be expensive and/or difficult to install (e.g., in a nuclear power plant, inside moving components of motors and engines). Due to the growing interest from the industrial community, several international standards have been developed recently for industrial WSNs (IWSNs), including ZigBee, WirelessHART, and ISA 100 [16].

Industrial wireless sensor networks are essentially wireless sensor networks that have been specifically adapted to industrial applications. As such, many of the techniques developed for WSNs in general can also be applied to IWSNs. However, IWSNs also have a number of additional requirements such as:

- Reliable and real time data delivery. An IWSN should be able to respond to the changing networking conditions proactively, and in a timely manner [17], [53].

- Stringent dependability. Faults leading to wireless system failures may result in economic losses, environmental damage, or pose a threat to the safety of people nearby [40].

- Stringent Quality-of-service (QoS). The data reported to the sink node must represent what is actually occurring in the industrial environment [16].

- Support for a large number of sensor nodes. An IWSN may include over 100 sensor/actuators in a physical area of a few meters radius, in a discrete manufacturing environment [53], and the network should be able to handle them.

- Energy-efficient data communication. An IWSN should use an appropriate data transmission scheme that conserves energy as much as possible, as nodes are usually powered by batteries [31].

- Ability to operate in industrial environments. An IWSN should be able to communicate reliably in industrial deployment, even with complicated layout, stationary and moving obstacles, and radio interferences [17], [39].

- Built-in fault tolerance and secure communication. An IWSN should be capable of handling failures in communication links and sensor nodes, as well as, be resistant to possible security threats [39].

There has been a steady growth of industrial applications for wireless sensor networks in recent years. However, the pace at which IWSN systems are being adopted in industries has been impeded to some extent by the system availability and reliability requirements for industrial applications [45]. The widespread adoption of WSNs in industries requires that the deployed networks be able to meet the specified reliability requirements, in terms of safety, security and availability, with a high degree of effectiveness. The reliability of an IWSN is challenged by communication failure due to poor quality radio frequency (RF) links, high background noise levels, low signal strength, longer path lengths, obstacles, and multipath fading [51]. Industrial devices such as motors, pumps, actuators, electrical switches, and relays can create significant electromagnetic interference (EMI) for wireless transmission channels. EMI from other sources in the same radio band may corrupt the transmitted signals or even make it impossible to decode them at the receiver. Changes of the industrial environment due to variations in temperature, pressure, humidity, presence of heavy equipments, and variations in radio conditions can lead to uncertainties, packet loss and transmission delay [51]. For an IWSN to become widely acceptable in industrial applications, these issues and their effects on the reliable operation of the system have to be resolved. Three critical requirements for a reliable IWSN have been identified in [1] as:

- Safety: Safety of humans, environment, and property in any industrial setting is extremely important. Even components that are not safety-critical need to be designed and implemented with care. Equipment should be designed such that if a problem should occur, it can be detected, and the process is put into a safe state. Recent work in safety-critical communication has been reported in [2].

- Security: The data transmitted over an IWSN may not be confidential. However, authentication, integrity, and non-repudiation remain important security objectives. Furthermore, adversarial intrusions can result in harmful, or even disastrous situations in an industrial setting. Security issues have been addressed in [12], [39].

- Availability: The availability of an IWSN is extremely important since even transient faults can cause significant loss in production. In this context, the use of self-healing, self-configuring mesh networks have been proposed because of its inherent redundancy. Mesh networks can also improve the network performance, balance the load and extend network coverage; however, network capacity, fading and interference have been reported as major challenges in mesh networks [18].

Availability of an IWSN largely depends on the fault tolerance capability of the system and the network itself. Fault tolerance ensures that a system remains operational, preferably without any interruption, even in the presence of faults in some underlying components such as network nodes, links, or node/link subsystems. Thus, fault tolerance enhances the reliability, availability, and consequently dependability of the system [25].

An IWSN may fail for various reasons, including radio frequency interference, de-synchronization, battery exhaustion, or dislocation. A fault can come from various sources. For example, from sensors and network components, software and hardware, environmental conditions, or due to conflict in timing of other legitimate activities [43]. In the network protocol stack, faults can occur in any layer, and a faulty node/link can disrupt the communication in the entire network. For providing enhanced availability, an IWSN system must be able to resume its normal operations, i.e., *recover*, in the event of certain minor faults in its underlying components. However, in a fault tolerant IWSN design, different types of faults may need to be addressed separately, and one single solution is unlikely to be found for all faults. This chapter has focused on fault tolerance techniques proposed in the network *protocol stack* to handle potential faults that can disrupt data communication in an IWSN.

The rest of the chapter is organized as follows. In Section 7.2, several typical sources of faults in IWSNs, along with techniques for detecting and handling them, are outlined, and some specific types of network level faults are discussed. The fault tolerance mechanisms for managing network faults in three major IWSN standards are discussed in Section 7.3. In Section 7.4, several promising fault tolerant WSN design approaches that can be easily adapted to improve network level fault tolerance of an IWSN are presented. The chapter concludes with some remarks in Section 7.5.

7.2   Faults in IWSNs

A resilient IWSN system must be able to perform two basic functions, *fault detection* and *fault recovery*. Accordingly, in order to design a fault-tolerant system [43], it is necessary to i) understand the sources of various faults in the network, ii) detect whether a specific component is or will become faulty, and iii) take appropriate steps to prevent or recover from the detected fault. These issues have been discussed in this section.

7.2.1   Sources of Faults in IWSNs

A fault in an IWSN can occur in individual nodes, at the network layer, or even in the sink node itself. Individual sensor nodes have many hardware and software components that may be vulnerable to failure. For example, exposure to environmental stress such as contact with liquid, or unexpected shock impacts may lead to failures, as reported in [29]. Also, depletion of battery power can often lead to incorrect readings [46]. Another common source of error is software related [48]. Software errors can have a disastrous effect on the network, particularly if the defective node is also responsible for aggregating/forwarding data from other nodes.

In IWSNs, communication links between nodes are highly volatile, which often results in reduced data rates [44] and constant changes in routing paths

[38]. If some nodes have mobility, this may result in certain nodes becoming unreachable due to topology changes. Another source of link failure is due to radio frequency interference from other devices and the coexistence conflicts with other nearby networks.

Faults in the sink node mean a potential for loss of all the data gathered by the sensors in the network. Sink failure has been reported in [29], where the sink was in a remote location and powered by solar cells. After the solar panels had been unexpectedly covered with snow, the sink node suffered from a power failure.

7.2.2 Fault Detection in IWSNs

The goal of fault detection is to detect "abnormalities" by verifying that the services being provided are functioning properly and to predict, if possible, whether they will continue to function in the near future [43]. In many cases, it is possible for a node to perform a self-diagnosis to check for any possible anomalies. For example, the work in [20] has introduced fault tolerance into WSNs by monitoring the status of each wireless sensor node. The focus has been on the detection of physical malfunctions of sensor nodes caused by impacts or incorrect orientation. Similarly, battery depletion can be detected if nodes can measure the battery output [34]. Incorrectly generated values from a faulty sensor node can also be detected using group detection. Typically, it is expected that sensors from the same *region* will generate similar values as additional/redundant sensors can usually be deployed in the regions in order to obtain finer-grained information [24]. Hence, a fault probability can be calculated based on the information from the neighbors of a node [24]. Another approach, based on quartile method, has been proposed in [47] for fault tolerant sensing. The approach can select correct data based on data discreteness so that actors can perform appropriate actions.

Another fault detection approach, used in hierarchical *cluster-tree* based networks, determines that a gateway (or a cluster head) has failed if no other gateway can communicate with it. This type of group detection requires regular exchange of status updates among nodes. Hierarchical detection can be implemented by creating a *detection tree*, where each node forwards the status of its child nodes to its parent node [36]. This approach is scalable but consumes network resources.

7.2.3 Fault Recovery in IWSNs

Fault recovery techniques enable IWSNs to continue operating even in the presence of certain faults. Most IWSNs have some built-in redundancies, which can be exploited to improve the overall reliability and availability of the network through *replication*. Replication based approaches can be classified as *active* or *passive*. In active replication, all requests are processed by all replicas. If an anomalous value is detected from a node, the reading from that node can simply be discarded [20]. In terms of the network layer, multi-path routing can be used to replicate routing paths [32].

In passive replication, the primary component receives all requests and processes them. A backup component takes over only when the primary one fails [43]. If there are multiple backup replicas, an important issue is to determine which one shall start operating as the *service provider* to replace the services that had been the responsibility of the failed component. This can be achieved through self-election, group-election, or hierarchical election. In this context, a learning and refinement module has been proposed in [37], which enables an adaptive and self-configurable fault tolerance solution based on changes in the network conditions.

7.2.4 Network Faults in IWSNs

An IWSN must ensure that the data from individual nodes in the network can reach the sink node. Usually, successful data communication involves forwarding data over a set of network nodes and links using some defined protocols. However, this communication can be disrupted by faults from several networking components. From a network communication perspective, a fault can be defined as *an adverse occurrence at any network component which may render corrective operation of the network.* Network faults can be *transient* (e.g., caused by noise or electromagnetic interferences) or *permanent* (usually as a result of hardware malfunctions) [40].

A network fault can occur in different layers of the protocol stack. For example, a fault can occur in the MAC layer due to an incorrect clock synchronization [10], in the physical layer due to a strong electromagnetic interference, or in the network layer due to a routing path becoming unusable as an intermediate node has depleted its battery power. Whenever possible, faults should be addressed locally without modifying the data transmission scheme, as the latter involves higher complexity and consumes additional network resources. Examples of these include *retransmission* of data in case of missing acknowledgements, and use of *forward error correction* techniques in case of noisy links. In this context, use of *chirp spread spectrum* technology [21] and *adaptive forward error correction* mechanism [11] have been studied for industrial WSN. However, this may not always be possible. When the fault cannot be dealt with locally, the network level fault management schemes are used in a fault tolerant IWSN to detect and handle faults with minimum disruption to the functionality of the network.

Faults in a network component (e.g., node, link) can be detected by the neighboring nodes when they fail to receive any transmissions from the failed node (or over the failed link). Such faults can be handled by the network protocol stack (e.g., in MAC layer, in the network layer) to provide a fault tolerant IWSN, which is the focus of this chapter. Therefore, in the subsequent discussion, *faults* and *fault tolerant* networks have been used in the sense of failures in the network components (e.g., nodes, links), and fault tolerance in the protocol stack (e.g., MAC layer, network layer), respectively.

In the following section, three major IWSN standards, along with their fault tolerance techniques, are discussed. Several approaches that enhances the built-in fault tolerance mechanisms of these standards are also presented.

7.3 Fault Handling in IWSN Standards

Over the past few years, several industrial standards have been developed for IWSNs that include ZigBee, WirelessHART, and ISA 100 [16]. These standards define basic fault tolerance/fault recovery mechanisms for IWSN, as discussed in the following sections.

7.3.1 ZigBee Networks

ZigBee [52] is a standard for a suite of high level communication protocols. It provides wireless communication and networking solution for low data rate and low power consumption applications such as home and industrial automation. ZigBee executes on small, low-power digital radios based on the IEEE 802.15.4 standard, which specifies two bottom layers of the network protocol stack, namely, the physical layer (PHY) and the medium access control layer (MAC). The ZigBee protocol defines the higher layers of the protocol stack, i.e., the network (NWK) layer and the application (APL) layer [14]. The ZigBee devices operate in the unlicensed radio frequency bands centered at 2.4 GHz, 915 MHz, and 868 MHz ISM band, respectively. The ZigBee protocol is designed to provide reliable data communication through hostile radio frequency (RF) environments.

ZigBee specifies three types of networking devices, *coordinator*, *router*, and *end device*. The coordinator is responsible for initializing, maintaining, and controlling the network. A router is responsible for routing messages, and an end device usually performs the sensing task. ZigBee specification enables three network topologies: *star*, *mesh*, and *cluster-tree* [5]. Figure 7.1 illustrates examples of each topology along with the device roles in each topology. In the star topology, a coordinator device is the central node. In a mesh topology, a coordinator acts as a central node; however, each node can communicate with any other node. A cluster-tree network consists of clusters, and typically, a router acts as a cluster head in each cluster. A coordinator takes the role of the master. The coordinator is linked to a set of routers and end devices, which become its *children*. The routers can be linked to a set of other routers and end devices, which then become its children (and hence, the *grand children* of the coordinator). This process can continue for multiple levels. Therefore, the topology forms a *tree* rooted at the coordinator (hence the name tree-

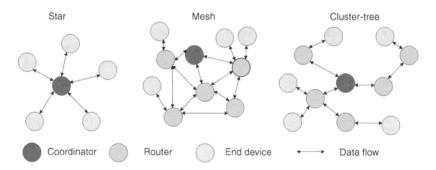

FIGURE 7.1
Star, mesh, and cluster-tree topology of ZigBee network.

topology), and there is a single routing path between any pair of nodes. The network is formed by *parent-child* relationships (a child can have only one parent at a time), and each new device is required to associate itself as a child of an existing cluster head/coordinator [5], [14]. An end device does not have a child node.

A ZigBee network is initiated by a coordinator, and any device that wants to join the network can do so by specifying its role (e.g., router, end device), and performing the *association procedures*, defined in the standard. If accepted, it is allocated a 16 bit network address. Once the devices are associated with a network, they can send/receive data to/from the coordinator. However, a child device may lose connectivity with its parent, due to either failure of the wireless link (e.g., due to electromagnetic interference (EMI), introduction of obstacles between the nodes), or failure of the parent device (e.g., failure of hardware, battery, software) [5]. To address this issue, IEEE 802.15.4/ZigBee protocol stack specifies fault-tolerance through standard mechanisms.

When a ZigBee network uses mesh topology, multiple paths interconnect each pair of nodes. The connections are updated dynamically through a built-in mesh routing table. Also, nodes have the capability of self-discovery, and can reconfigure routing paths if some nodes leave the network. Therefore, the mesh network is relatively stable under changing conditions and/or to the failure of a single link/node.

However, a single-point failure may impact the network operation, depending on the role of the failed device, when a ZigBee network uses a cluster-tree topology. Usually, failure of an end device does not impact the overall operation of the network, because it has no child node, and hence, does not forward data for any other nodes. On the other hand, failure of the coordinator will require electing a new root for the tree and re-configuring the entire network. Failure of a cluster head device will disconnect all children of the failed device from the network [6].

A child device that has lost connectivity to its parent (either due to the device failure or link failure) is called an "orphan." The process of associating an orphan device to the network is called the *orphan process* [14]. A child device uses MAC layer defined constants for *maximum allowable beacon frame* (from its parent) losses, and *maximum re-transmissions* to determine if it has been orphaned. Once it determines that it has been orphaned, the child device initiates the orphan process, which is defined in IEEE 802.15.4/ZigBee standards as its basic fault-tolerance mechanism.

The orphan process begins by an *orphan scan procedure*, where the orphan device performs a physical channel scan using the same list of channels of the association process and sends a MAC directive known as *orphan notification* [14]. The parent device, if it receives the notification, replies with a *coordinator realignment* frame. The orphan device, if and when it receives the realignment frame, stops the channel scan procedure, updates its information, and the realignment between the parent and the child is accomplished. The goal of the realignment procedure is to return the orphan device to its current parent [14]. On the other hand, if the orphan device completes the channel scan but cannot find its parent, it starts a new association mechanism, where the device once again scans all channels and searches for a suitable parent. Once found, the device starts *association procedures* after finishing the synchronization with the new parent. An orphan device cannot transmit or receive any messages while scanning, synchronizing and associating, and hence remains inaccessible for the normal network operation [5].

The orphan process is key to the fault recovery in ZigBee networks. However, this process introduces delays during which the orphan device remains inaccessible. As mentioned earlier, in a standard orphan process, the child first searches for the parent to which it was connected, so that it can return to it. In the process, both devices exchange *orphan commands* that require competing for the channel access for transmitting data frames, beacons, and acknowledgements. Such communications introduce traffic delay and additional energy consumption, especially under high channel utilization conditions [5], [14]. Moreover, if an orphan device fails to reconnect to its parent, the association process must begin for obtaining a new coordinator, which could introduce additional delays [14]. When an orphan device is the parent of some other devices, a long inaccessible time may imply many more orphaned children and grandchildren devices in the network. Therefore, several approaches have been proposed recently to improve the standard orphan mechanism of IEEE 802.15.4/ZigBee protocol. Some of these approaches are discussed below.

Reactive and Proactive approaches for orphan process [5]: This work has focused on reducing the inaccessible time that may be experienced by an orphan device. Two different approaches have been proposed, namely, a *reactive* approach and a *proactive* approach. Both approaches make use of a Parent Adoption Quality Indicator (PAI), which has been defined based on several metrics such as the link quality indicator, the depth of the candidate parent in the tree, traffic load, and the energy indicator. While choosing a new parent, the orphan uses the PAI to assess the adoption potential of the set of available new parents.

The reactive approach allows a child node to react *quickly* to a total link failure with its parent by re-associating to a new parent. This is achieved by enhancing the standard orphaned device realignment procedure to ensure that the orphan device performs the channel scan procedure only once, whether to realign itself with its current parent or to associate with a new parent. In a reactive re-association mechanism, while the device performs an orphan scan procedure to search for its parent, it simultaneously searches for other potential parents. In each scanned logical channel, an orphan notification is sent, and the device also listens to other potential parent devices while waiting for a realignment command. If the device receives its parent realignment command, it re-associates itself with its previous parent. Otherwise, if the child device discovers a potential parent, it calculates and stores the value of the PAI indicator for each of them. After scanning all channels, if an association with the previous parent cannot be established, the device selects the best potential parent based on the PAI indicator, and starts a re-association procedure.

The proactive approach is beneficial when a child device experiences *frequent* problems (i.e., temporary or transient) with its current parent, and switching to a new parent appears to be a better option. It allows the device to plan a re-association to a more reliable parent in advance, and hence, to avoid the device re-association procedures. This can be seen as a preventive measure, where a child device may change a parent to avoid the total loss of connectivity or substantial degradation of the link with its current parent. For this approach to work, a node has to periodically sample the PAI of its current parent. If the PAI falls below a certain threshold (which indicates a degradation in connectivity), some additional consecutive samples are taken and compared with the threshold. If the degradation is confirmed, the device then searches for a more suitable parent (in terms of the PAI) during the *inactive period* (a period when nodes switch to a *sleep* mode to save energy). If such a parent is found, the device associates itself to the new parent and disassociates from the current one.

Optimized Orphan Algorithm (OOA) [14]: The OOA algorithm also aims to decrease the amount of delay during the orphan process. Additionally, it reduces the associated power consumption, which is very important for an IWSN that runs on battery. The algorithm allows prioritizing the orphan commands with respect to any other communications. The orphan device keeps a backup list of other possible coordinators. If the device does not find its former coordinator, it simply selects one from the list, avoiding the other standard processes.

Problems with the orphan process have also been addressed in [15], [22], [30], and [35]. An orphan device needs to obtain a 16-bit network address from the coordinator or a router to join a network. However, a problem may arise if the corresponding coordinator/router runs out of addresses while unused address spaces still remain in the network. The schemes proposed in [15] allow borrowing of addresses from other nodes (which have spare addresses), when a node's sub-address space is insufficient. The limitation on both the number of child nodes and the depth of the network has been considered in [30], and an approach to reduce the orphan routers has been proposed. An on-demand scalable address assignment algorithm based on segmentation of address spaces has been proposed in [35]. In [22], the orphan problem has been addressed where the objective has been to reduce the passive scan time in any topology, with an aim for reducing energy-consumption.

The standard orphan process provides fault tolerance in ZigBee networks. However, it has limitations. The approaches discussed above enhance, in various ways, the standard process to overcome these limitations, and hence, extends the fault tolerant capabilities of ZigBee networks.

In a fault tolerant network, a failure of one device should not bring down the entire network. However, one of the pitfalls of ZigBee is that the failure of the coordinator can potentially affect the entire network. The star topology fails completely. Being the root node, a failure in the coordinator also renders a tree topology nearly inoperational. The mesh topology is robust against failures of a single device. However, this may also fail if *indirect binding* is used, as under such configuration, the binding table describing the necessary routing information is stored at the coordinator [23].

A failure of the coordinator device is typically handled by assigning the role of a coordinator to a surviving device, while the application is still running [13]. However, some disadvantages of this approach are reported in [23] as:

- It cannot be used in all topologies.

- A dynamic role assignment might change the hardware address through which the device can be addressed. This requires the application on top of the ZigBee stack to ensure that a failure is handled properly, and the addressing and the application are not disturbed.

- It is not completely *transparent* to device applications and to device users.

A transparent solution to handle failures of the coordinator device is presented in [23]. The approach can be used with any network topology. The idea is to mirror the critical device, and to ensure that the backup also copies the MAC address and the ZigBee key, as these are required to enable a fully-functional ZigBee stack. In addition, network-specific tables, such as routing, binding, and neighborhood tables are also copied, since ZigBee stack needs these tables to maintain the network and to organize communication. In case of a failure of the coordinator device, the backup device switches its role to the coordinator, together with the MAC address and the ZigBee key. The method is claimed to be completely transparent to the user, where the system only sends a report to the user that there has been a failure. However, a few disadvantages of the approach have been noted, which include increased network traffic due to the *ping* messages, and the administrative traffic as the coordinator has to send its entire network settings when there is a change in the network structure.

This approach can be seen as an example of fault tolerance by replication, as it adopts the concept of creating a backup copy and uses it in the event of faults in the primary device.

7.3.2   WirelessHART Networks

WirelessHART, released in September 2007, is the first open wireless standard for industrial process control [41]. It is a secure networking protocol operating in the 2.4 GHz ISM band. WirelessHART is based on the physical layer of IEEE 802.15.4; however, it specifies new data link, network, transport, and application layers.

WirelessHART employs a central network manager that configures the network and schedules all activities in the network. The network manager is a key entity that looks after the overall management, scheduling, and optimization of the network. It is responsible for initializing and maintaining communication parameter values and allowing devices to join and leave the network. It also manages all dedicated and shared network resources and collects and maintains network health monitoring and fault diagnosis. The network devices are required to report traffic statistics and network uses to the network manager such as signal levels, packet loss rate for each neighbor, and discovery of new neighbors [33], [42].

In a WirelessHART network, each device can act as a router (forward packets on behalf of other devices) and can be the source and the sink of packets. In the network layer, WirelessHART supports mesh networking technology that provides redundant paths, which permits routing messages through different routes for avoiding broken links, interference, and physical obstacles. This flexibility offers the capability of self re-organizing and self healing of a WirelessHART network [42].

In a WirelessHART network, mainly the *graph routing* is used, although *source routing* has also been defined [41] in the standard.

- Graph Routing: A graph is a collection of paths connecting the network nodes and is identified by a graph ID. The network manager creates the paths in each graph and sends this information to each individual network device. For communication, a specific graph ID (based on the destination) is written by the source node in the network header. All intermediate devices on the path from the source to the destination are aware of the neighbors to which the packets may be forwarded as they are pre-configured with graph information [41]. To ensure reliable data forwarding, each intermediate device is given no less than two nodes in the graph that they can use to send the packets to. Such redundancy facilitates transmission retries in the event of a fault [33], and adds fault tolerant capability to the WirelessHART network.

- Source Routing: This is a supplement to the graph routing and is used primarily in the network commissioning phase for diagnostics and testing. To send a message to its destination, a source device includes an ordered list of intermediate nodes in its header. As the message is routed, each intermediate device receives the next network device address from the list and determines the next hop until the destination device is reached. In addition, the device at a point of failure must notify the network manager for remedy actions so that the message can be delivered.

WirelessHART uses mesh topology that offers redundant paths between any pair of source and destination nodes. However, not all paths have the same reliability. The reliability of communication in a standard WirelessHART network can be affected by the harsh industrial environment. If the reliability of communication for all the routing graphs can be evaluated, the network manager may use that information to set up most reliable routes. In [33], one such strategy, based on Finite-state Markov Model, has been proposed. It mitigates the limitations imposed by such unpredictable environmental variations through reliability evaluation and prediction technology. Such evaluation is feasible as the WirelessHART network is typically a static network with mesh topology, as opposed to the ZigBee, which can be considered as a relatively dynamic mesh network. The idea in [33] is to evaluate the success probability of communication graphs, which can be used by the system manager for calculating the routing schedule in the network. It can choose the highest success probability graphs for each node. Using this approach, the system manager is able to determine the most reliable routing graph by not only comparing single links but also considering the whole network. In addition, the proposed approach can alert the network operator if the success probability of some communication links falls below a preset value so that necessary corrective actions can be initiated immediately.

WirelessHART has been designed to provide security, reliability, safety, and timeliness in industrial wireless sensor networks. It combats faults through the provision of redundant paths between a gateway and field devices. It uses AES-128 ciphers and keys at both MAC and network layers for secure communication. WirelessHART provides reliability and co-existence with other IEEE 802.15.4-based devices by introducing channel hopping on top of the IEEE 802.15.4 physical layer.

7.3.3 ISA100.11a Networks

ISA100.11a standard, defined by the Instrumentation, Systems, and Automation Society (ISA), ISA100 committee, aims at providing a reliable and secure wireless system for industrial applications. It addresses the non-critical process applications that can tolerate delays up to 100 ms. ISA100.11a has many aspects in common with WirelessHART, such as network topology, TDMA access method, superframe structure, and routing strategy. However, ISA100.11a aims to provide more comprehensive and flexible service with a larger set of options.

ISA-SP100.11a adopts the existing physical layer of the IEEE 802.15.4-2006 and operates in the unlicensed 2.4-GHz ISM radio band only. It supports channel hopping options to improve robustness against interferences. Furthermore, it can blacklist some frequency bands from the frequency hopping and use adaptive frequency hopping to deal with crowded frequency bands, which can be used to deliver a reliable network [49]. The routing strategy of the ISA100.11a is very similar to that of WirelessHART, both adopting the graph routing and the source routing schemes. It also adopts the centralized network structure with mesh topology. Due to this similarity, ISA100.11a offers robustness, fault tolerance and reliability in the similar fashion as in the WirelessHART network.

Fault tolerant stabilizability in the ISA100 networks has been investigated in [9]. ISA100 supports industrial control systems that require wireless communication between sensors, actuators, and computational units. Industrial control usually involves conveying information from sensors to the controller and from the controller to actuators through multiple communication hops. Such communication needs to address issues such as fading, time-varying throughput, communication delays, and packet losses. Additionally, analysis of stability, performance, and reliability of wireless networked control systems requires addressing issues such as scheduling and routing with real time communication protocols. The ISA100 standard specifies mechanisms for scheduling; however, schedules and routing remain a challenge for the network designers. A mathematical model incorporating the effect of scheduling and routing on the control system has been presented in [9]. The approach is based on the concept of multi-hop control network (MCN). The MCN model allows modeling multi-hop control networks that implement the ISA100 standard mechanism. However, it allows modeling general routing and scheduling communication protocols, which specify TDMA, FDMA, and/or CDMA access to a shared communication resource, for a set of communication nodes interconnected by an arbitrary radio connectivity graph [9]. The proposed methodology designs scheduling and routing of a communication network, which preserves controllability and observability, for any set of failures configurations that preserve connectivity within the scheduling period between the controller and the plant. The approach provides fault tolerant stability and can be used in WirelessHART networks as well.

In this section, three standard fault tolerant IWSNs have been discussed, and their fault tolerance mechanisms have been outlined. In the following section, several promising approaches for fault tolerant, cluster-tree based IWSN design are reviewed.

7.4 Fault-Tolerant IWSN Design

The network topology is the backbone for any network based system. The topology for an IWSN application should be chosen based on a number of considerations such as connectivity, adaptability, mobility, scalability, responsiveness, and reliability [51]. A hierarchical, cluster-tree based network topology has been identified as an important network model in industrial settings [8], [49]. In this network model, sensor nodes form clusters and forward data to their respective cluster heads. The cluster heads form a tree topology and forward their collected data to the sink node (also known as the *base station*), which is the root of the tree. An example of a hierarchical, cluster-tree based WSN is shown in Figure 7.2. As discussed in Section 7.3, major industrial standards also provide support for this topology. Several fault tolerant design techniques have been proposed in the literature for this network model. Some of them are presented in this section.

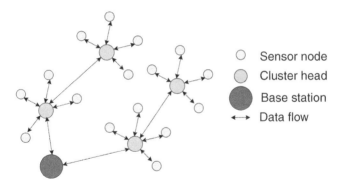

FIGURE 7.2
A hierarchical, cluster-based WSN.

7.4.1 Fault Tolerant Routing

IWSNs require a real-time and reliable data delivery, along with an energy-efficient transmission scheme. The inherent vulnerabilities of wireless links in a harsh industrial environment, and typically high failure rates of sensor nodes, can lead to undesirable disruptions of the normal operation of an IWSN. A node/link failure along a routing path may cause the entire route to fail, preventing a network packet from reaching the destination, or delaying it beyond an acceptable time limit. Therefore, a fault tolerant routing protocol must be used in an IWSN that is reconfigurable and energy efficient. Fault tolerant routing approaches proposed in the literature can be classified as *retransmission* based, and *replication* based [25].

- Retransmission based: In this approach data packets are retransmitted to the sink for a predetermined number of times. The sink node transmits an acknowledgement back to the source when a data packet is received, indicating a successful transmission. If the acknowledgement is not received by the sender before a timeout, the data packet will be retransmitted (a predetermined number of times). This method ensures that lost packets are recognized and eventually reach the sink. However, multiple transmissions increase the network traffic and may lead to increased delay in delivery and packet loss due to collisions. Furthermore, more memory space is needed in the sensor node to buffer the packet until it receives an acknowledgment from the destination.

- Replication based: In this approach, multiple copies of the same packet are transmitted to achieve a higher reliability. One technique is multi-path routing, where a set of alternate paths between the source nodes and the sink are determined at the expense of increased energy consumption and traffic generation. An extreme example is to use the *flooding* (packets are disseminated to all nodes in the network) to achieve reliability.

A fault tolerant routing scheme for cluster-tree based network is proposed in [7]. Some special nodes, called *relay* nodes, have been used to act as cluster heads and assigned the responsibility of routing the data to the sink. The objective of the proposed approach is to compute an alternate path to forward data in the event of a single relay node failure such that the energy consumption in the network is minimized. Such minimization, in turn, maximizes the lifetime of the network when the nodes in the network are powered by batteries. An integer linear program (ILP) formulation has been presented to find the optimal routing schedule for any single fault scenario. The proposed scheme considers the load on each relay node, and determines an alternate routing schedule in such a way that i) the updated paths avoid the failed relay node, and ii) the lifetime of the network is maximized. This route selection process is performed off-line, and the backup routing schedule for each possible fault is stored at either the individual relay nodes or at the sink node (which can be downloaded to the surviving relay nodes when a fault is detected).

Fault tolerant routing has also been addressed in a number of other papers, including those in [25] and [50]. In [25], a fault tolerant routing, called Routing with Error Reporting Protocol (RERP), is developed. It is a proactive approach that uses energy and past performance of the link as a metric for selecting the best route. The protocol also enables error reporting through exchange of query/response messages, which can be used for diagnostic purposes. There are three such error reporting messages: i) the link failure message, ii) the destination unreachable message and iii) the critical battery message. In addition, each node maintains a backup path, and whenever a sensor node fails to deliver a packet due to the non-availability of a packet delivery path, it switches over to the backup path. This provides for fault tolerance, thus increasing the reliability and availability of the system.

In [50] a dynamic jumping real-time fault-tolerant routing protocol (DMRF) has been proposed for time-critical WSN applications. Each node utilizes the remaining transmission time of the data packets, and the state of the forwarding candidate node set to choose the next hop dynamically. The DMRF has two data transmission modes: *hop-by-hop* mode and *jumping transmission* mode. Once a node failure, or network congestion or empty region occurs, the transmission mode switches to jumping mode, which can reduce the transmission time delay, ensuring that the data packets are sent to the sink node within the specified time limit. The jumping transmission mode, along with a feedback mechanism, is used to guarantee real-time and fault-tolerant characteristics. Furthermore, the average energy consumption of each node in the network is balanced when selecting the subsequent hop to prolong the lifetime of the entire network.

7.4.2 Fault Tolerant Node Placement and Clustering

In a cluster tree network, using cluster heads to route data through multi-hop paths, one of the best approaches for fault tolerant routing is to have redundant paths between each source-destination pair of nodes. To achieve this, it is necessary to ensure that there are sufficient (redundant) relay nodes (acting as cluster head and/or router) in the network so that it is possible to provide redundant edge/node disjoint paths between each pair of source-destination nodes. However, it is desirable to minimize the number of redundant nodes for several reasons: to reduce the cost of installation and maintenance of additional nodes, and to optimize the use of network resources. This requires that the relay nodes be placed in the network following some "placement" strategy. A fault tolerant placement strategy must ensure that in the event of any failure(s) in relay node(s),

i) each sensor node belonging to the cluster of a failed relay node should be able to send its data to another fault-free relay node, and

ii) data from all fault-free relay nodes are able to reach the base station successfully.

In [8], such a fault tolerant node placement approach has been presented that considers the possibility of faults in multiple relay nodes in a cluster-tree based, two-tiered network. Fault tolerance has been defined for both tiers of the network, namely, the *lower* and the *upper* tiers. The lower tier deals with the communication between the sensor nodes and the relay nodes (intra-cluster), whereas, the upper tier deals with the communication among the relay nodes (and the sink). The concept of the network being k_s-survivable and k_r-survivable in the lower and upper tier, respectively, has been introduced as follows:

• The lower tier network is k_s-survivable if each sensor node can communicate with at least k_s, $(k_s \geq 1)$ relay node(s). This means that each sensor node can still transmit its data to at least one relay node, even if there are up to k_s - 1 relay node(s) (that the sensor node can communicate with) fail.

- The upper tier network is k_r-survivable if each relay node can communicate with at least k_r, $(k_r \geq 1)$ other suitable relay nodes, each having a valid path to the sink.

The desired level of redundancy, i.e., the values of k_s and k_r will depend on the intended application, and can be adjusted for each specific IWSN. An integrated ILP formulation has been presented that meets the specified fault-tolerance requirements of the application.

The problem of relay node placement in fault tolerant network has also been addressed in several recent papers, for example, [3], [4], and [27]. In [27], a two-step approximation algorithm has been presented to obtain a 1-connected (in the first step) and a 2-connected (in the second step, by adding extra back-up nodes to the result of the first step) sensor and relay node network. In [3], relay node placement problem in 3D space has been considered, where the objective has been to maximize the network connectivity. Scenarios with different probabilities of node/edge failure have been investigated. In [4], partitioning of a network due to node failure has been considered, and a scheme for relay node placement on a 3D-grid for reestablishing connectivity using the remaining functional nodes has been proposed.

In a cluster based WSN, *fault tolerant clustering* is the ability of the children of a failed cluster head node to join other surviving cluster heads. Fault tolerant clustering problem has been addressed in [19], and a mechanism for recovering sensor nodes that belong to a failed cluster head has been proposed. This approach does not require a full-scale re-clustering. The approach works in two-phase, *detect* and *recover*. Fault detection requires the cluster heads exchanging message vectors periodically, indicating the status of other cluster heads. These vectors are used by each operational cluster head to identify the failed cluster heads. For the fault recovery, the scheme depends upon the backup information created during the cluster formation phase. During this phase, cluster heads identify all sensors within their radio range and partition these sensors into primary and backup cluster members. During the recovery phase, sensors in the backup sets are reassigned to the primary set of the cluster heads.

Another fault tolerant mechanism has been proposed in [26]. Due to the broadcasting nature of wireless transmissions, many nodes in the vicinity of a sender node can overhear its packet transmissions, which can be utilized in the fault tolerant mechanisms in WSNs. In this approach, all nodes within a cluster are involved in the processing and monitoring of links to the cluster head, and overhear transmissions from neighboring cluster heads. When cluster members detect that the fault is a permanent failure of the cluster head, they act cooperatively to select a new cluster head. When the cluster members detect that the fault is a medium error, they subsequently transfer themselves to the neighboring clusters or relay nodes. Both detection and recovery are performed locally, without requiring any global knowledge of the network.

7.5 Conclusions

In this chapter, several approaches proposed to handle faults at the network level in IWSNs have been discussed. The failure of a network node due to the harsh industrial environments, and faults in communication links due to electromagnetic interference from other sources are of particular importance for a fault tolerant IWSN. It can be concluded that existing fault tolerant mechanisms are capable of handling faults up to a certain degree; however, these mechanisms can be further augmented and new approaches may be investigated for enhancing the fault tolerance in IWSNs.

Three widely accepted industrial standards for IWSNs, ZigBee, WirelessHART, and ISA 100, define some basic fault tolerance mechanisms. However, these basic fault management schemes often have significant limitations, and there is a critical need to develop more effective techniques that can handle different types of failures and can be utilized in a wide range of applications. A number of recent advances in the area of fault tolerant WSNs design can be readily extended and/or adapted for IWSNs. Some proposed approaches that appear particularly suited for IWSNs include the design of fault-tolerant and energy-efficient routing techniques and the development of hierarchical, cluster tree based topologies, with built-in redundancy, to improve network availability, reliability, and fault tolerance.

The design of fault-tolerant IWSNs for industrial applications is an important and growing area of research. This study has concluded that, although significant progress has been made in this area, much work still remains to be done. There exists tremendous opportunity for the development of new models and strategies for fault tolerant IWSNs.

Acknowledgment:
A. Bari and J. Jiang would like to acknowledge the financial support for this work from the Natural Sciences and Engineering Research Council of Canada (NSERC); the Ministry of Research and Innovation (MRI), Ontario, Canada; the University Network of Excellence in Nuclear Engineering (UNENE), and ISTP Canada.

A. Jaekel would like to acknowledge the financial support from the NSERC for this work.

References

[1] J. Akerberg, M. Gidlund, and M. Bjorkman. Future research challenges in wireless sensor and actuator networks targeting industrial automation. In *Proceedings of the 9th IEEE International Conference on (INDIN 2011)*, pages 410–415, 2011.

[2] J. Akerberg, F. Reichenbach, and M. Bjorkman. Enabling safety-critical wireless communication using wirelesshart and profisafe. In *Proceedings of IEEE Conference on Emerging Technologies and Factory Automation (ETFA2010)*, pages 1–8, 2010.

[3] F. M. Al-Turjman, H. S. Hassanein, and M. A. Ibnkahla. Connectivity Optimization for Wireless Sensor Networks Applied to Forest Monitoring. In *Proceedings of IEEE International Conference on Communications (ICC 2009)*, pages 1–6, 2009.

[4] F. M. Al-Turjman, H. S. Hassanein, and S. M. A. Oteafy. Towards augmenting federated wireless sensor networks. *Procedia Computer Science*, 5(2011):224–231, 2011.

[5] S. B. Attia, A. Cunha, A. Koubaa, and M. Alves. Fault-Tolerance Mechanisms for Zigbee Wireless Sensor Networks. In *Proceedings of the 19th Euromicro Conference on Real-Time Systems (ECRTS 2007)*, 2007.

[6] S.B. Attia. *Fault-Tolerance Mechanisms for ZigBee Cluster-Tree Wireless Sensor Networks*. Master's thesis, Ecole Superieure des Communications de Tunis, 2007.

[7] A. Bari, A. Jaekel, and S. Bandyopadhyay. Energy Aware Fault Tolerant Routing in Two-Tiered Sensor Networks. In *Proceedings of International Conference on Distributed Computing and Networking (ICDCN 2011). LNCS 6522*, pages 293–302, 2011.

[8] A. Bari, A. Jaekel, J. Jiang, and Y. Wu. Design of fault tolerant sensor networks satisfying coverage, connectivity and lifetime requirements. *Computer Communications*, 35(3):320–333, 2012.

[9] M.D. Benedetto, Di D'Innocenzo, and E. E. Serra. Fault Tolerant Stabilizability of Multi-Hop Control Networks. In *Proceedings of IFAC World Congress*, pages 5651–5656, 2011.

[10] B. Bengi Aygun and V.C. Gungor. Wireless sensor networks for structure health monitoring: recent advances and future research directions. *Sensor Review*, 31(3):261–276, 2011.

[11] B.E. Bilgin and V.C. Gungor. Adaptive error control in wireless sensor networks under harsh smart grid environments. *Sensor Review*, 32(3):4–4, 2012.

[12] Y.-S. Choi, Y.-J. Jeon, and S-H. Park. A study on sensor nodes attestation protocol in a Wireless Sensor Network. In *Proceedings of Advanced Communication Technology (ICACT 2010)*, volume 1, pages 574–575, 2010.

[13] A. Coers, H.-C. Muller, and R. Kokozinski. An approach to reduce the risk of network dropouts through dynamic role assignments. In *Proceedings of 1st European ZigBee Developers' Conference*, 2007.

[14] A.-J. Garcia-Sanchez, F. Garcia-Sanchez, and J. Garcia-Haro. Enhancements in the Orphan Process for Wireless Personal Area Networks: Real Implementation Scenarios. In *Proceedings of 4th International Conference on Software Engineering Advances (ICSEA 2009)*, 2009.

[15] D. Giri and U.K. Roy. Address Borrowing in Wireless Personal Area Network. In *Proceedings of IEEE International Advance Computing Conference (IACC 2009)*, pages 181–186, 2009.

[16] V.C. Gungor and G.P. Hancke. Industrial wireless sensor networks: Challenges, design principles, and technical approaches. *IEEE Transactions on Industrial Electronics*, 56(10):4258–4265, 2009.

[17] V.C. Gungor, B. Lu, and G.P. Hancke. Opportunities and challenges of wireless sensor networks in smart grid. *IEEE Transactions on Industrial Electronics*, 57(10):3557–3564, 2010.

[18] V.C. Gungor, D. Sahin, T. Kocak, S. Ergt, C. Buccella, C. Cecati, and G.P. Hancke. Smart grid technologies: Communication technologies and standards. *IEEE Transactions on Industrial Electronics*, 7(4):529–539, 2011.

[19] G. Gupta and M. Younis. Fault-Tolerant Clustering of Wireless Sensor Networks. In *Proceedings of Wireless Communications and Networking*, volume 3, pages 1579–1584, 2003.

[20] S. Harte and A. A. Rahman. Fault Tolerance in Sensor Networks Using Self-Diagnosing Sensor Nodes. In *Proceedings of IEE International Workshop on Intelligent Enviroment*, pages 7–12, 2005.

[21] A. Kadri, R.K. Rao, and J. Jiang. Low power chirp spread spectrum signals for wireless communication within nuclear power plants. *American Nuclear Society Journal of Nuclear Technology*, 166:156–167, 2009.

[22] M. Kohvakka, J. Suhonen, M. Kuorilehto, Hnnikinen M., and T.D. Hmlinen. Network signaling channel for improving ZigBee performance in

dynamic cluster-tree networks. *EURASIP Journal on Wireless Communications and Networking,* Hindawi Publishing Corporation, 2008:1–15, 2008.

[23] R. Kolln and A. Zimmermann. Transparent coordinator failure recovery for ZigBee networks. In *Proceedings of IEEE Conference on Emerging Technologies & Factory Automation (ETFA 2009),* pages 1–8, 2009.

[24] B. Krishnamachari and S. Iyengar. Distributed Bayesian algorithms for fault-tolerant event region detection in wireless sensor networks. *IEEE Transactions on Computers,* 53:241–250, 2004.

[25] K. Kulothungan, J.A.A. Jothi, and A. Kannan. An adaptive fault tolerant routing protocol with error reporting scheme for wireless sensor networks. *European Journal of Scientific Research,* 60(1):19–32, 2011.

[26] Y. Lai and H. Chen. Energy-Efficient Fault-Tolerant Mechanism for Clustered Wireless Sensor Networks. In *Proceedings of the 16th International Conference on Computer Communications and Networks (ICCCN 2007),* pages 272–277, 2007.

[27] H. Liu, P. Wan, and W. Jia. Fault-Tolerant Relay Node Placement in Wireless Sensor Networks. In *Proceedings of of COCOON,* 2005.

[28] K.S. Low, W.N.N. Win, and M.J. Er. Wireless Sensor Networks for Industrial Environments. In *Proceedings of Computational Intelligence for Modelling, Control and Automation, 2005 and International Conference on Intelligent Agents, Web Technologies and Internet Commerce,* volume 2, pages 271–276, 2005.

[29] K. Martinez, P. Padhy, A. Riddoch, H. Ong, and J. Hart. Glacial environment monitoring using sensor networks. In *Proceedings of REALWSN,* 2005.

[30] M.S. Pan, C.H. Tsai, and Y.C. Tseng. The orphan problem in zigbee wireless networks. *IEEE Transactions on Mobile Computing,* 8(11):1573–1584, 2009.

[31] P. Pangun, C. Fischione, A. Bonivento, K.H. Johansson, and A. Sangiovanni-Vincent. Breath: An adaptive protocol for industrial control applications using wireless sensor networks. *IEEE Trans. on Mobile Computing,* 10(6):821–836, 2011.

[32] S. Pratheema, K.G. Srinivasagan, and J. Naskath. Minimizing end-to-end delay using multipath routing in wireless sensor networks. *International Journal of Computer Applications,* 21(5):20–26, 2011.

[33] W. Quan and W. Ping. A Finite-State Markov Model for Reliability Evaluation of Industrial Wireless Network. In *Proceedings of the 6th*

International Conference on Wireless Communications Networking and Mobile Computing (WiCOM 2010), pages 1–4, 2010.

[34] D. Rakhmatov and S.B. Vrudhula. Time-to-Failure Estimation for Batteries in Portable Electronic Systems. In *Proceedings of the 2001 International Symposium on Low Power Electronics and Design*, pages 88–91, 2001.

[35] Z. Ren, P. Li, and J. Fang. Segmentation-Based On-Demand Scalable Address Assignment for ZigBee Networks. In *Proceedings of IEEE Vehicular Technology Conference (VTC Fall)*, pages 1–5, 2011.

[36] S. Rost and H. H. Balakrishnan. A health monitoring system for wireless sensor networks. In *Proceedings of SECON*, 2006.

[37] I. Saleh, H. El-Sayed, and M. Eltoweissy. A fault tolerance management framework for wireless sensor networks. In *Proceedings of Innovations in Information Technology*, pages 1–5, 2006.

[38] T. Schmid, H. Dubois-Ferriere, and M. Vetterli. Sensorscope: Experiences with a wireless building monitoring sensor network. In *Proceedings of REALWSN*, 2005.

[39] S. Shin, T. Kwon, G-Y. Jo, Y. Park, and H. Rhy. An experimental study of hierarchical intrusion detection for wireless industrial sensor networks. *IEEE Trans. on Industrial Informatics*, 6(4):744–757, 2010.

[40] I. Silva, A.G. Guedes, P. Portugal, and F. Vasques. Reliability and availability evaluation of wireless sensor networks for industrial applications. *Sensors*, 2012(12):806–838, 2012.

[41] J. Song, S. Han, A.K. Mok, D. Chen, M. Lucas, and M Nixon. WirelessHART: Applying Wireless Technology in Real-Time Industrial Process Control. In *Proceedings of the 14th IEEE Real-Time and Embedded Technology and Applications Symp (RTAS 2008)*, pages 377–386, 2008.

[42] J. Song, S. Han, X. Zhu, A. Mok, D. Chen, and M. Nixon. Demo Abstract: A Complete WirelessHART Network. In *Proceedings of ACM SynSy 2008*, pages 381–382, 2008.

[43] L.M. Souza, H. Vogt, and M. Beigl. A Survey on Fault Tolerance in Wireless Sensor Networks. *Website: http://digbib.ubka.uni-karlsruhe.de/volltexte/documents/11824*, l.a. 2012.

[44] R. Szewczyk, A. Mainwaring, J. Polastre, J. Anderson, and D. Culler. An analysis of a large scale habitat monitoring application. In *Proceedings of the 2nd International Conference on Embedded Networked Sensor Systems*, pages 214–226, 2004.

[45] J.H. Taylor and J. Slipp. An integrated testbed for advanced wireless networked control systems technology. In *Proceedings of the 36th Annual Conference on IEEE Industrial Electronics Society (IECON 2010)*, pages 2101–2106, 2010.

[46] G. Tolle, J. Polastre, R. Szewczyk, D. Culler, N. Turner, K. Tu, S. Burgess, T. Dawson, P. Buonadonna, D. Gay, and W. Hong. A macroscope in the redwoods. In *Proceedings of the 3rd International Conference on Embedded Networked Sensor*, pages 51–63, 2005.

[47] C. Tuan, Y. Wu, W. Chang, and W. W. Huang. Fault tolerance by quartile method in wireless sensor and actor networks. In *Proceedings of International Conference on Complex, Intelligent and Software Intensive Systems*, pages 758–763, 2010.

[48] G. Werner-Allen, K. Lorincz, M. Ruiz, O. Marcillo, J. Johnson, J. Lees, and M. Welsh. Deploying a wireless sensor network on an active volcano. In *Proceedings of IEEE Internet Computing*, volume 10, pages 18–25, 2006.

[49] A. Willig. Recent and emerging topics in wireless industrial communications: A selection. *IEEE Trans. on Ind. Informat.*, 4:102–124, 2008.

[50] G. Wu, C. Lin, F. Xia, L. Yao, H. Zhang, and B. Liu. Dynamical jumping real-time fault-tolerant routing protocol for wireless sensor networks. *Sensors*, 10:2416–2437, 2010.

[51] G. Zhao. Wireless sensor networks for industrial process monitoring and control: A survey. *Network Protocols and Algorithms*, 3(1):46–63, 2011.

[52] ZigBee. Company website: http://www.zigbee.org.

[53] R. Zurawski. Wireless Sensor Network in Industrial Automation. In *Proceedings of Embedded Software and Systems (ICESS 2009)*, 2009.

8

Network Architectures for Delay Critical Industrial Wireless Sensor Networks

Nazif Cihan Tas

Siemens Corporation, Corporate Research and Technology

CONTENTS

8.1 Introduction

The last two decades have seen an increasing attention in the wireless sensor networks (WSN) domain, with thousands of publications, dedicated conferences, real-life deployments and testbeds [16]. This vast interest is based on two main aspects: WSNs provide a very convenient, attractive, and inexpensive programming environment for developing and deploying distributed applications; and the increasing need for designing and implementing novel algorithms for this new hardware platform, with its unique requirements much different than the traditional networks. The WSN concept is also one of the major enabling technologies for the exciting Internet of Things paradigm, which allows the devices to interact and cooperate with each other to reach common goals [17].

Until recently, the usage of WSNs mainly revolved around low-duty monitoring applications for e.g., military and environmental surveillance, and non-real-time control applications for e.g., remote management of home devices and integrated building automation [16]. With the recent advances in the wireless communication technology, we have seen a new deployment area for WSNs: industrial settings. These networks are identified as industrial wireless sensor networks (IWSNs) in the literature and expose unique requirements, not common in other WSN deployments [22].

In this chapter, we elaborate on such IWSN requirements specific to industrial settings, focusing on the communication challenges caused by the harsh deployment environments that are common in industrial settings, and the strict application requirements imposed by the real-time nature of the industrial processes.

Showing that the traditional wireless network standards, such as IEEE 802.11 [7] and IEEE 802.15.4 [8], fall short on supporting packet transmissions with strict deadlines (as we call the Delay Sensitive Networks or DSNs), we propose a cross-layer architecture where the application using the wireless medium associates each packet it is pushing down the communication stack with a delay requirement. For this purpose, we further propose three new mechanisms: ER (Early Retirement), MAC+, and PHY+. These mechanisms utilize the deadline associated with each packet for further transmission, backoff decision, and redundancy optimization.

The rest of this chapter is organized as follows: In Section 8.2, we define the common IWSN applications and settings, elaborating on the usage of wireless technology in industrial settings. In Section 8.3, we describe a typical distributed control system and illustrate the communication challenges associated with such a system. In Section 8.4, we describe our proposed mechanisms for real-time applications with strict delay requirements, and report on the observed simulation results. Finally, we close our chapter with the conclusions and discussions in Section 8.5.

8.2 Industrial Applications and Settings

Requirements for IWSNs are radically different than other types of networks as the typical network deployment settings and the applications differ greatly than the traditional networks. Before going in details about these unique requirements, it is essential to have a look at some key application scenarios for industrial environments.

8.2.1 IWSN Applications

The term *industrial* is regarded as a very general term spanning the domains from energy to transportation and the technological advances from pure-software design to complex hardware implementation. From wireless sensor networks' view, however, an industrial setting is usually of respect to a dedicated setting (such as a manufacturing plant, or another type of automation control system) and involves heavy or mid-heavy machinery, automated robots and similar machines. In such settings, the applications and their requirements differ heavily from the rest of the domains, which we will discuss in the upcoming sections. The typical applications that are deployed in industrial settings can be categorized in two main areas: monitoring applications and cable replacement applications.

8.2.1.1 Monitoring Applications

These applications are similar to the general use cases of sensor networks for monitoring a certain phenomenon. Monitoring applications usually involve tens of nodes that are strategically placed on pre-meditated locations to monitor a certain device, process, or physical property. Such applications ensure correct operation and preferably act on out-of the bound or abnormal behavior of the monitored entity to avoid unnecessary disruption and potentially hazardous situations, performing predictive and proactive maintenance. Examples of such systems include motor conditioning and fault detection for electrical systems [24, 41, 42], real-time status monitoring of weaving machinery [30], and other factory machinery [50].

Typical monitoring applications might involve complex signal processing and classification techniques, and might require collecting data from various types of sensors and executing high-level inter-dependency analysis and data correlation. As the faults might damage the sensor hardware as well, a thorough fault/fail analysis has to be made for potential applications, considering several types of fail scenarios and their potential after effects.

Monitoring applications can be further divided into three categories [39]:

1. **Periodic Monitoring:** In this kind type of monitoring, the state of the equipment (or similar phenomenon) is periodically monitored

through frequently transmitted status messages. These messages can involve direct sensor readings or semi-processed or combined multi-sensor readings. This class of monitoring is also called cyclic data collection [33].

2. **Event-Based Monitoring:** This type of monitoring relies on on-board pre-processing techniques which continuously watch the sensor readings and compares them against to a pre-defined criteria or a condition. If (and only if) a violation of such a condition is determined, event or alarm messages are transmitted. This type of monitoring better utilizes the transmission medium and thus is more suitable for large scale deployments. Further, as radio transmissions are costly, it saves valuable battery power and extends the lifetime of the network. In contrast, as they are required to perform more complicated operations such as on-board validation checks, the wireless devices are expected to be armed with better processing power. Moreover, the deployment of the IWSNs has to be performed in a more diligent way as the devices are required to be pre-programmed with normal operating conditions and abnormal behavior detection criterion of the equipment they are monitoring. This class of monitoring is also called acyclic data collection [33].

3. **Store and Forward Applications:** This type of application targets radio transmission related challenges by enabling the devices to store the data when the radio medium is not available, and transmit their data in bulk once it becomes available.

8.2.1.2   Cable Replacement Applications

Another group of industrial applications directly targets at bridging the communication gap between different types of devices wirelessly and enabling easy and efficient communication between different network entities through the wireless sensor network technology. The main theme behind this group of applications is to replace the costly wires and cumbersome wire installations with easy-to-use and flexible wireless links.

As a typical industrial network can be composed of a complex multi-level hierarchical system, the cable replacement can take place at several different architectural levels. These levels include the connectivity between the control-center and controllers such as the programmable logic controllers (PLCs) at the highest level, PLC-to-PLC and other similar device-to-device communication at the mid-level; and finally, the communication between the sensors, actuators and the rest of the network at the lowest level. This structure will be discussed further in Section 8.3.

Potential high-level use cases for cable replacement are remote control/configuration/information exchange [48] and inventory management [43]. More control oriented applications, such as controlling the speed of an induction motor using a wireless position feedback [36], wireless fieldbus replacement [19] and Bluetooth-based WISA (Wireless Interface for Sensors and Actuators) sensor/actuator systems are also reported [20].

8.2.2 Why (and Why Not) Wireless in an I(W)SN?

The development of wireless sensors are expected to improve the production efficiency by 10% [6].Wireless technologies promise a cost reduction of approximately 70% of initial cost and 80% annual maintenance costs [15] compared to the wired installations. A wireless system has several advantages over a wired system, such as easiness of use and increased productivity. Wireless technology is especially suitable for harsh, hard-to-reach settings where the chemical and physical processes might harm or prevent installing wires [49].

Communicating with wires has the limitations of constantly increasing high installation and maintenance costs, high failure rate of connectors together with difficulty of troubleshooting the connectivity issues; whereas the wireless communications have clear advantages such as lower installation and maintenance costs, ease of replacement and upgrading, reduced connector failures, greater physical mobility and freedom, capability of utilizing extremely small form factor sensors without the bulky wires and faster commissioning [14].

These advantages encouraged several industrial companies to install wireless sensing technologies in their businesses. One of the notable deployments is the BP's remote monitoring of LPG tank fill levels using battery-powered ultrasonic sensors that transmit information by radio signal to a low Earth orbit satellite. Using this network BP reports over 33% efficiency improvement [31]. In a similar effort, General Motors save 10%-20% on production costs through monitoring the health of their manufacturing equipment such as stamp releases, conveyer belts, and other types of machinery with a mesh-based wireless sensor network [23].

Even though the advantages and promises of using wireless technology in the manufacturing plants and industrial environments are encouraging, the penetration rate of the IWSNs into the industry is rather slow. The resistance for the acceptance of the wireless communication in the industry is rooted in three main concerns:

1. **Lack of one-fits-all wireless standard:** The plethora of today's available communication technologies creates an overwhelming effect on the designers and plant owners. ZigBee, Bluetooth, WirelessHART, and ISA100 are among the promising technologies of tomorrow's wireless industrial networks [22], among which the last two technologies become the most competitive [29]. However, there is currently no silver-bullet wireless technology which is proven to be successfully deployable in all the industrial settings.

2. **Data security concerns:** As the wireless communication takes place directly on the air, it is susceptible to security threats such

as eavesdropping and spurious packet injection. For the industrial environments where loss of production can cost millions of dollars and malfunctioning equipment can cost precious human lives, ensuring secure and reliable communication is crucial for keeping the entire industrial ecosystem at safe operation. Naturally, the plant engineers desire uninterrupted and correct operation with the same level of security as they are accustomed to with the wired systems, about which they are reluctant to achieve with wireless technology [34].

3. **Perceived reliability issues:** Industrial environments might be challenging for wireless communications because of the interference generating equipment (e.g., heavy machinery, motors, microwave and ultrasonic equipment, etc.) and the geometry of the locations (e.g., large, thick metallic and concrete walls). Reliable and fault-tolerant communication requirements are among the main reasons of lack of enthusiasm towards the wireless technology for industrial deployments [14]. Contrary to the common belief, many real-world applications might benefit from replacing the wires with antennas. For instance, Proctor & Gamble was able to improve the communication reliability through replacing the slip rings with an 802.11-based wireless network that was designed to optimize its existing EtherNet/IP network [34].

8.3   A View on a Distributed Control System

An industrial automation and control system is a complex structure that supports various functionalities for fast, flexible, and easy management, configuration, and operation. A typical automation system, such as shown in Figure 8.1, might contain different types of devices including sensors/actuators, controllers, human-interface devices, and regular PCs. Logically, there are three levels in an automation system [4]:

- **Management level:** enables the holistic view of the system, supporting operations such as remote diagnostics and maintenance of controls and devices, and storage of process and device information.

- **Control level:** manages the controllers and similar devices for changing and saving device parameters.

- **Field/Device level:** enables fast, reliable communication between the sensors/actuators and the rest of the automation system.

The main communication technologies evolved in the industry to support the functionalities of the automation systems are industrial Ethernet, fieldbus,

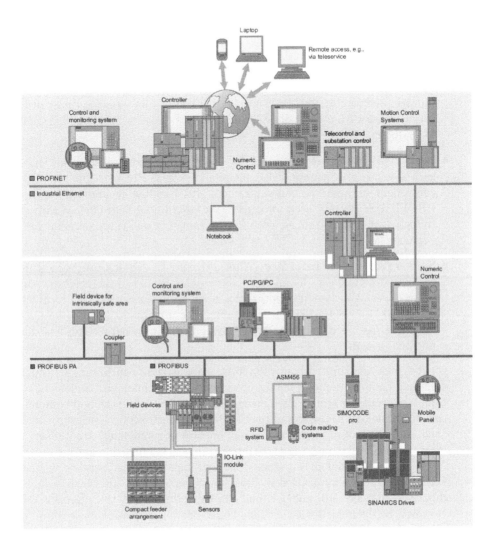

FIGURE 8.1
A typical distributed control system, captured from [13] with permission.

and sensor/actuator IO interfaces. These technologies can be summarized as follows:

- **Industrial Ethernet:** The traditional Ethernet technology is designed for offices and IT departments. Industrial Ethernet technologies, such as PROFINET, address the tougher conditions imposed in industrial applications. PROFINET targets various industrial specific requirements such as data-intensive parameter assignment and synchronous I/O signal transmission. PROFINET's flexible network topology support, increased availability through redundancy solutions and intelligent diagnostic concepts, and deterministic and isochronous transmission capabilities for extreme time-critical process data with a jitter of even less than $1\mu s$ makes it an attractive choice for industrial Ethernet communication [24].

- **Industrial Fieldbus:** A fieldbus is a network for connecting field devices such as sensors, actuators, field controllers (such as PLCs), regulators, drive controllers, and man-machine interfaces [46]. Fieldbus technologies such as PROFIBUS [11] are used for connecting distributed field devices with extremely fast response times, with the guarantee of real-time communication behavior [47].

- **Sensor/Actuator Interface:** Alternative to the configuration in which the actuators and sensors might be directly connected to the fieldbus, other low-cost, easy-to-assembly options such as Actuator Sensor Interface (AS-interface) [1] and IO-Link [3] also exist. These technologies provide faster and error-free installation, and easier commissioning owed to their less complicated designs.

A typical distributed control system, such as shown in Figure 8.1, might contain a combination of different communication technologies. Industrial automation arena has seen excited discussions about choosing the best communication technology to use in industrial settings [21]. However, such discussions (and the existence of plethora of communication technologies) are unlikely to cease, as each technology finds its interest in this competitive market.

All of the technologies discussed above typically utilize installed wires as their physical communication medium. Wireless product alternatives for industrial Ethernet [10] and fieldbus [35] can be found in the market today. However, as the field/device level might require very strict response times (on the order of sub-ms levels for certain applications such as motion control [38]), reliable communications for such fast drive controls (with time constraints in the $1ms$ range) arguably cannot be handled by current wireless technologies [2].

8.3.1   Challenges in IWSNs

The challenges related with IWSNs are previously reported in an excellent work before [22] and include concerns about the scarce hardware and software resources, security related issues to avoid attacks and intrusion, integration issues with Internet, large-scale deployment topics, and link quality issues. In this chapter, we focus on communication related challenges coupled with the strict application requirements imposed by the nature of the control systems.

The challenges related with deploying wireless technologies in industrial automation systems originated from two aspects: critical and strict system application requirements, and harsh nature of the deployment environments.

The manufacturing plants usually contain physical equipment such as large obstructive machinery which disrupts the line-of-sight transmissions, and rotating machines which generate electromagnetic interference [32]. Furthermore, such plants typically contain large steel and/or concrete walls which make the radio transmission more challenging due to multipath interference [40].

As the manufacturing plants are complex systems consisting of various types of subsystems, each utilizing their own communication technologies, the IWSNs face the issue of coexistence with other applications and processes, which also utilize the radio frequency technology. Coexistence between different technologies has been analyzed before from several perspectives [2, 26], and methods for designing networks that decrease the likelihood of interference between different unlicensed band wireless technologies are presented [25].

In industrial applications, loss of production is particularly undesired and can be very costly. Hence it is tremendously important to have communication systems which are reliable, predictable, and fault tolerant [48]. Maintaining the reliability in industrial applications is especially challenging because of the physical characteristics mentioned above. There has been some suggestions in the literature in order to overcome such negative effects caused by the harsh physical characteristics. For instance, it has been observed that interference from industrial noise sources to be less significant for communication systems operating above 1 or 1.6 GHz when measurements were made at distances in excess of four meters from the noise sources [26], and utilization of higher frequency bands, such as 5GHz IEEE 802.11a, is suggested to overcome the difficulties caused by the noise sources [18]. Even though the correct choice of the physical alternative alleviates the significance of the problem, it does not solve it entirely as the key performance values for a successful industrial automation system depend not entirely on the physical properties, but also the transmission order and frequency of the devices in the network.

An automation system might be composed of several applications running concurrently and communicating simultaneously. Each such application has its own specific requirements and optimal operating conditions. Real-time (RT) behavior is an essential component of most of the automation system applications. RT response requires that a system has a clearly defined time response which is guaranteed under all operating conditions [38]. According to the criticality of the communication, there are three main RT classes [37, 38]:

- **Soft-RT:** Response times less than $100ms$ are desirable. Common applications include factory floor and process automation, and monitoring and diagnostics.

- **Hard-RT:** Response times less than $10ms$ are desirable. Common applications include mobile operators and safety.

- **Isochronous RT:** Response times less than $1ms$ are desirable. Common applications include motion control and fast drive controls.

As the requirements for Hard-RT and Isochronous RT are extremely challenging, we focus on the Soft-RT systems in the rest of this document and propose techniques to increase the wireless communication efficiency for processes which have strict deadlines to keep in order to operate correctly.

8.4   Delay Sensitive Networks

As aforementioned in previous sections, in automation control systems where direct physical control is performed, application latency plays a critical and important role for the reliability of the system. In control systems, the device response times are bound to strict limits in order to continue proper operation. Consequently, communication latency plays an important role in ensuring the correct functioning of the entire ecosystem. In such systems, clearly, the communication latency should be kept under firm boundaries as the packets that arrive at their destinations later than their deadlines will be of no use and ignored by the target application immediately, wasting valuable wireless medium capacity. We refer to such systems as Delay-Sensitive Networks (DSN) following their dependency on the firm packet deadlines [1].

In traditional communication protocols, such as IEEE 802.11 [7] and IEEE 802.15.4 [8], each packet is handled equally. As these standards were designed mainly for uncritical operations, such as daily web usage of common Internet users in the case of IEEE 802.11 and periodic monitoring of an asset or a region in the case of IEEE 802.15.4, the traffic belonging to these networks do not carry packets with extremely critical deadlines, i.e., the communication can tolerate much higher levels of delay. Even though these standards have some QoS extensions for more quality critical applications, such as IEEE 802.11e [9], these methods fall short in handling the priority on the packet level, by classifying the traffic into discrete number of priority levels and giving higher access probabilities to the higher priority traffic. Our proposed mechanisms, on the other hand, manage the transmission procedure per-packet basis by considering the deadline of the packet and the immediacy of the deadline.

In the rest of this section, we propose a cross-layer architecture where the

---

[1] A preliminary version of the study that is presented here was reported before in [45]. In this chapter, we extend this previous methods by introducing a new mechanism, MAC+.

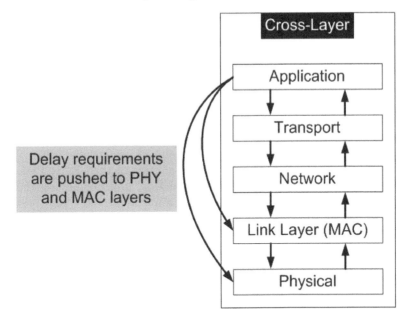

FIGURE 8.2
Delay Sensitive Networks: Cross-layer approach.

application using the wireless medium associates each packet it is pushing down the communication stack with a delay requirement as in Figure 8.2. For this purpose, we propose three new mechanisms: ER(Early Retirement), MAC+, and PHY+ which utilize the deadline associated with each packet for further transmission, backoff decision, and redundancy optimization respectively.

8.4.1 Proposed Mechanisms

Our proposed mechanisms operate on the PHY and MAC layers, and make use of the delay information per packet provided by the application layer. For this purpose, we assumed that the PHY and MAC layers are extended with additional interfaces following a cross-layer approach [44, 28] as illustrated in Figure 8.2. The readers are assumed to have a basic understanding of the CSMA/CA algorithms as described in [8, 7].

The first two of our mechanisms were introduced in detail in one of our previous works [45]. For completeness we briefly summarize these mechanisms and encourage the reader to consult the original work for more details.

8.4.1.1   Early Retirement (ER)

In a DSN where the packets are marked with strict deadlines, correct and timely delivery of the packets before their deadlines is a major concern. If a packet is not successfully received by its designated destination before its deadline, it is regarded as lost and any transmission attempts for this packet would waste precious bandwidth. If the deadline information attached to each packet is known a priori before transmission, the transmitter can avoid any potentially unsuccessful transmissions by predicting the packet's reception time and comparing it with the packet's deadline, and ultimately discarding the packet if a timely reception is unlikely. We name this method Early Retirement (ER) as the packets retire before the actual MAC protocol discards them as a result of a successful transmission or after a maximum retry limit reach.

ER predicts the potential reception time of the packet before the first transmission attempt, and right after any unsuccessful attempt. This action ensures timely disposal of the packets if the next retry of the packet is scheduled (randomly) to a time later than the packet deadline. If the next packet retry is scheduled to be attempted later than the deadline, ER automatically discards the packet, retrieves the next packet from the queue and starts the new packet transmission, acting as if the previous packet's transmission attempts reached the retry limit. Please consult to our previous publication [45] for more information about the ER mechanism.

There are three main advantages of ER: 1) as the packets which do not have any hope of being received timely are discarded in advance, the average delay of successfully transmitted packets decreases. In other words, the received packets always have low transmission delays. 2) ER makes the system transmit more number of packets in a given amount of time, which consequently reduces the queuing delays. 3) ER, in some scenarios, can improve the throughput since the average transmission time of a packet is reduced, as will be illustrated in the simulation sections.

ER's principle is simple: do not attempt to transmit a packet if there is no chance that it will be successfully received. This principle ensures that the packets which are about to be transmitted have at least non-zero probability of being successfully received, but it also indirectly interferes with the medium access protocol in effect. The medium access protocol arbitrates each user's medium access and resolves packet collisions, if there are any. The traditional Binary Exponential Backoff (BEB) mechanism, for instance, deploys a probabilistic approach and assigns random transmission times to each user competing for the medium. Upon any collision, BEB ensures that the next transmission times are randomly chosen from a larger probability space, through exponentially increasing the contention window size. ER, on the other hand, behaves in just the opposite manner, and it avoids the window size growing larger. Hence, in networks where the number of users is large, ER might have worsening effects as shown in [45]. However, as will be shown in the next sections, ER still can improve performance in such large networks when it is used in conjunction with the other proposed methods.

8.4.1.2 Variable Redundancy Error Correction (PHY+)

Our second mechanism, PHY+, is a variable complexity error correction technique which favors the packets closer to their deadlines to have a better chance of being received compared to the packets that are far away from their deadlines. PHY+ inserts arbitrary amount of redundancy in each packet right before the transmission, considering how close the packet is to its deadline.

Notice that PHY+ is not a channel coding technique per se and its novelty is that it enables the transmitters to inject just the right amount of redundancy in the channel coding algorithms so that the critical packets which are closer to their deadlines are less vulnerable to channel errors.

For simplicity, we assumed a very rudimentary channel coding technique to demonstrate the strength and value of PHY+. Through this technique, each packet is transmitted, not only once, but several times on the medium in a back-to-back fashion, as if the entire chunk of the transmission is a single packet. This technique follows the idea of Partial Packet Recovery (PPR) [27], in which the receivers can still decode the partially collided packets through strategically replicated PHY headers in the packet. PPR assumes a temporal and/or transmission size diversity, which leaves parts of the collided packets still decodable upon collision. PHY+ is similar to PPR, however, it repeats the entire packet (including the PHY headers) multiple times in order to maximize the likelihood of a repeated packet to be still decodable after a collision. Notice that as the packets are being sent in bursts and regarded as single packets, the medium access methods in use are not triggered between repeated transmissions which explains the *PHY* in PHY+.

PHY+ assigns *redundancy factors* to each of the stations according to their delay requirements. The redundancy factor (r) of a packet specifies the level of redundancy for that packet, which is the number of times the packet will be repeated in our example channel coding technique. Thus, $r = 1$ means that there is no requirement for this packet and it has to be sent only once; $r = 2$ suggests that the packet has to be repeated twice, and so on so forth. Intuitively, if two packets collide, the packet that was repeated more number of times (i.e., with higher redundancy factor) would still be decodable.

Clearly, PHY+ increases the transmission times as each packet is being repeated several times, which might negatively affect the average delay and throughput because of the medium time wasted. However, as shown in [45], when the collisions are unavoidable, as in the case of highly congested networks, PHY+ can resolve the collisions for the benefit of the most delay critical packet.

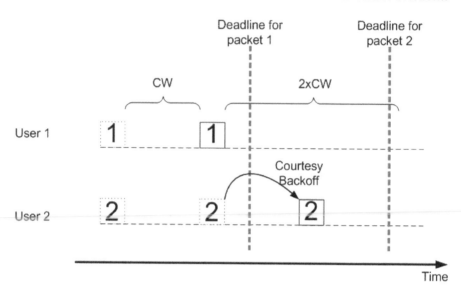

FIGURE 8.3
Courtesy backoff example.

8.4.1.3   Enhanced MAC for Delay Sensitive Networks (MAC+)

Our third method, MAC+, focuses on enhancing the collision resolution schema by considering the delay requirements of each potential colliding packet. In our schema, the users might choose not to transmit even though the BEB mechanism permits a transmission. These backoffs, as they leave their place for more important users available intentionally, are called *courtesy backoffs*.

In Figure 8.3, we give an example for courtesy backoffs schema. In this figure, there are two users ready to transmit their packets, User 1 and User 2 respectively. Their corresponding packets waiting to be transmitted, packet 1 and packet 2, are scheduled to be transmitted at the same time in this scenario. In a typical scenario where there is no information about the delay requirements of each packet to be transmitted, the collision would be unavoidable and the users would retransmit their packets by adjusting their contention window sizes to overcome the dynamicity of the environment. However, such a schema would totally ignore any delay-related requirements and would treat all the packets same.

In our proposed MAC+ schema, the users calculate their probability of transmission at each attempt by checking their temporal proximity to the deadline scheduled. For instance, in our example, let's assume that packet 2 can tolerate longer delays than packet 1. Each user, once they are given the opportunity to transmit, calculate the probability of transmission (p_1 and p_2

in this case), and since packet 2's deadline is far away, the probability that the first user transmits is higher ($p_2 < p_1$). In this case, User 1 will transmit its packet whereas User 2 will not transmit but behave as if the transmission was executed and there was a collision. Hence, this schema gives more priority to the packets with closer deadlines. Notice that if two users have close deadlines than the system behaves exactly the same as the conventional method, i.e., $p_1 = p_2 = 1$.

MAC+ is expected to increase performance most in scenarios where the number of the users in the system is high; as in such scenarios, the probability of collision and the expected number of collisions are also high. In uncongested traffic scenarios, on the other hand, MAC+ cannot offer much as the collision rate is expected to be low. Notice that MAC+ essentially *increases* the average delay as it schedules less important packets to be transmitted further in the future. Ultimately, as their deadline approaches, each packet's turn will come, and gain more importance. In a DSN, increased delay does not automatically cause any performance degradation as the DSNs do not possess minimal delay requirements but strict deadlines for correct reception. Hence, a transmission is regarded as a successful transmission as long as it is received by the destination before its deadline passes, and a sooner reception does not yield any further benefits.

MAC+, essentially, provides another dimension of randomness on top of the randomness supplied by the BEB mechanism. This second dimension favors the closer-deadline packets and lets the further-deadline packets not be transmitted even though the local BEB mechanisms mandate certain transmission schedules.

8.4.2 Simulations

We have implemented a discrete-time network simulator using Matlab and adopted our proposed mechanisms in order to compare their performance against the traditional 802.15.4 standard[2], similar to the experiments we have conducted in our previous work [45]. We have investigated the effects of the network size and the criticality of the communication through varying the number of users in the system and the mean delay requirements per user, respectively. For the latter, we assumed a uniform distribution of delay among users with half mean difference from both sides, i.e., if mean delay is τ, then each user randomly picks a delay requirement from the interval $(\tau/2, 3\tau/2)$ with uniform distribution. Furthermore, we assumed that a packet transmission is completed in four slots which follows from the TinyOS [5] default packet length of 36 bytes and 802.15.4 [8] default slot length of 20 symbols ($320\mu s$).

As in [45], at each transmission, PHY layer computes the r using the following equation at time T:

---

[2]In the rest of this document, we use the terms IEEE 802.15.4 and 802.15.4 interchangeably.

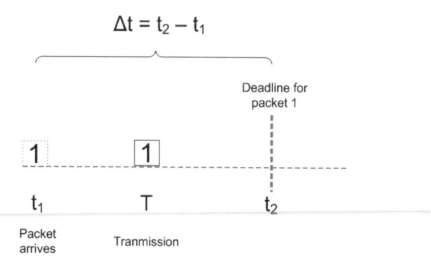

FIGURE 8.4
Calculating r in PHY+.

$$r_T = \left\lceil \frac{T - t_1}{\frac{\tau}{maxR}} \right\rceil \tag{8.1}$$

where t_1 is the time the packet arrived, t_2 is the deadline of the packet and $\tau = \Delta t = t_2 - t_1$ as can be seen in Figure 8.4. In this formula, $maxR$ denotes the maximum value that the redundancy level can take and it is an application parameter which has to be decided by the system administrator. However, in our simulations, we have conducted several experiments by varying this value in order to see its effect on the performance as well.

In our schema, at each transmission, MAC layer computes the probability of transmission, p, using the following equation at time T:

$$p = c + \frac{(T - t1) \times (1 - c)}{(t2 - t1)} \tag{8.2}$$

where t_1 is the time the packet arrived, t_2 is the deadline of the packet as above. In this formula, c denotes the minimum value that the probability of transmission can take and it is an application parameter which has to be decided by the system administrator.

In our experiments, we performed simulations for all seven combinations of the three methods proposed. However, because of limited space, we only report on the four most promising cases of PHY+, ER PHY+, MAC+, and ER MAC+. For each of these cases, we created three systems varying number of users between 5, 20, and 50; and for each of these three systems we looked

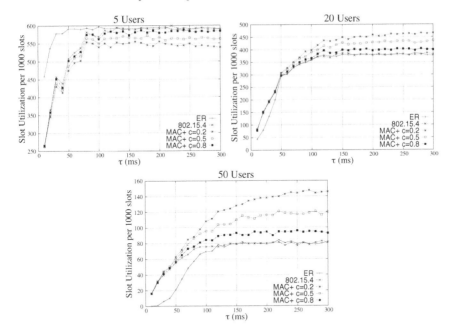

FIGURE 8.5
Slot utilization using MAC+.

at the performance by increasing the average delay requirement (τ) between 5ms and 300ms.

As the notion of success in DSNs differs from the traditional networks, in our simulations we measure a modified metric which captures the criteria of packets meeting or missing their associated deadlines. For this purpose, when we are measuring the throughput, we not only check if the packet is received by the designated destination, but we also check if the deadline of this packet is met. In all the simulations conveyed in the rest of this section, we report on the *Slot Utilization per 1000 slots* metric which reflects the number of slots spent for transmission of packets that were received successfully *and* on time. This metric is equivalent to conventional throughput metric extended with the packet deadline consideration, thus we use these two terms interchangeably.

As a second performance criterion, we investigate the average delay of all packets received correctly, including the ones that were not on time but received anyways. In essence, the average delay is not a vital metric as the success of a packet transmission depends solely on the packet deadline and correct reception. Hence, if a packet is received correctly before its transmission deadline, it is regarded as successful transmission; and average delays less than the packet's deadline do not necessarily contribute towards packet's success. However, we include the average delay analysis in our results in order

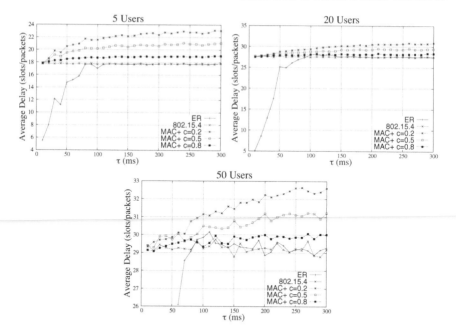

FIGURE 8.6
Average delay using MAC+.

to give the reader insight about the effects of the proposed mechanisms on the delay behavior.

8.4.2.1 ER and PHY+ Simulations and Discussions

Our previous work [45], reports on the performance results of the ER and PHY+ mechanisms extensively. For convenience, we summarize these findings here and suggest the reader to consult to the original document for further details.

- **Low number of users, high delay requirements** [3]:

  In a wireless network in which the number of users is not large, the contention on the channel and consequently the expected number of collisions become insignificant. As PHY+ relies on its intelligent collision resolution mechanism for the benefit of the delay critical packets, it is not effective in low contention situations. On the contrary, as PHY+ increases the medium usage times of each packet by essentially repeating the packet several times,

---

[3]This is in reality not a target scenario for a typical DSN since the delay critical IWSNs impose low delay requirements and are typically not very small in size. However, for completeness, we report on our performance findings under these circumstances as well.

it increases the average delay per packet, decreases the number of packets successfully received in a given time interval, and consequently decreases the throughput.

ER, on the other hand, becomes equivalent to the original 802.15.4 CSMA/CA mechanism, as the high delay requirements avoid the packets timing out.

- **High number of users, high delay requirement:**

  In a wireless network with high number of users, the probability of packets colliding increases with the contention on the medium. In a traditional wireless network, in case of collisions, the collided packets are lost and each source retries to transmit their collided packet in a future time hoping that this time the packets will not collide. PHY+, on the other hand, resolves the collisions through its variable redundancy error correction mechanism and performs well in high contention situations. Indeed, PHY+ promises over 200% slot utilization increase when the packet deadlines are considered [45].

  Clearly, the selection of $maxR$ has a considerable effect on the system as a very low $maxR$ makes PHY+ nonfunctional with several packets being transmitted with the same redundancy factor; and a very high $maxR$ makes the redundancy levels (and thus the number of times the packets are transmitted) soaring which consequently increases the average delay.

  Similar to the previous case, ER performs equivalent to the original 802.15.4 mechanism with deadlines further away than the BEB retry schedules.

- **High number of users, low delay requirement:**

  As in the previous high number of users case, with the high contention on the channel, PHY+ mechanism performs well. ER, on the other hand, has an adverse effect on the system as it discards the packets earlier than the BEB mechanism mandates. In a high contention situation with packets with low delay requirements, ER avoids the contention window size to grow and invalidates the effectiveness of the BEB mechanism. This causes the system to suffer high number of collisions and decreases the throughput. However, notice that, with the aggressive timeout mechanism, ER facilitates lower delays for the packets which could be transmitted successfully, with a very high cost of decreased throughput.

  When we use both of the methods in conjunction (referred as ER PHY+), the performance of the wireless system improves considerably since these two methods complement each other very well: ER pushes the system to be more competitive with its timeout mechanism while PHY+ utilizes the channel competition for increased number of successful and on-time packets with its redundancy mechanism.

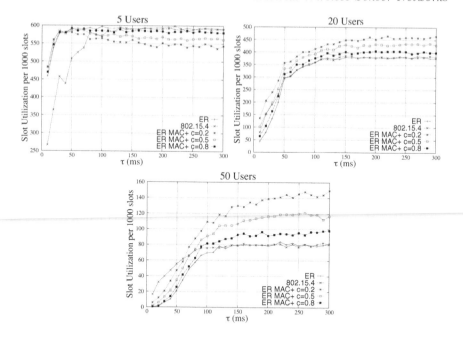

FIGURE 8.7
Slot utilization using ER MAC+.

- **Low number of users, low delay requirement:**

  PHY+ is not effective in low contention scenarios in which collision probability is low. Moreover, increased redundancy also increases average delay without the benefit of collision resolution. ER, on the other hand, avoids the contention window size growing large, and is effective in decreasing the average delay in situations where contention is low.

8.4.2.2   MAC+ Simulations and Discussions

In this subsection, we report on the performance results of MAC+ in Figure 8.5 and Figure 8.6; and on ER MAC+ in Figure 8.7 and Figure 8.8.

- **Low number of users, high delay requirement** [4]**:**

  For small number of users without very strict delay requirements, our methods do not promise much improvement. As discussed before, our ER mechanism is effective only with low delay requirements and for larger such values it is ineffective and equivalent to the original 802.15.4. Similarly, MAC+ mechanism aims at resolving possible collisions before they happen and it

---

[4]This is in reality not a target scenario for a typical DSN since the delay critical IWSNs impose low delay requirements and are typically not very small in size. However, for completeness, we report on our performance findings under these circumstances as well.

does not comprise any significant performance improvements in situations where number of users is low. Moreover, MAC+ increases average delay by pushing some of the packets to be transmitted further in future even though the channel is available. This effect, in addition, makes several slots to be wasted idle even though the possibility of collision is low. Hence, for low number of user and high delay requirement case, neither of our methods is helpful on improving the performance of the wireless system.

- **High number of users, high delay requirement:**

  For high number of users without very strict delay requirements, the expected number of collisions is very high because of the highly competitive environment created by increased contention. Hence, by avoiding collisions by giving higher precedence to more critical users, our MAC+ method increases throughput considerably. Thus, MAC+ extension has a significant effect on increasing the number of successful on-time packets and subsequently the throughput. However, as MAC+ method pushes the packets further, average delay per successful packets is much higher than the MAC+-disabled cases.

  As the delay requirements are not very strict in this scenario, ER method does not have a significant effect on the performance of the system and produces very similar results to the original 802.15.4.

  Also immediate from the bottom two graphs of Figure 8.5 and Figure 8.6 is that the choice of c parameter has a big effect on the throughput and delay of the system. As the c value increases, the system behaves more like the original 802.15.4 and the throughput gain decreases, however the average delay decreases also since the number of courtesy backoffs decreases. In a well-managed system, we trust that a system administrator can choose a proper c value according to the needs of his/her wireless system.

- **High number of users, low delay requirement:**

  For high number of users with low delay requirements, ER affects the throughput and the slot utilization metrics unfavorably. The reason for this is that ER does not let the contention window increase by dropping the packets much earlier once it sees that their deadlines will not be met. As the BEB mechanism uses the contention window for arbitrating the medium access times, ER decreases the throughput values considerably, especially in large networks. We believe that unless the system needs extremely low average delays and can trade very low throughput for the sake of this low delay requirement, using ER alone is not a desirable option in situations where the number of users contending for the channel is large.

  MAC+, on the other hand, outperforms 802.15.4 in terms of slot utilization. This outcome is gained by the courtesy backoffs executed by MAC+ as the probability of collisions is decreased by deliberately missing transmissions if the delay deadline is far off in the future.

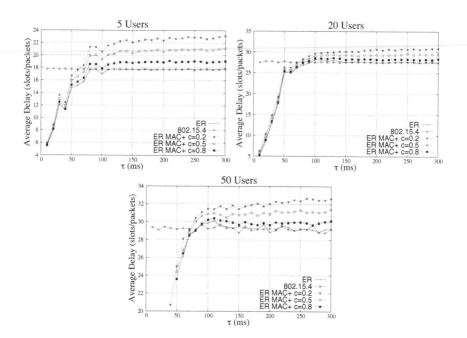

FIGURE 8.8
Average delay using ER MAC+.

TABLE 8.1
Slot utilization growth when the proposed methods are in use.

		Number of users	
		Low (5)	High (50)
Delay Bounds	Low (15 ms)	ER (20%) ER MAC+ (20%)	ER PHY+ (250%)
	High (100 ms)	Outside the scope	PHY+ (200%) MAC+ (90%)

Lastly, we point out that the aforementioned effect of c parameter is valid in this case as well and lowering this parameter increases throughput with the cost of increasing the delay. However, notice that this effect is not very immediate for extremely low delay requirements, but becomes clearer as the delay requirement starts increasing.

• **Low number of users, low delay requirement:**

In a small network in which the users have very strict delay requirements, using both ER and MAC+ schemas in conjunction outperforms the cases where these methods are used separately. The reason for this behavior is twofold: 1) ER, by eliminating unnecessary transmissions, increases throughput efficiency, increases number of successful packets received and decreases average delay (leftmost graph in Figure 8.5). 2) MAC+ increases throughput by deploying courtesy backoffs, especially useful for eliminating early collisions. Together, ER MAC+ performs very well by deploying courtesy backoffs and eliminating transmission attempts for the packets which cannot meet their respective deadlines, especially clear in the rightmost graph in Figure 8.7.

As the probability of collision is lower in smaller networks, the adverse effects of ER become more negligible in such networks. Similarly, MAC+ (with proper low c value) cannot harm a small size wireless network as the probability of collision after a courtesy backoff is insignificant. Since the users execute courtesy backoffs according to their proximity to the deadlines, the throughput is improved when compared with the original 802.15.4 mechanism.

8.4.2.3 General Discussions

Previous section presented clear performance improvements on the successful packet reception rates in DSNs, when the proposed mechanisms are in use. Following our simulation results, we created a selection matrix in Table 8.1 which identifies the best methodology combination to choose for each setting. For each of these choices, the delta improvement values are also reported. Notice that even though several other combinations also outperform the traditional 802.15.4 case, this table reports only on the best improvement combinations.

For low number of users with low packet delay requirements, using ER (for smaller networks) or ER MAC+ (for intermediate size networks) improves the slot utilization by 20%. For the cases with high number of users, on the other hand, due to the successful collision resolution and avoidance strategies of our proposed mechanisms, the improvement is much higher. For low delay requirement, using ER and PHY+ together improve the performance by 250%; and for high delay requirement, using PHY+ improves the performance by 200%. Similarly, MAC+ promises a 90% performance improvement in large networks with high delay requirements.

8.5   Conclusions

Typical industrial applications and deployment settings expose unique challenges that are not commonly found in other systems. In this chapter, we focus on such challenges related with the communication technology, specifically the strict delay requirements caused by the real-time behavior of such systems.

In response to these challenges, we propose three cross-layer mechanisms on PHY and MAC layers which address the problems associated with Delay Sensitive Networks where each packet created has a strict deadline to meet and the packets received after their deadlines are not useful even though they are decoded correctly. The proposed mechanisms utilize the deadline information associated with each packet for further transmission, backoff decision, and redundancy optimization.

Performed simulations showed that our proposed systems improve the packet success rate of the traditional CSMA/CA up to 250%, making them attractive solutions for systems with strict low delay requirements, such as the Soft-RT in automation industry.

Acknowledgements

The author would like to thank Dr. Yakup Genc and Vania Mesrob for their valuable contributions.

References

[1] AS-Interface. as-interface.net.

[2] Coexistence of Wireless Systems in Automation Technology. ZVEI-German Electrical and Electronic Manufacturers' Association, Technical Report.

[3] IO-Link. http://www.io-link.com/en/index.php.

[4] Planet Fieldbus: PROFIBUS and PROFINET. FESTO White Paper.

[5] TinyOS webpage: http://www.tinyos.net/.

[6] Report to the president on federal energy research and development for the challenges of the twenty-first century. Presidents Committee Of Advisors On Science And Technology, November 1997.

[7] *IEEE 802.11: Wireless LAN Medium Access Control (MAC) and Physical Layer (PHY) Specifications.*, 1999.

[8] *IEEE 802.15.4: Wireless Medium Access Control (MAC) and Physical Layer (PHY) Specifications for Low-Rate Wireless Personal Area Networks (LR-WPANs)*, 2003.

[9] *IEEE Part 11: Wireless LAN Medium Access Control (MAC) and Physical Layer (PHY) specifications, Amendment 8: Medium Access Control (MAC) Quality of Service Enhancements*, 2005.

[10] Industrial Wireless LAN: Industrial Features and Current Standards. SIEMENS SIMATIC NET White Paper, 2009.

[11] PROFIBUS System Description Technology and Application. White Paper, 2011.

[12] PROFINET System Description Technology and Application. White Paper, 2011.

[13] SIEMENS SIMATIC NET Industrial Communication Catalog IK PI. Product Catalog and White Paper, 2012.

[14] Advanced Manufacturing Office. Industrial wireless technology for the 21st century. Technical report, U.S. Department of Energy, 2002.

[15] Advanced Manufacturing Office. Wireless Success Story - Industrial Technologies Program (ITP). Technical report, U.S. Department of Energy, 2010.

[16] I.F. Akyildiz, W. Su, Y. Sankarasubramaniam, and E. Cayirci. Wireless sensor networks: a survey. *Computer Networks*, 38:393422, 2002.

[17] L. Atzori, A. Iera, and G. Morabito. The internet of things: A survey. *Computer Networks*, 54(15):2787–2805, 2010.

[18] R. Calcagno, F. Rusina, F. Deregibus, A. Vincentelli, and A. Sangiovanni, Bonivento. Application of wireless technologies in automotive production systems. In *ISR 2006 - ROBOTIK 2006: Proceedings of the Joint Conference on Robotics*, 2006.

[19] F. De Pellegrini, D. Miorandi, S. Vitturi, and A. Zanella. On the use of wireless networks at low level of factory automation systems. *Industrial Informatics, IEEE Transactions on*, 2(2):129–143, may 2006.

[20] D. Dzung, C. Apneseth, J. Endresen, and J.-E. Frey. Design and implementation of a real-time wireless sensor/actuator communication system. In *Emerging Technologies and Factory Automation, 2005. ETFA 2005. 10th IEEE Conference on*, volume 2, pages 10 pp.–442, sept. 2005.

[21] M. Felser and T. Sauter. The fieldbus war: history or short break between battles? In *Factory Communication Systems, 2002. 4th IEEE International Workshop on*, pages 73–80, 2002.

[22] V.C. Gungor and G.P. Hancke. Industrial wireless sensor networks: Challenges, design principles, and technical approaches. *Industrial Electronics, IEEE Transactions on*, 56(10):4258–4265, Oct. 2009.

[23] P. Hochmuth. Case Study: GM cuts the cords to cut costs. *TechWorld*, 2005.

[24] L. Hou and N.W. Bergmann. Induction motor condition monitoring using industrial wireless sensor networks. In *Intelligent Sensors, Sensor Networks and Information Processing (ISSNIP), 2010 6th International Conference on*, pages 49–54, Dec. 2010.

[25] I. Howitt. WLAN and WPAN coexistence in UL band. *Vehicular Technology, IEEE Transactions on*, 50(4):1114–1124, July 2001.

[26] I. Howitt, W. W. Manges, P. T. Kuruganti, G. Allgood, J. A. Gutierrez, and J. M. Conrad. Wireless industrial sensor networks: Framework for QoS assessment and QoS management. *ISA Transactions*, 45(3):347–359, 2006.

[27] K. Jamieson and H. Balakrishnan. PPR: Partial Packet Recovery for Wireless Networks. In *ACM SIGCOMM*, Kyoto, Japan, August 2007.

[28] V. Kawadia and P.R. Kumar. A cautionary perspective on cross-layer design. *Wireless Communications, IEEE [see also IEEE Personal Communications]*, 12(1):3–11, 2005.

[29] T. Kevan. What's delaying wireless adoption? *Sensors Magazine*, 2011.

[30] K. Khakpour and M.H. Shenassa. Industrial control using wireless sensor networks. In *Information and Communication Technologies: From Theory to Applications, 2008. ICTTA 2008. 3rd International Conference on*, pages 1–5, april 2008.

[31] J. King. BP Pioneers Large-Scale Use of Wireless Sensor Networks. *COM-PUTERWORLD*, 2005.

[32] S. Kjesbu and T. Brunsvik. Radiowave propagation in industrial environments. In *Industrial Electronics Society, 2000. IECON 2000. 26th Annual Conference of the IEEE*, volume 4, pages 2425–2430, 2000.

[33] K. S. Low, W.N.N. Win, and Meng Joo Er. Wireless Sensor Networks for Industrial Environments. In *International Conference on Computational Intelligence for Modelling, Control and Automation*, volume 2, pages 271–276, 2005.

[34] A. Lutovsky. Overcoming real and imagined barriers to wireless adoption. *Industrial Ethernet Book*, 68:34–35, 2012.

[35] P. Matkurbanov, S. Lee, and D.-S. Kim. A survey and analysis of wireless fieldbus for industrial environments. In *SICE-ICASE, 2006. International Joint Conference*, pages 5555–5561, Oct. 2006.

[36] S.K. Mazumder, R. Huang, and K. Acharya. Rotor Position Feedback Over an RF Link for Motor Speed Control. *Power Electronics, IEEE Transactions on*, 25(4):907–913, April 2010.

[37] P. Neumann. Communication in industrial automation.What is going on? *Control Engineering Practice*, 15(11):1332–1347, 2007.

[38] R. Pigan and M. Metter. *Automating with PROFINET*. Publicis Publishing, 2008.

[39] H. Ramamurthy, B.S. Prabhu, R. Gadh, and A.M. Madni. Wireless industrial monitoring and control using a smart sensor platform. *Sensors Journal, IEEE*, 7(5):611–618, May 2007.

[40] T.S. Rappaport. Characterization of UHF multipath radio channels in factory buildings. *Antennas and Propagation, IEEE Transactions on*, 37(8):1058–1069, Aug. 1989.

[41] F. Salvadori, M. de Campos, R. de Figueiredo, C. Gehrke, C. Rech, P.S. Sausen, M.A. Spohn, and A. Oliveira. Monitoring and diagnosis in industrial systems using wireless sensor networks. In *Intelligent Signal Processing, 2007. WISP 2007. IEEE International Symposium on*, pages 1–6, Oct. 2007.

[42] F. Salvadori, M. de Campos, P.S. Sausen, R.F. de Camargo, C. Gehrke, C. Rech, M.A. Spohn, and A.C. Oliveira. Monitoring in industrial systems using wireless sensor network with dynamic power management. *Instrumentation and Measurement, IEEE Transactions on*, 58(9):3104–3111, Sept. 2009.

[43] Xingfa Shen, Zhi Wang, and Youxian Sun. Wireless sensor networks for industrial applications. In *Intelligent Control and Automation, 2004. WCICA 2004. Fifth World Congress on*, volume 4, pages 3636–3640 Vol. 4, June 2004.

[44] V. Srivastava and M. Motani. Cross-Layer Design: A Survey and the Road Ahead. *IEEE Communications Magazine*, pages 112–119, December 2005.

[45] N.C. Tas, V. Mesrob, and Y. Genc. Wireless sensor networks in the control loop: Delay-sensitive networks. In *Consumer Communications and Networking Conference (CCNC), 2011 IEEE*, pages 575–579, Jan. 2011.

[46] J.-P. Thomesse. Fieldbus technology in industrial automation. *Proceedings of the IEEE*, 93(6):1073–1101, June 2005.

[47] E. Tovar and F. Vasques. Real-time fieldbus communications using profibus networks. *Industrial Electronics, IEEE Transactions on*, 46(6):1241–1251, Dec 1999.

[48] P.-A. Wiberg and U. Bilstrup. Wireless technology in industry-applications and user scenarios. In *Emerging Technologies and Factory Automation, 2001. Proceedings. 2001 8th IEEE International Conference on*, pages 123–131 vol.1, 2001.

[49] A. Willig, K. Matheus, and A. Wolisz. Wireless technology in industrial networks. *Proceedings of the IEEE*, 93(6):1130–1151, June 2005.

[50] P. Wright, D. Dornfeld, and N. Ota. Condition monitoring in end-milling using wireless sensor networks (WSNs). *Transactions of NAMRI/SME*, 36:177–183, 2008.

9

Network Synchronization in Industrial Wireless Sensor Networks

Carlos H. Rentel

R&D Engineer. Creare Inc. Hanover, New Hampshire, U.S.A.

CONTENTS

9.1 Introduction

The objective of a Network Synchronization approach is to align the time-scales of a network of clocks. The core of a clock is an oscillator, and in a simple implementation one could count the cycles of the oscillator signal to obtain time. More generally, the frequency of the oscillator signal is integrated to obtain phase, which is proportional to time. A network synchronization approach could attempt to equalize the frequency and phase of a group of geographically separated oscillator signals by physically controlling the frequency of the oscillators or virtually modifying the time read from a clock using software. Furthermore, the goal may be to synchronize to an international time reference such as the Coordinated Universal Time (UTC), or achieve relative synchronization among the clocks comprising a network regardless of time conventions. In either case the interest is on keeping track of when events of interest occur, or must occur, relative to one another, such as when is the time at which a node in a network is to transmit, or when a sensor detected a phenomenon or signal.

Network synchronization is an important feature for current and future industrial wireless sensor networks. Some of the current tasks requiring network synchronization can be summarized as follows:

- Time Division Multiple Access (TDMA): Time synchronization is indispensable for TDMA. An example of an industrial network utilizing TDMA is the WirelessHART standard, which is based on the Dust's network Time Synchronized Mesh Protocol (TSMP). Code Division Multiple Access (CDMA) making use of synchronous codes (e.g., Walsh sequences) also rely on network synchronization to achieve orthogonality among multiple user transmissions.

- Power-management: Having a common time-scale makes it possible for wireless sensor nodes to turn-off their transceivers and other sub-systems since they can wake-up at pre-determined intervals to avoid missing messages from neighboring nodes.

- Some security protocols rely on a common time-scale to operate properly. For instance, timestamped messages can be used as part of authentication in order to avoid replay attacks. However the network synchronization protocol messages must be secured as well to avoid attacks from malicious nodes. Secure network synchronization protocols are uncommon and represent a relatively new area of research.

- Localization techniques are based on the comparison among the time of arrival of multiple signals at geographically separated nodes.

- Sensor fusion merges spatial and temporal data from multiple sensors to obtain a better representation of the sensed phenomenon. A common time-scale plays a fundamental role when combining the data from multiple sources. Typical examples are beam-forming and tracking applications.

- Numerous industrial and manufacturing processes are in fact synchronous and rely on a pre-defined sequence of events that must occur within certain intervals. For instance, in a network of manufacturing robots along an assembly line. A wireless sensor network that is capable of network synchronization may integrate better to these applications by, for instance, ensuring messages are sent within appropriate time-limits.

Network synchronization can be classified based on the strategy used to organize time dissemination. The two most fundamental strategies are: 1) Master-slave, and 2) Mutual Network Synchronization; other strategies have been derived from these two, such as hybrid master-slave, and hierarchical variants

[4]. In the most fundamental master-slave approach there is a single (master) clock to which all the rest of the (slave) clocks synchronize to. The network topology can take any form, but it is typically modeled as a tree with the master clock at its root. In the mutual network synchronization approach every clock synchronizes to all of its neighbors. The mutual network synchronization approach has been investigated in the past within the context of wire line computer networks [7], and more recently within the context of wireless ad hoc networks [14]. Convergence and stability of the mutual network synchronization approach has been proved under certain scenarios. However, there are still open problems that make this an interesting research area.

A slightly different classification is based on the role of senders and receivers in the network. In sender-receiver synchronization one of two nodes sends a time-stamp while the other one receives it. The node that adjusts its clock based on the clock differences depends on the implementation. In receiver-receiver synchronization a sender will send a signal which is received by multiple receivers. The received signal serves as the point in time to trigger receivers into exchanging and comparing their times. Other classifications of network synchronization are possible, such as hardware versus software-based synchronization, local versus global synchronization, multi-hop versus single-hop, on-demand (post-facto) versus continuous, among others. Performance metrics used to compare network synchronization approaches in wireless sensor networks include: precision, accuracy, energy efficiency, scalability, stability, fault-tolerance, mobility support, and more recently, security. Security is an important aspect of any network algorithm and network synchronization is not exempted, but security will not be treated in this chapter, for more details see, for instance, [3] where security is investigated in the context of network synchronization

Currently, network synchronization is the exception across deployed Industrial Wireless Sensor Networks. The Time-sync Protocol for Sensor Networks (TPSN) is one of the exceptions and will be described in this chapter along with other protocols deemed fundamental in wireless sensor networks. Network synchronization will play an increasingly important role as new application demands and more efficient methods are discovered tailored to the particulars of industrial requirements. There are numerous good references that summarize network synchronization protocols in general wireless sensor networks, such as [19], [18], and [15]. In this chapter we will focus on protocols for Industrial Wireless Sensor Networks, and spend more time on other aspects that have been less common in the survey literature. In Section 9.2 we will review the physics of clocks and their models. Section 9.3 will focus on network synchronization protocols. We will briefly review four protocols namely TPSN, RBS, Random Time Source, and an Industrial Wireless Sensor Network approach based on Kalman filters that has been implemented and tested in industrial environments [2]. We will also describe in some level of detail the fundamentals of mutual network synchronization and two particular protocols in this family referred to as Clock Sampling Mutual Network

Synchronization (CSMNS) [14] and the Diffusion algorithm in [11]. Our goal is to provide the reader with a different perspective to the one encountered in the existing survey literature on this subject, and also to stimulate further research. Section 9.4, summarizes some important results from the estimation of network synchronization parameters.

9.2   Clocks

In this chapter we review physical clock behavior to gain an understanding on the problem faced by a network synchronization approach.

A clock is comprised of an oscillator and a counter or integrator. For convenience assume the signal out of the oscillator $s(t)$ is a general sinusoidal signal as follows

$$s(t) = A(t)\sin(\emptyset(t)) \tag{9.1}$$

$A(t)$ is a time-varying amplitude, which can be considered constant for our purposes, and $\emptyset(t)$ is a time-varying phase. Note that ideally $\emptyset(t) = 2\pi f t$. However, that is only true for a perfect clock with perfectly stable and constant frequency. Rather, real physical clocks deviate from that ideal, and their frequency is a function of time. Therefore, in general when we integrate the oscillator's frequency to obtain phase we have

$$\int_0^t f(t)dt = \frac{1}{2\pi} (\emptyset(t) - \emptyset(0)) \tag{9.2}$$

The time representation produced by a clock using the oscillator is the phase $\emptyset(t)$ multiplied by a scaling factor. For the lack of a better estimate, assume the scaling factor is $1/(2\pi f_n)$, where f_n is the nominal frequency of the oscillator, then

$$T(t) = \frac{1}{2\pi f_n} (\emptyset(t) - \emptyset(0)) \tag{9.3}$$

Where $T(t)$ is the time process of the clock (a function of real time), and $T(t) \neq t$ because $f_n \neq f(t)$. We could use instead a measured frequency to obtain time (rather than a nominal frequency), but this measure, although potentially better, will also deviate from the true frequency of the oscillator in practice. If one could precisely measure the true frequency $f(t)$ of the oscillator then we could obtain real time t (i.e., divide Eq. (9.2) by $f(t)$ on both terms and solve for t). However, this is not possible in practice because the frequency measurements depend also on accurate time measurement, which rely on imperfect clocks. The representation of time by a clock is therefore, tied to its frequency. The more stable and constant this frequency is the more precise the clock will be. Oscillators depend on components whose parameters depend on manufacturing tolerances, environmental factors (e.g., temperature and pressure), and age, hence their frequency variations. In what follows we describe an accepted model for the instantaneous frequency of oscillators.

A typical way of modeling an oscillator frequency is as follows [4], [1]

$$f(t) = f_0 + f_d(t) + f_r(t) \tag{9.4}$$

$$f_0 = f_n + \Delta f \tag{9.5}$$

where Δf is a constant deviation from the nominal frequency due to initial calibration errors, or manufacturing tolerances. $f_d(t)$ is the deterministic time-dependent frequency component,

$$f_d(t) \cong D f_n t \tag{9.6}$$

Where D is the linear fractional frequency drift rate. Finally

$$f_r(t) = \frac{1}{2\pi} \frac{d\theta(t)}{dt} \tag{9.7}$$

$f_r(t)$ are frequency variations due to random phase variations $\theta(t)$. Combining (9.5), (9.6), and (9.7) into (9.4)

$$f(t) = f_n + \Delta f + D f_n t + \frac{1}{2\pi} \frac{d\theta(t)}{dt} \tag{9.8}$$

Integrating (9.8) with respect to t (see Eq. (9.2)) we obtain

$$\emptyset(t) = 2\pi \left(f_n + \Delta f \right) t + \pi D f_n t^2 + \theta(t) + \emptyset_0 \tag{9.9}$$

Where $\emptyset_0 = \emptyset(0) - \theta(t)$. Dividing (9.9) by $2\pi f_n$ we obtain

$$T(t) = t + \frac{\Delta f}{f_n} t + \frac{D}{2} t^2 + \frac{\theta(t) + \emptyset_0}{2\pi f_n} \tag{9.10}$$

In the ideal case phase increases linearly with time, that is

$$\emptyset_d(t) = 2\pi f_n t + \emptyset(0) \tag{9.11}$$

and

$$T_{ideal}(t) = t + \frac{\emptyset(0)}{2\pi f_n} \tag{9.12}$$

The time of a real clock, however, has a time drift effect due to $(1 + \frac{\Delta f}{f_n})t$, a frequency drift effect due to $\frac{D}{2} t^2$, and a noise effect due to $\frac{\theta(t) + \emptyset_0}{2\pi f_n}$ as observed in Equation (9.10) Frequency drift terms affect the frequency of the oscillator in a longer time-scale than do the noise and time-drift terms. However, as the demand on precision and accuracy increases, these long time-scale terms become more relevant. This is particularly true today for time references obtained with highly precise and stable atomic clocks. For the current demands of wireless network synchronization accuracies, it is sufficient to simplify (9.10) to

$$T(t) = \beta t + T(0) \tag{9.13}$$

Where $\beta = 1 + \frac{\Delta f}{f_n}$, which is referred as the skew of the clock.

All these imperfections are different from one clock to the next, and they vary over time. Therefore network synchronization is needed to keep time of every clock as close to one another as possible. Additionally, drifts and skew terms make it impossible for the network synchronization approach to rely on a single (one time) correction. Frequent error tracking and correction is necessary.

Clock performance is typically measured by its stability and accuracy. Stability is referred to the ability of the clock to generate a constant time interval over real time. There are short-term and long-term measurements of stability. Accuracy refers to the agreement of the clock with a time reference, such as UTC. Clock stability performance is typically given in parts per million (ppm). For instance, the clocks driving certain functions of the IEEE 802.15.4 PHY are required to be within ±40ppm, which means the clocks can deviate up to 80 microseconds per second in the worst case. Stability over temperature is also a performance frequently encountered, for instance 1ppm/°C, which states the stability of the clock as a function of temperature (i.e., its temperature coefficient). Note, that in an embedded device, such as a sensor node, the clock can also be affected by its power source voltage/current variations, electronic noise, and offsets.

9.3 Network Synchronization Protocols for Industrial Wireless Sensor Networks

Numerous Network Synchronization protocols have been proposed for Wireless Sensor Networks in general. In the particular case of Industrial Wireless Sensor Networks the IEEE 802.15.4 standard specifies mechanisms for contention-free access that require network synchronization for proper operation. However, the synchronization method is left open as a vendor differentiation feature. Table 9.1 shows few companies producing wireless sensor networks with network synchronization. As of this writing, available solutions with network synchronization are limited. However, more solutions are likely to appear in the market due to the increased maturity of wireless sensor software and hardware, new applications, and the importance of network synchronization as a mechanism to improve reliability and security.

Low cost and simplicity is an important requirement for Industrial Wireless Sensor Networks. Several network synchronization protocols have been proposed and implemented focusing on minimizing cost and complexity. In this regard a well-known network synchronization approach is the one used

TABLE 9.1

An example of Available Industrial Networks with Network Synchronization

Company	Country of Origin	Description
Dust Networks	USA	TPSN
VTT	Finland	Multi-hop synchronization
MicroStrain	USA	Master-slave beaconing

by Dust networks, referred to as the Time-sync Protocol for Sensor Networks (TPSN), and used in the WirelessHART protocol standard. TPSN will be described in the next section, followed by RBS as a good representation of simple Network Synchronization protocols that have either been implemented or are deployed in industry. In the case of MicroStrain shown in Table 9.1, their mXRS system uses a beaconing synchronization protocol integrated with the IEEE 802.15.4 standard. This is a master-slave synchronization approach that keeps a group of distributed wireless sensing nodes synchronized to within tens of micro-seconds for the implementation of TDMA.

Reliability is another important requirement in industrial environments, and Wireless Sensor Network protocols and technologies must increase their reliability if they are to be successfully adopted in place of wired networking technologies. All protocols in an Industrial Wireless Sensor Network must be reliable and fault-tolerant. Reliability concerns in network synchronization have been investigated and several methods for Industrial Wireless Sensor Networks have been proposed. We will describe the Random Time Source protocol which attempts to improve reliability and fault-tolerance in industrial environments by utilizing different time sources selected randomly over time. Additionally, towards the end of this section we will present Mutual Network Synchronization fundamentals and two example protocols in this category, the diffusion algorithms in [11], and Clock Sampling Mutual Network Synchronization (CSMNS) in [14], which also attempt to increase reliability via the utilization of a diverse number of time sources.

9.3.1   Timing-Sync Protocol for Sensor Networks (TPSN)

TPSN is used by Dust Networks and WirelessHART standard. It is a sender-receiver approach [8]. Two nodes try to synchronize by exchanging time-stamps and compute the offset between their clocks. The sender initiates the exchange and adjusts its clock to that of the receiver. The original TPSN algorithm does not estimate clock skew, which requires more frequent re-synchronization events. However, joint skew and offset estimation is possible [13]. Figure 9.1 shows the sequence of messages exchanged by a sender node i and a receiver node j; multiple time-stamp exchanges are possible to obtain a better estimate of the offset between the two clocks as shown.

The sender node starts preparing to create the synchronization message to

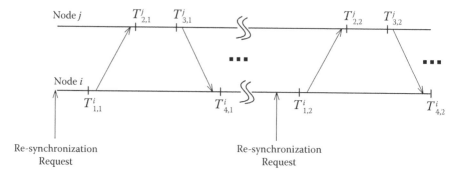

FIGURE 9.1

Sender-Receiver time-stamp exchanges in TPSN.

be transmitted at re-synchronization request time. Ideally the synchronization message should carry a time-stamp with a time T (a function of real time t) as close to the time of transmission as possible to avoid uncertainties in transmission processing and channel access.

In the time interval from time-stamp creation to transmission there is a delay that may be due to channel access and other transmission processing delay uncertainties. The time at which the time-stamp is created (on transmission) or obtained (on reception) is shown in Figure 9.1 with small vertical lines above or below the corresponding time-stamp T.

Node i creates its time-stamp $T_{1,1}^i$ and transmits it after a delay to node j. Node j receives the time-stamp sent by node i, and after a reception processing delay time-stamps this message with $T_{2,1}^j$ and prepares to send its own time-stamp. Node j creates its time-stamp $T_{3,1}^j$ and sends it to node i along with $T_{2,1}^j$, which in turn time-stamps it at $T_{4,1}^i$ after a reception processing delay. Node i ends up with all time-stamps needed to compute its offset with respect to node j, therefore in TPSN the sender (node i in Figure 9.1) synchronizes to the receiver (node j). This process repeats at every re-synchronization event.

The relationships between the time-stamps exchange by the nodes at the kth exchange event is as follows

$$T_{2,k}^j = T_{1,k}^i + O + \gamma_t^i + d + \gamma_r^j$$
$$T_{4,k}^i = T_{3,k}^j - O + \gamma_t^j + d + \gamma_r^i \qquad (9.14)$$

Where O is the offset of the two clocks, d includes the transmission and propagation delays, γ_r^i is the time-stamp reception delay from the time of message reception to decoding of time-stamp at node i, and γ_t^i is the time-stamp transmission delay from the time of time-stamp creation to actual time-stamp transmission. Eq. (9.14) assumes the propagation and transmission delays d are both the same, this is reasonable only if the transmission rates in both directions are equal, which is typically the case. Node i can compute an estimate of O and adjust its clock accordingly as

$$O = \frac{\left(T_{2,k}^j - T_{1,k}^i\right) - \left(T_{4,k}^i - T_{3,k}^i\right)}{2} \tag{9.15}$$

If the γ delays are assumed equal the estimation of the clock's offsets is excellent. If these delays are made very small their influence on the final estimation error can also be reduced. Typically, these delays are modeled as random variables. If the delays γ are lumped into random variables Γ_{sr} and Γ_{rs} for the sender-to-receiver and receiver-to-sender transmissions respectively then (9.14) becomes

$$\begin{aligned} T_{2,i}^j &= T_{1,i}^i + O + d + \Gamma_{sr} \\ T_{4,i}^i &= T_{3,i}^i - O + d + \Gamma_{rs} \end{aligned} \tag{9.16}$$

The estimation of the offset becomes a statistical parameter estimation problem. As explained before the original TPSN estimates offset only. A method to estimate skew of the clocks is also possible as, for instance, in the Tiny-sync and Mini-sync approaches [17] that also use a two-way time-stamp exchange approach.

So far we have described how two nodes synchronize in TPSN. In a network, the synchronization of more than two nodes is achieved by building a spanning-tree where every node synchronizes to its parent, and everyone is synchronized to the root node. The creation of the spanning-tree is performed first in a so called level discovery phase, and once the spanning-tree is formed and every node knows its position in the tree, the synchronization phase starts, where every node synchronizes to its parent node in the way explained previously. Several algorithms exist for the creation of spanning-trees and TPSN uses a simple one, which is suitable for energy-constraint wireless sensor networks. For a more detailed description of the TPSN network-wide synchronization approach see [8] or [10].

9.3.2 Reference Broadcast Synchronization (RBS)

RBS [12] is a receiver-receiver synchronization approach. The fundamental idea is that a sender node transmits a message or pulse (not necessarily time-stamped) and the receivers use the reception of that message to exchange the times at which they received the sender's message. This allows the receiver nodes to synchronize to one another. The same approach is used in the CesiumSpray synchronization approach [21], which is inspired by the *a posteriori* agreement internal synchronization algorithm [20] RBS extends this approach to a multi-hop network. Figure 9.2 shows the message exchanges performed in RBS. It is easy to see that the offset between the two nodes can be found by computing the difference between the two time-stamps in every node. A node does not adjust its clock directly instead it updates a table with the

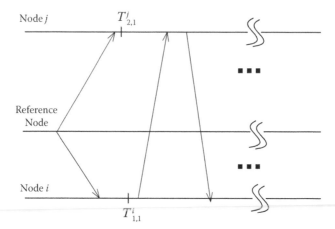

FIGURE 9.2
Receiver-Receiver time-stamp exchanges in RBS.

parameters of its neighbors, which is used to translate the time from a neighbor to its own. In order to synchronize a multi-hop network nodes that share multiple broadcast domains (referred to as gateways) are used as translators between two nodes in different broadcast domains; see [12] for more details. A hybrid approach that combines sender-receiver and receiver-receiver methods is proposed in [18] and referred to as TSync.

9.3.3 Random Time Source Protocol

The Random Time Source Protocol was proposed in [22] to improve the reliability of network synchronization approaches in industrial environments. The network is modeled as a tree and every child node synchronizes to a parent source node using a linear regression technique.

The precision and survivability of the synchronization approach depends on the parent source selected by the child nodes. Industrial wireless environments are plagued with interference and communication outages. Therefore, all the children of a parent node (or time source) will be out of synchronization if any problem occurs in the time source itself or the link connecting the time source to its children. In the Random Time Source Protocol, a node in the network employs several potential nodes as time source and randomly selects one for time synchronization every time period. The rules of the Random Time Source Protocol are as follows:

- A given node selects its time source based on link quality. Link quality is verified periodically to guarantee a good time source is selected regularly. A list of potential time sources is kept in every node and potential time source nodes are deleted if the link quality decreases below an established threshold.

- A re-synchronization process is performed in the event that a node fails to synchronize to all of its potential time sources.

Random time source selection can reduce the probability of resynchronization and improve the robustness of the synchronization mechanism. The authors in [22] implemented this protocol in a real platform and demonstrated time synchronization accuracies to within $5\mu s$.

The synchronization approaches in Section 9.3.5 utilize a similar approach in which multiple time sources are present to improve reliability and fault-tolerance. These approaches improve reliability and fault-tolerance in the event of time-source failure.

9.3.4 Kalman-Based Industrial Wireless Sensor Network Synchronization

The authors in [2] propose a method of synchronization that utilizes a Kalman filter to correct time deviations between the nodes of an Industrial Wireless Sensor Network. The main motivation of this work was to provide network synchronization among the nodes of a system capable of determining the wireless communication conditions in an industrial environment. That is, the quality of wireless network links (measured in terms of packet success rate, outage probability, etc.) was determined a priori with a number of synchronized wireless nodes capable of acquiring wireless channel quality data that was off-line processed to assess the wireless network channel conditions. Wireless sensor deployment strategies were then recommended from the results of the off-line analysis.

The wireless nodes needed to be synchronized for the correct time-order of events, and the network synchronization approach was required to survive harsh Radio Frequency (RF) industrial environments. The proposed network synchronization approach is able to maintain synchronization to within tens of micro-seconds in an RF environment with very low packet success rate (as low as 0.1% in some cases). Each node implements a Kalman filter that compares the time at which a packet is actually received with the time at which the packet should have ideally been received. Both clock rate and offset are adjusted in every node.

9.3.5 Mutual Network Synchronization

The mutual network synchronization approach is also referred to as peer-to-peer network synchronization. In this approach, all nodes try to synchronize to one another with no root or master clock. Every clock exchanges its time information with its neighbors and distributed algorithms in every node are designed to make all the times converge towards a common time. In its purest

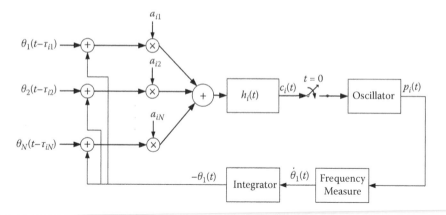

FIGURE 9.3
Node phase adjustment in mutual network synchronization.

form there is really no need for network protocols that arbitrate the order of the message exchange, initiators, or references as in TPSN or RBS. There is not assumed or purposely built network topology to disseminate time as in, for instance, the spanning-tree used in TPSN or the Random Time Source Protocol. All that is required is network connectivity and a mean to exchange time among the clocks in no particular order. A mutual network synchronization approach as described promises many advantages including fault-tolerance, reduced energy consumption, and mobility support due to simpler overhead protocols. However, mutual network synchronization approaches can become unstable if parameters change beyond design considerations. In what follows, we first present a fundamental result regarding stability and the final convergence frequency, then we present two approaches that use mutual network synchronization.

An analysis of the mutual network synchronization approach was done in [7] and we follow this approach hereafter. We assume the clocks exchange their instantaneous phases. Figure 9.3 shows the clock in a node, and the control used to adjust its phase based on the delayed phases of its neighbors.

The received phases are compared to node's i phase and the difference is linearly combined to form the input to the $h_i(t)$ filter. The output of $h_i(t)$ is then

$$c_i(t) = h_i(t) * \sum_{j=1}^{N} a_{ij} \left[\theta_j \left(t - \tau_{ij} \right) - \theta_i(t) \right] \tag{9.17}$$

Where $*$ denotes convolution. The output of the filter is used to control the oscillator's frequency as follows,

$$\dot{\theta}_i(t) = c_i(t) + f_{n,i} \tag{9.18}$$

$f_{n,i}$ is the nominal frequency of the oscillator in node i. Note that for simplicity we are assuming the nominal frequency of the clock to be constant. We are interested in using the previous model to infer the properties of the filter needed to make the system stable and converge to a final frequency common to all the oscillators. We assume the filter is causal, stable, and has a positive DC gain (i.e., $H_i (s = 0) \equiv \lambda_i > 0$). Without loss of generality we have

$$\sum_{j=1}^{N} a_{ij} = 1 \qquad (9.19)$$

Note that the a_{ij} terms are zero if there is no connection between nodes i and j. Furthermore, it is assumed that the filter is connected to the oscillator at time $t = 0$ in all nodes. The latter assumption can be relaxed by, for instance, connecting one node at a time (not all at once) see for instance [5], and [14], or by broadcasting a signal across the entire network that will be used solely to command the nodes to close the switch at once.

If the oscillators have been free running before the switch is closed at $t = 0$, then the phase for $t < 0$ is

$$\theta_k(t) = f_{n,k}t + \theta_k(0) \quad t < 0, \quad k = \{1, \ldots, i, \ldots, N\} \qquad (9.20)$$

Equation (9.17) can be written as follows for $t \geq 0$

$$\begin{aligned} c_i(t) &= h_i(t) * \sum_{j=1}^{N} a_{ij}[\theta_j(t - \tau_{ij})u(t - \tau_{ij}) \\ &+ \theta_j(t - \tau_{ij})[u(t) - u(t - \tau_{ij})] - \theta_i(t)]t \geq 0 \end{aligned} \qquad (9.21)$$

The term $\theta_j (t - \tau_{ij}) [u(t) - u (t - \tau_{ij})]$ is the phase signal of node j and in the link at the time the switch is closed (which corresponds to a free running oscillator at node j). That is,

$$\theta_j (t - \tau_{ij}) [u(t) - u (t - \tau_{ij})] = \begin{cases} f_{n,j} \cdot (t - \tau_{ij}) + \theta_j(0) & \text{for} \quad 0 \leq t \leq \tau_{ij}, \\ 0 & \text{Otherwise} \end{cases} \qquad (9.22)$$

Substituting (9.21) into (9.18) and taking the Laplace transform yields

$$\begin{aligned} s\Theta_i(s) &= H_i(s) \sum_{j=1}^{N} a_{ij} \left[\Theta_j(s)e^{-s\tau_{ij}} + \Theta_j^-(s)e^{-s\tau_{ij}} - \Theta_i(s)\right] \\ &+ \theta_i(0) + \frac{1}{s} f_{n,i} \end{aligned} \qquad (9.23)$$

Where $\Theta_j(s)e^{-s\tau_{ij}}$ is the Laplace transform of $\theta_j (t - \tau_{ij}) u (t - \tau_{ij})$, $H_i(s)$ and $\Theta_i(s)$ are the Laplace transform of $h_i(t)$ and $\theta_i(t)$ respectively, and $\Theta_j^-(s)e^{-s\tau_{ij}}$ is the Laplace transform of (9.22), which is given by

$$L\{\theta_j (t - \tau_{ij}) [u(t) - u (t - \tau_{ij})]\} = \int_0^{\tau_{ij}} (f_{n,j} \cdot (t - \tau_{ij}) + \theta_j(0)) e^{-st}dt \qquad (9.24)$$

After a change of variables $\rho = t - \tau_{ij}$ (9.24) results in

$$\Theta_j^-(s)e^{-s\tau_{ij}} = \int_{-\tau_{ij}}^{0} (f_{n,j} \cdot \rho + \theta_j(0)) \, e^{-s\tau_{ij}} e^{-s\rho} d\rho = \int_{-\tau_{ij}}^{0} \theta_j(t)e^{-s\tau_{ij}} e^{-st} dt$$

(9.25)

Equation (9.23) can be written as

$$s\Theta_i(s) = H_i(s) \sum_{j=1}^{N} \hat{a}_{ij}\Theta_i(s) - H_i(s)\Theta_i(s) + \mathrm{K}$$

(9.26)

Where $\hat{a}_{ij} = a_{ij}e^{-s\tau_{ij}}$, and

$$\mathrm{K} = \theta_i(0) + \frac{1}{s}f_{n,i} + H_i(s) \sum_{j=1}^{N} \hat{a}_{ij} \int_{-\tau_{ij}}^{0} \theta_j(t)e^{-s\tau_{ij}} e^{-st} dt$$

(9.27)

It can be shown that the last term in K diminishes to zero as time approaches infinity. Solving for $\Theta_i(s)$ in (9.26) yields

$$\Theta_i(s) = \frac{H_i(s)}{s + H_i(s)} \sum_{j=1}^{N} \hat{a}_{ij}\Theta_i(s) + \frac{\mathrm{K}}{s + H_i(s)}$$

(9.28)

The condition for (9.28) to be stable is that [7]

$$\left| \frac{H_i(s)}{s + H_i(s)} \right| < 1$$

The authors in [7] also found the final frequency at which all oscillators converge is finite. This frequency is not necessarily the average, but in fact can be below the lowest frequency of all oscillators. The frequency of all oscillators can tend towards zero as well, which can bring the system down to zero frequency. However, by proper delay estimation it can be shown that the final frequency reaches the average of all frequencies. Stability of mutual network synchronization under disturbances has also been proved and the reader is referred to [7] and [5] for the details.

We now present a more recent example of a mutual network synchronization algorithm referred to as Clock-Sampling Mutual Network Synchronization (CSMNS). Rather than transmitting analog instantaneous phase signals, what would be the consequence of transmitting digital timing information in the form of time-stamps? What happens if the links are not static, instead nodes move and connections and neighbors change? Or perhaps links are wireless, which are subject to degradation, obstacles, multipath fading, etc.? How does mutual network synchronization perform in that case? The work in [14] tried to answer these questions.

The nodes in CSMNS exchange timestamps periodically, and every node utilizes the scheme shown in Figure 9.4 to adjust both frequency and time

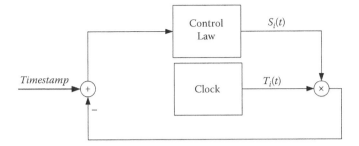

FIGURE 9.4
Control scheme in every node for CSMNS.

offsets. Every node corrects its clock by multiplying it by a correction factor $s_i(t)$, which transforms (9.13) into

$$T(t) = s_i(t)\beta t + s_i(t)T(0) \quad \text{for the } i^{\text{th}} \text{ node.} \tag{9.29}$$

The scheme in Figure 9.4 automatically and recursively adjust both frequency and time without the need to either estimate or directly control the frequency of any clock.

The authors in [14] proved stability and convergence of this procedure in a simple scenario, and showed via simulation stability and convergence in a more complex scenario where nodes move under different mobility models. Figure 9.5 shows one of these results using the Random Waypoint (RWP) mobility model for two different networks with 100 nodes and 500 nodes Figure 9.5 shows the maximum deviation observed in any two clocks over time. In this case the maximum error observed converges to a value between $20\mu s$ and $30\mu s$.

A similar approach was proposed by the authors in [11]. This approach is referred to as diffusion algorithms. Rather than letting the clocks converge by simply exchanging times each node asks its neighbors about their clocks and it computes an average. The average time is then sent back to the neighbors in order for them to update their clocks.

9.4 Parameter Estimation in Network Synchronization for Industrial Wireless Sensor Networks

The offset, link delay, and skew of the clocks may need to be estimated for the network synchronization algorithms to work properly. The key is for these estimation algorithms to maintain good accuracy in the face of limited computational resources in a wireless sensor network. TPSN and RBS estimate clock offset and skew in a simple way to minimize computational complexity.

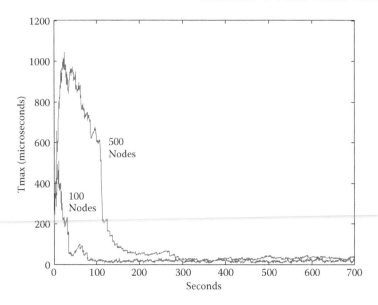

FIGURE 9.5
CSMNS for mobile networks of 100 and 500 nodes.

Other methods have been proposed for wireless sensor networks that jointly estimate offset and skew of the clocks. Some examples of the work in this area can be found in [9] and [16].

9.5   Conclusions

We have presented an overview of network synchronization for Industrial Wireless Sensor Networks. We have described important protocols that have been successfully implemented and others that show promise for the near future. With this chapter our goal was to summarize some of the most important fundamentals in the area, and also stimulate the reader to find more information on his or her own. Network synchronization is an evolving field with a critical importance in the improvement of reliable data transfer, event timing, and security. More solutions will become available as the wireless sensor network implementation methods mature, and network synchronization approaches become less resource intensive.

References

[1] Allan D. W., "Clock characterization tutorial," *Proceedings of the 15th Annual Precise Time and Time Interval (PTTI) Applications and Planning Meeting*, 1983.

[2] Armstrong B. S. R., Pereira L., and Rentel C. H., "Synchronization System and Method for Wireless Communication Nodes." U.S Patent No. 7,912,164 B2. March 2011.

[3] Boukerche A. and Turgut D., Secure time synchronization protocols for wireless sensor networks, *IEEE Wireless Communications*, 14 (2007), 64-69.

[4] Bregni S., "Clock stability characterization and measurement in telecommunications," *IEEE Trans. On Instrumentation and Measurement*, Vol. 46, No. 6, December 1997, pp. 1284-1294.

[5] Brilliant M. B., "The determination of frequency in systems of mutually synchronized oscillators," Bell Syst. Tech. J., Vol. 45, Dec. 1966, pp. 1737-1748.

[6] Dai H. and Han R., "TSync: A lightweight bidirectional time synchronization service for wireless sensor networks." *ACM SIGMOBILE Mobile Computing and Communications Review*, Vol. 8, No. 1, 2004, pp. 125-139.

[7] Gersho A. and Karafin B. J., "Mutual synchronization of geographically separated oscillators," Bell Systems Tech. J., Vol. 45, December 1966, pp.1689-1904.

[8] Ganeriwal S., Kumar R., and Srivastava M., "Timing synch protocol for sensor networks," *In Proceedings of 1st International Conference on Embedded Network Sensor Systems*, Los Angeles, CA, November 2003, pp138-149.

[9] Jeske D. R., "On the maximum likelihood estimation of clock offset," *IEEE Trans. On Communications*, Vol. 53, 2005, pp. 53-54.

[10] Karl H. and Willig A., "Protocols and architectures for wireless sensor networks," Wiley, 2005.

[11] Li Q. and Rus D., "Global clock synchronization in sensor networks," *IEEE Trans. On Computers*, Vol. 55, 2006, pp.214-226.

[12] Nelson J., Girod L., and Estrin D., "Fine-Grained network time synchronization using reference broadcast," *Proceedings of the 5th Symposium on Operating Systems Design and Implementation (OSDI'02)*. December 2002, pp. 147-163.

[13] Noh K. L., Chaudhari Q., Serpedin E., and Suter B., "Novel clock phase offset and skew estimation using two-way timing message exchanges for wireless sensor networks," *IEEE Trans. On Communications*, Vol. 55, 2007, pp. 766-777.

[14] Rentel C. H. and Kunz T., "A mutual network synchronization method for wireless ad hoc and sensor networks," *IEEE Trans. On Mobile Computing*, Vol. 7, No. 5, May 2008, pp. 1-14.

[15] Sadler B. M. and Swami A., "Synchronization in sensor networks: an overview," in *Proceedings of IEEE Military Communications Conference*, Washington DC, 2006, pp. 1-6.

[16] Serpedin E. and Chaudhari Q. M., "Synchronization in wireless sensor networks. Parameter estimation, performance benchmarks, and protocols," Cambridge University Press 2009.

[17] Sichitiu M. L. and Veerarittiphan C., "Simple, accurate time synchronization for wireless sensor networks," in *Proceedings of IEEE Wireless Communications and Networking Conference (WCNC)*, New Orleans, LA, March 2003, pp. 1266-1273.

[18] Sivrikaya F. and Yener B., " Time synchronization in sensor networks: a survey," *IEEE Networks*, 2004, p. 45-50.

[19] Sundararaman B., Buy U., and Kshemkalyani A. D., "Clock synchronization for wireless sensor networks: a survey," Ad-Hoc Networks, Vol. 3, 2005, pp. 281-323.

[20] Veríssimo P. and Rodrigues L., "A posteriori agreement for fault-tolerant clock synchronization in broadcast networks". In *Digest of Papers, The 22nd International Symposium on Fault-Tolerant Computing*, Boston – USA, July 1992. IEEE. INES AR/65-92.

[21] Veríssimo P., Rodriguez L., and Casimiro A, " Cesium Spray: A precise and accurate global clock service for large-scale systems," Journal of Real-time Systems, Vol. 12, No. 3, May 1997, pp. 243-294.

[22] Wang F. et al., "A Reliable Time Synchronization Protocol for Wireless Sensor Networks," *Proceedings of the 3rd International Symposium on Computer Science and Computational Technology (ISCSCT)*, Jiaozuo, P.R. China, August 2010, pp. 9-13.

10

Wireless Control Networks with Real-Time Constraints

Alphan Ulusoy

Division of Systems Engineering, Boston University

Ozgur Gurbuz and Ahmet Onat

Faculty of Engineering and Natural Sciences, Sabanci University

CONTENTS

10.1 Introduction

The trend for today's industrial control systems is to design them as distributed real-time systems for reliability, cost effectiveness, ability for fault diagnosis, and upgradability. Industrial control systems and networks, traditionally realized through wired communication systems/protocols suffer from maintenance costs due to numerous installations of expensive cables. Utilization of wireless technologies in industrial control systems is an urgent need.

However, one cannot simply combine a general-purpose wireless network and a real-time system and expect satisfactory performance, due to the conflicting natures of the following two entities: A real-time system, such as an industrial control system, does not expect the data traveling in the system to be lost or delayed, whereas packet losses and delays are only typical in a wireless network because of its stochastic nature. Since wireless communication channels are inherently prone to data loss and corruption resulting in unpredictable delays, it is necessary to implement methods that will bound the latency of data delivery between the communicating nodes. This is especially important when a feedback loop is closed over the network as in a *networked control system* (NCS) [3, 2, 13, 7], because unbounded delays directly affect the phase delay, and thus the stability of the control loop.

The transmission latency is typically unbounded in general communication networks, which aim for high average throughput. The delay characteristics are caused by two main reasons; *transport losses*, caused by the physical characteristics of the channel such as fading, electromagnetic interference, etc., which enforce retransmissions, and *protocol delays*, which may delay the node from starting the transmission without bound. The obvious method of bounding the transmission latency is to implement a communication protocol and a physical transport layer which have bounded protocol delays and low probability of corruption, respectively. Modern industrial networks typically revolve around this idea, resulting in industrial wireless sensor networks (IWSNs) and standards such as [16, 33, 28, 23, 13]. ZigBee [33] is a specification for a suite of high level communication protocols using small, low-power digital radios based on an IEEE 802 standard for personal area networks, IEEE 802.15.4 [16], targeting applications such as industrial control and monitoring, building and home automation, energy system automation and embedded sensing [12]. WirelessHART [28, 30], an extension of the Highway Addressable Remote Transducer (HART) protocol, utilizes a time synchronized, self-organizing, and self-healing mesh architecture that uses IEEE 802.15.4 radios for process monitoring and control applications. IETF's 6LoWPAN [23] also works on IEEE 802.15.4 radios to provide Internet Protocol support and connectivity to industrial wireless devices and networks. ISA-100.11a standard [13] was developed to provide reliable and secure wireless operation for non-critical monitoring, alerting, supervisory control, open loop control, and closed loop control applications. The standard defines the protocol suite, system management, gateway and security specifications for low-data-rate wireless connectivity with fixed, portable, and moving devices supporting very limited power consumption requirements. The application focus is to address the performance needs of applications, such as monitoring and process control, where latencies on the order of 100 ms can be tolerated, with optional behavior for shorter latency.

An alternative approach to IWSNs is to design the overall control system taking the limitations of the underlying communication methodology into consideration. The so called *co-design* methods have received much attention recently, and they aim to optimize or guarantee the performance of the control by compensating for the weaknesses of the communication methodology in another part of the system: in the control design. This can be achieved by balancing the control sampling time with communication bandwidth requirement based on an optimization criterion [22, 21, 5], or by regulating the data transmission rate for communication by using plant models where data is only transmitted if the plant state prediction errors are greater than a margin. The bandwidth thus gained can be used for retransmissions, or error correction [26, 19]. Another common approach is to employ a plant model to calculate control predictions and transmit all of them at once to the actuator node. This redundancy allows the plant to run in open loop for short durations where the control signal does not arrive on-time [8, 25, 24]. Similar methods applying the field of model predictive control also exist in [20].

This chapter focuses on the joint networked control problem for designing and operating a system with real-time constraints over a wireless network, employing simple, off-the-shelf, IEEE 802.11 (WiFi), and IEEE 802.15.4 (ZigBee) radio technologies. We consider the wireless networked control problem as an example real-time wireless industrial application, evaluate the effects of the underlying wireless network on the performance of the system and propose ways to improve it. More specifically, we compare the performance of a basic wireless networked controlled system (b-WNCS) with that of a Wireless Model Based Predictive Networked Control System (WMBPNCS) [31, 25] under various operating conditions. The b-WNCS that we consider is a conventional closed-loop control system with the exception that the nodes communicate with each other wirelessly. The WMBPNCS, on the other hand, is a control system that is specifically designed to operate under random packet delays and losses [31]. The operating conditions that we consider include three sources of wireless channel errors that typically occur in industrial settings: ambient wireless traffic, block fading, and fast fading; and three wireless network technologies: IEEE 802.11, IEEE 802.11 with Cooperative Medium Access Control Protocol (COMAC), which were initially presented in [31], and IEEE 802.15.4. We establish through extensive experiments that WMBPNCS outperforms b-WNCS in all cases that we consider. Nevertheless, we see that overall system performance remains limited by the quality of the underlying wireless channel. We show that the adverse effects of ambient wireless traffic can be virtually eliminated through a modification of several medium access control parameters. We also demonstrate that the insensitivity to changes in the reference signal caused by severe fast fading can be significantly reduced through the use of a cooperative protocol such as IEEE 802.11 with COMAC as we have also shown in [31]. Finally, in this chapter we show that WMBPNCS does not necessarily need a sophisticated high throughput wireless network such as IEEE 802.11 and WMBPNCS can even be operated over IEEE 802.15.4 in the presence of other WMBPNCSs.

The rest of this chapter is organized as follows: In Section 10.2, we provide some background on wireless access, discuss various channel errors that occur

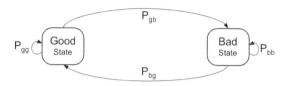

FIGURE 10.1
Gilbert/Elliot channel model.

in typical industrial settings, and present the wireless networked control problem as an example wireless industrial application with real-time constraints. In Section 10.3, we review the wireless network technologies that we consider. In Section 10.4, we introduce the Wireless Model Based Predictive Networked Control System (WMBPNCS). In Section 10.5, we present the results of our experiments and evaluate the performances of both the WMBPNCS and the b-WNCS when operated using different wireless network technologies under various wireless channel conditions. We conclude with final remarks in Section 10.6.

10.2 The Wireless Control Network and the Industrial Setting

The characteristics of a wireless channel typically depend on a multitude of factors, and controlling all such contributing factors simultaneously is not possible in a test-bed of limited size and range. Thus, one typically uses various channel models in order to emulate a previously characterized wireless channel in a controlled and reproducible way. In the following, we discuss these various sources of channel errors, namely *ambient wireless traffic*, *block fading*, and *fast fading*, and explain how they have been incorporated into our experiments to introduce errors to an otherwise ideal link.

Wireless channel is of broadcast nature, and *ambient wireless traffic* may interfere with a node's transmissions. As the nodes retry after each failed transmission attempt, their packets may suffer latencies long enough to cause them to miss their deadlines. In the applicable experiments, we have utilized a test-bed node to generate disrupting wireless traffic and evaluated its effects on the performance of the system.

Wireless channel errors occur typically in bursts followed by practically error-free periods rather than occurring uniformly randomly [27]. This type of fading is referred to as *block fading* where the duration and separation of these error bursts are longer than the symbol time, the duration in which the channel must preserve its characteristics for satisfactory operation. We consider the

Gilbert/Elliot model given in Figure 10.1 [6] to model the bursty packet loss characteristics of the wireless channel. At any given time, the characteristics of the emulated channel are determined by the *good* and *bad* states of the model, in which packets are lost according to packet loss probabilities P_{loss}^g and P_{loss}^b respectively.

The next state of the model is determined after each packet by state transition probabilities P_{gb} and P_{bg}. Since state transition probabilities are typically small, the channel state remains unchanged for some time after a transition is taken imitating bursts of packet loss when the model is in the *bad* state and periods of almost error free transmission when the model is in the *good* state.

In [32], the authors give the measurement results of a wireless link in a realistic industrial setting. We use their results to derive the parameters of our model as $P_{gb} = 0.0196$, $P_{bg} = 0.282$, $P_{loss}^g = 0$, and $P_{loss}^b = 1$ and use this model in the applicable experiments to model the characteristics of a channel with block fading.

Fast fading occurs when the characteristics of the channel change faster than the symbol time. In an industrial setting with numerous obstacles and no direct line of sight between the transmitter and the receiver, multi-path fading causes rapid fluctuations in the received signal strength which result in increased number of retransmissions, increasing packet latency and packet loss. The fluctuations that we consider can be modeled with the Rayleigh distribution [11, 27] which is essentially an exponential distribution with mean $\bar{P}_{rx}$, the average received signal power. For a distance aware Rayleigh fading model, in the applicable experiments, we use an exponential random variable (Y) scaled by P_t/d^α where P_t is transmission power, d is the distance between the transmitting and receiving nodes and α is the path loss exponent.

Wireless networked control systems (WNCS) where sensors, controllers, and actuators communicate over a wireless network are suitable for scenarios that require spatial distribution. They offer significant advantages in terms of reliability, commissioning, and maintenance, especially for complex systems. In a WNCS (Figure 10.2), sensor nodes are responsible for periodically measuring plant outputs and communicating this data to controller nodes over the network. Controller nodes use plant outputs to calculate the control signals and communicate them to actuator nodes. Ultimately, actuator nodes apply the control signals to the plant.

WNCSs are *delay-sensitive* systems that require dedicated wireless real-time networks to ensure proper operation. However, the overhead of designing and installing such a network often hinders the commissioning of the control system and discourages its use. One ideally wants to be able to operate a WNCS over a proven and low-cost general-purpose wireless network. Unfortunately, this is not possible without additional measures as the stability of the system is compromised by unbounded delays and random packet losses that are typical of such general-purpose networks. Without loss of generality, end to end latency of a WNCS can be broken into 5 main components (Figure 10.2): internal latencies of the sensor (τ_s), controller (τ_c), actuator nodes (τ_a), sensor

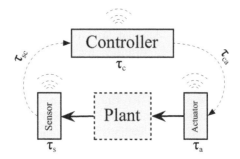

FIGURE 10.2
Latency components of a typical wireless networked control system (WNCS).

to controller communication latency (τ_{sc}) and controller to actuator communication latency (τ_{ca}). Here, τ_s, τ_c, and τ_a refer to the computational and functional latencies that a particular node introduces and are bounded for hard real-time computer systems. Remaining delay components τ_{sc} and τ_{ca} refer to the latencies induced by the communication medium of choice and are not bounded for general-purpose wireless networks where medium access is based on random back-off times and multiple retransmissions may be required depending on the state of the channel as we discuss next.

10.3   Wireless Network Alternatives

In the following subsections we provide some background on wireless network alternatives that we consider in this chapter. IEEE 802.11 and IEEE 802.15.4 are selected as off-the-shelf radio technologies and IEEE 802.11 with COMAC is considered as an enhancement via cooperation. For more details, we refer the interested reader to the references cited in this section.

10.3.1   IEEE 802.11

IEEE 802.11 medium access control (MAC) [17] is based on the *distributed coordination function* (DCF), which provides collision avoidance and resolution mechanisms to coordinate the transmissions of wireless nodes. The collision avoidance is provided via a back off period that takes place after the channel is checked to be free for a DCF inter frame space ($DIFS$). The back off period is a random outcome obtained from the contention window (CW), and collisions are avoided since different stations are likely to choose different outcomes. CW is started from CW_{min} and in case of collisions, it is doubled after each collision up to CW_{max}, so that a larger contention window is employed before each retry for resolving collisions. Another feature of DCF is virtual carrier sensing, which is provided by the exchange of Request to Send (RTS) and Clear to Send (CTS) frames. RTS/CTS frames as well as data frames involve a duration field that notifies the remaining time until the end of the current packet exchange. All the nodes that hear the RTS/CTS and/or data frames update their network allocation vector (NAV) with the duration information from those packets as they infer that the medium is busy for that period of time.

For a wireless node, DCF operation is as follows: Before each transmission, the node checks its NAV and if it is zero, the node senses the channel and makes sure that it remains idle for DIFS duration. If the node's NAV is non-zero or the medium becomes busy within DIFS, the node defers transmission until the channel becomes free. After the deferral period, the node waits for an additional amount of time determined by its back off timer for collision avoidance. The node begins transmission after its back off timer expires, if the channel remains free for the back off period (if not, the transmission is postponed until the next time the channel is free). If the receiving the node can decode the DATA frame successfully, it sends an ACK frame after one short inter frame space (SIFS). When an ACK is not received, the transmitted packet has failed either due to channel errors or a collision. A node whose packet is lost doubles its CW and selects its back off timer from the new CW before each retransmission. This scheme is repeated until all contentions are resolved and all nodes receive their ACKs. A node wishing to protect its transmission with virtual carrier sensing mechanism sends an RTS frame to the destination node before transmitting the DATA frame. The destination node replies with a CTS frame. When the source node receives the CTS reply, it disseminates the DATA frame which contains the actual payload and waits for the ACK frame. An ACK frame from the destination node concludes the frame exchange.

10.3.2 IEEE 802.11 with Cooperative Medium Access Control Protocol (COMAC)

Multi-path fading can cause significant degradation in the received signal strength over wireless channels, hence the capacity and quality of wireless communication networks are seriously degraded. Diversity techniques try to alleviate the effects of fading by generating and/or combining independently fading copies of the transmitted signal at the receiver [11], and cooperative diversity is one of the recent techniques to provide antenna diversity across cooperating nodes of a wireless network. In cooperative communications, the wireless broadcast advantage is exploited to disseminate the data to the cooperating nodes and via the cooperative transmissions, the receiver is provided with multiple copies of the original signal emanating from geographically separated transmitters (cooperating nodes); thus creating diversity at the receiver.

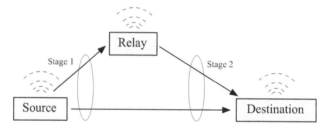

FIGURE 10.3
Two stages of cooperative communication.

At the receivers of cooperating nodes, Maximal Ratio Combining (MRC) is implemented, so that signals received from different nodes (branches) are combined in such a way that output Signal to Noise Ratio (SNR) is maximum, equal to the sum of SNRs of individual branches. This results in improved SNR at the receiver, higher link reliability, hence reduced number of retransmissions.

In order to facilitate cooperative communication over an IEEE 802.11 based air interface, the COMAC protocol has been proposed in [9], where all relevant frame formats and frame exchange procedures are defined and the improvement of cooperative communications is shown with higher packet success rates, hence higher throughput and lower latency values. In COMAC, cooperative communication is initiated by a C-RTS (Cooperative RTS) frame sent from the source node to the destination node for reserving the medium for one COMAC exchange. The destination node replies with a C-CTS (Cooperative CTS) frame. Overhearing the C-RTS and C-CTS frames, a relay node sends an ACO (Available to Cooperate) frame to indicate its intention for cooperation. COMAC involves a two stage frame exchange as illustrated in Figure 10.3, where the source disseminates the data frame in stage-1 to the destination. The data is overheard and decoded also by the relay, and in stage-2 the source and the relay simultaneously send a copy of the data frame to the destination. After combining and successfully decoding the received data packets, the destination ends the cooperative transaction with a *C-ACK* (Cooperative ACK) frame. Use of C-RTS and C-CTS frames along with two-stage cooperative exchange incurs an overhead on the order of several microseconds, however the amount of improvement in the packet success rate overcomes this overhead as studied in detail in [9, 10].

10.3.3   IEEE 802.15.4

IEEE 802.15.4 (ZigBee) protocol is proposed for low rate wireless networks with nodes with low computational resources, which is particularly suitable for low-cost industrial applications. IEEE 802.15.4 standard [16] defines the physical and medium access control layers for those low cost, low rate personal area networks and ZigBee alliance [33] defines the network layer specifications for different network topologies while also providing a framework for programming applications to run over IEEE 802.15.4. We refer to this protocol stack as IEEE 802.15.4 (ZigBee) protocol.

At the physical layer, the IEEE 802.15.4 standard [16] defines three frequency bands: 2450MHz (with 16 channels), 915MHz (with 10 channels), and 868MHz (1 channel), all using the Direct Sequence Spread Spectrum (DSSS) access mode. The 2450MHz band employs Offset Quadrature Phase Shift Keying (O-QPSK) for modulation with a data rate of 250 kbps, while the 868 and the 915MHz bands make use of Binary Phase Shift Keying (BPSK), supporting 20 and 40 kbps, respectively. At the MAC layer the standard defines two types of nodes, namely, Reduced Function Devices (RFDs) and Full Function Devices (FFDs). FFDs are implemented with a full set of MAC layer functions, which enables them to act as a network coordinator or a network end-device. When acting as a network coordinator, FFDs send beacons that provide synchronization, communication, and coordination, while RFDs can only act as end-devices that are equipped with sensors/actuators like transducers, light switches, lamps, etc.

In a ZigBee network, two types of topologies, star topology and the peer-to-peer topology can be implemented. In the star topology, a master-slave model is adopted, where a FFD acts as the network coordinator, and the other nodes can serve as RFDs or FFDs that communicate with the coordinator. In the peer-to-peer topology, a FFD can talk to other FFDs within its radio range and it can relay messages to other FFDs outside of its radio coverage through an intermediate FFD, forming a multi-hop network. In this work, we consider the peer-to-peer type of topology over a single hop.

In terms of medium access, IEEE 802.15.4 [16] is based on CSMA/CA. The network coordinator starts a super frame with a beacon, which provides synchronization and describes the super frame structure with active and inactive portions. The active portion is divided into Contention Access Period (CAP) where nodes compete for channel access using a slotted CSMA-CA protocol, and a Contention Free Period (CFP), where nodes transmit without contending for the channel in Guaranteed Time Slots (GTS) assigned and administered by the network coordinator. When an end-device needs to send data to a coordinator, it must wait for the beacon and later contend for channel access. On the other hand, the coordinator stores its messages to end-devices and announces pending delivery in the beacon. When an end device notices that a message is available, they request it explicitly during the CAP. In the second option, communication is based on unslotted CSMA/CA, operating similar to IEEE 802.11's access with no deferral period. This mode employs a back off interval selected randomly from a contention window, prior to sensing the medium. A network device can put its radio to sleep to conserve energy immediately after the reception of acknowledgement packet if there are no more data to be sent or received.

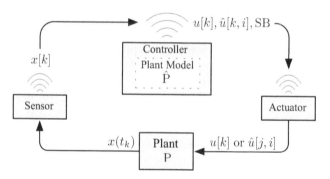

FIGURE 10.4
Operation of the Wireless Model Based Predictive Networked Control System.

10.4 Wireless Model Based Predictive Networked Control System (WMBPNCS)

In this section we introduce the Model Based Predictive Networked Control System (MBPNCS) and discuss its operation in detail. Wireless MBPNCS (WMBPNCS) is a special case of MBPNCS incorporating a wireless communication network. MBPNCS is a time-triggered discrete-time control system specifically designed to provide resilience to indeterministic bursty packet losses observed in the communication networks [31, 25]. A typical MBPNCS is made up of five components; the sensor, controller and actuator nodes, a communication network, and the actual plant P as shown in Figure 10.4. The network is generic, and does not necessarily provide real-time guarantees. We assume a broadcast network protocol in general, but other protocols with associated topologies may also be used. The system is time based where the nodes are time synchronized. They determine a deadline for arrival of expected data, after which a data packet is deemed lost or late. The sensor node periodically transmits plant readings (state or outputs) to the controller node over the network. The controller node contains a model of the plant, $\hat{P}$, which is used to predict future control signals as a means of tolerating intermittent packet losses. These control signals are then transmitted to and appropriately applied to the plant by a state machine in the actuator node.

In MBPNCS, the sensor node periodically transmits the measured instantaneous plant state $x[k]$. The controller node contains the control method and the plant model $\hat{P}$, and calculates the control signal $u[k]$ for the current sample and a predefined number of predicted future control signals $\hat{u}[k,i]$, $i = 1, 2 \ldots, n$. A control signal packet is formed using $u[k]$ and the predictions. Since an acknowledgement packet can be lost like any other packet, MBPNCS is designed not to require any explicit acknowledgement signals in contrast with various previously proposed systems, e.g., [14]. This is one of its

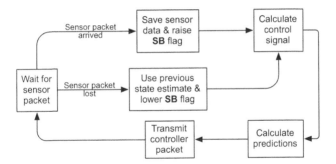

FIGURE 10.5
Operation of the controller node.

main novelties. In case of a communication failure between the sensor and the controller, the controller always assumes that the control signal of the previous time-step has been successfully received by the actuator and has been applied to the plant. (Although this assumption is not correct in the general case, a mechanism inside the actuator which will shortly be described, ensures this.) Based on this assumption, the controller produces either of two control signals at a given time-step before continuing with the calculation of future control signal predictions: If sensor data $x[k]$ is available, then the controller calculates the closed-loop control signal $u[k]$. However, if the sensor packet is lost on its way to the controller, then the controller uses the first state prediction $\hat{x}[k-1,1]$ of $\hat{P}$ from the previous time step for the predicted control signal $\hat{u}[k]$. In the latter case, the predicted control signal for the current time step $\hat{u}[k]$ and subsequent control signal predictions $\hat{u}[k,i]$, $i = 1 \ldots n$ are valid only if the previous control signal has been applied to P as expected by the controller. Consequently, whenever a controller packet is lost, further controller packets become obsolete until the next time the controller is *synchronized* with the plant (slightly abusing the term) by receiving a sensor packet. In order to differentiate such cases, a *sensor based (SB)* flag is also stored in control signal packets indicating whether the control signals in a given controller packet are unconditionally valid, i.e., they are based on sensor output $x[k]$, or the control signals in a given controller packet are valid only if the previous controller packet was successfully received by the actuator, i.e., they are based on the state predictions of $\hat{P}$. Figure 10.5 illustrates the operation of the controller node.

The mechanism as stated above, used in the actuator node to cope with the synchronization of the states of the actual plant and its model, is realized as the state machine shown in Figure 10.6. It has two states, one corresponding to intervals when the control signals enqueued in the actuator are coherent with the state of the plant P (*synchronized state*) and when the plant state assumption during their calculation was wrong (*interrupted state*) as discussed in [25, 31]. When the actuator is in the *synchronized state*, i.e., no controller

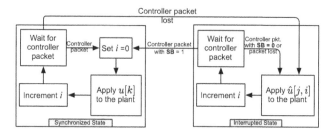

FIGURE 10.6
Operation of the actuator node.

packets have been lost since the last transition to this state was made, $u[k]$ of each packet is applied to P regardless of the condition of the SB flag until a controller packet is lost at time-step $j + 1$ when the actuator state machine makes a transition to the *interrupted state*. In the *interrupted state*, the actuator ignores the incoming controller packets and applies the predictions of the last controller packet received in the *synchronized state* consecutively $(\hat{u}[j, i], i = 1 \ldots n)$. The actuator state machine returns to the *synchronized state* when a *sensor based* controller packet is received with a set SB flag. After $\hat{u}[j, n]$ is applied, the actuator output is held constant until a controller packet with a set SB flag is received.

By using MBPNCS, stability of the networked control system is guaranteed while avoiding acknowledgement signals on the network. The stability of the system depends on the length of the prediction horizon n and the modeling errors in $\hat{P}$. The stability conditions of MBPNCS, directly applicable to WMBPNCS, are treated in [25] where the stability problem of MBPNCS is treated as a special case of Theorem 2 in [24] assuming full state measurement. Since in the following results we use output feedback, stability is not guaranteed. However, simulations and experiments show that the proposed method is stable under a wide range of operating conditions. The interested reader is referred to [24, 29, 15] for a broader theoretical discussion on the stability of model-based networked control systems in case of communication failures and to [25] for the specific case of MBPNCS.

10.4.1   The Plant and the Control Algorithm

In order to evaluate the performance of WMBPNCS, we consider the position control of a DC motor, as in our previous work [31]. The continuous-time dynamics of the plant are given by

$$\dot{x}(t) = Ax(t) + Bu(t)$$
$$y(t) = Cx(t) \tag{10.1}$$

with the following state, input, output matrices, and state vector

$$A = \begin{bmatrix} 0 & 1 & 0 \\ 0 & -b/J & K_t/J \\ 0 & -K_v/L & -R/L \end{bmatrix}, B = \begin{bmatrix} 0 \\ 0 \\ 1/L \end{bmatrix}$$

$$C = \begin{bmatrix} 1 & 0 & 0 \end{bmatrix}, x = \begin{bmatrix} \theta \\ \dot{\theta} \\ i \end{bmatrix}$$

(10.2)

where $b = 2.1 * 10^{-6}$ is the damping coefficient, $J = 6.28 * 10^{-6}$ is the rotor moment of inertia, $K_t = 0.11854$ is the torque constant, $K_v = 0.11789$ is the speed constant, $L = 3.1613 * 10^{-3}$ is the terminal inductance, $R = 11.8$ is the terminal resistance, θ is the position, $\dot{\theta}$ is the angular speed and i is the current of the motor. After obtaining the relevant parameters of the plant, we discretize the continuous-time model of the plant using zero-order hold for a sampling rate of 100 Hz and obtain the following discrete-time dynamics

$$x[k+1] = \bar{A}x[k] + \bar{B}u[k]$$
$$y[k] = Cx[k]$$

(10.3)

where $\bar{A}$ and $\bar{B}$ are discretized state and input matrices. Next, we design a full state feedback controller of the form $u[k] = G_r r[k] - Kx[k]$ with a bandwidth of 20.750 Hz resulting in the following closed-loop dynamics.

$$x[k+1] = (\bar{A} - \bar{B}K)x[k] + \bar{B}G_r r[k]$$
$$y[k] = Cx[k]$$

(10.4)

Finally, an observer may also be used in order to estimate the motor speed and current from measured motor position. We consider a no-delay full state Luenberger observer [1] with the following state and output reconstruction error dynamics.

$$\hat{x}[k] = (I - LC)\bar{A}\hat{x}[k-1] + (I - LC)\bar{B}u[k-1] + Ly[k]$$
$$\tilde{y}[k] = (I - CL)C\bar{A}\tilde{x}[k-1]$$

(10.5)

Since $C = [1\ 0\ 0]$, choosing $L = [1\ L_2\ L_3]'$ yields $\tilde{y}[k] = 0$ meaning that output can be estimated without error. After eliminating one equation this way, we implement a reduced order observer with a bandwidth of 80 Hz.

10.5 Case Studies

In this section we compare the performance of the WMBPNCS with that of a basic WNCS (b-WNCS), considering IEEE 802.11 based wireless networks, using our experimental results from [31], and also considering an alternative IWSN technology, namely IEEE 802.15.4. As discussed previously, WMBP-NCS is a real-time control system specifically designed to be robust to random delays and losses that are typical of wireless communication, whereas the b-WNCS that we consider consists of a time-triggered sensor node and event-triggered controller and actuator nodes that operate as follows: Upon the arrival of a sensor packet, the controller computes the control signal and sends it to the actuator node. The actuator updates the control signal applied to the plant after each new controller packet. If a sensor packet is lost or delayed, the controller does not generate any controller packets and the actuator node keeps applying the same control signal. In the experiments, controller performance of an NCS is determined by its percentage root mean square of error ($eRMS$) given by

$$Percentage\ eRMS = \sqrt{\frac{\sum_{k=1}^{n}(\theta[k] - r[k])^2}{\sum_{k=1}^{n} r[k]^2}} \tag{10.6}$$

where $r[k]$ and $\theta[k]$ are reference and plant positions at time step k. In the $eRMS$ experiments using IEEE 802.11 and COMAC, we use the same reference signal, a 0.5 Hz step input with an amplitude of *2 radians*. In the $eRMS$ experiments using IEEE 802.15.4, we use a 1 Hz step input with an amplitude of *1 radians*. For evaluating the time responses of the systems, we use a sawtooth reference with a slope of *4 radians/s*. We have performed each experiment 10 times and taken the average of the results to eliminate extremes. In the following, we present the results of our experiments and simulations obtained by operating both of the systems over three different types of wireless access: IEEE 802.11, IEEE 802.11 based COMAC and IEEE 802.15.4.

10.5.1   Performance Using IEEE 802.11

In this section, we compare the performance of WMBPNCS with that of b-WNCS under block fading, i.e., bursts of packet loss, and ambient wireless traffic, as initially presented in [31]. b-WNCS implements the same state-feedback control algorithm as WMBPNCS but lacks the model based predictive functionality. Thus, b-WNCS keeps the plant input unchanged in case of packet loss.

As given in [31], WMBPNCS and b-WNCS evaluated in our experiments are realized using Advantech PCM-9584 industrial computer boards, CNET CWP-854 wireless NICs, a Mesa 4i30 quadrature counter daughter board, and a Kontron 104-ADIO12-8 ADC/DAC daughter board. The plant that we consider is a Maxon RE-35 DC motor. We use the Debian GNU/Linux distribution as the operating system patched with the Xenomai real-time development framework for real-time support. The sensor and actuator nodes of the system are implemented in the same computer to reduce the number of components required. During the experiments, both systems use the IEEE

802.11 standard as the wireless access scheme at a transmission rate of 2 Mb/s with quadrature phase-shift keying (QPSK). Number of predictions calculated by the controller node of WMBPNCS is set to 50. We implement the bursty channel model, i.e., Gilbert/Elliot Model, in the nodes of the system to evaluate the effects of packet loss in a controlled way. The parameters of this model are derived from the results presented in [32] and are as follows: $P_{gb} = 0.0196$, $P_{bg} = 0.282$, $P_{loss}^g = 0$, $P_{loss}^b = 1$. In order to observe the effects of ambient wireless traffic on controller performance, we consider a constant bit rate traffic generated at a rate of 750 UDP packets/s with a payload of 50 bytes and a duration of $648\mu s$ per packet. The size of the traffic packets is chosen such that the majority of packet loss is caused by the randomness of the IEEE 802.11 backoff mechanism as opposed to their durations. Then, to evaluate the improvement provided by modified MAC parameters, we perform experiments using both stock and modified MAC parameters. Standard values for MAC parameters are $DIFS = 50, CW_{min} = 31, CW_{max} = 1023$, whereas the modified values that we consider are $DIFS = 30, CW_{min} = 0, CW_{max} = 3$. In the experiments, we keep the MAC parameters of the traffic generator at their stock values and no packet loss model is employed in the traffic generator for maximum interference.

First, we provide the results from [31] for the controller performance of b-WNCS under bursts of packet loss and ambient wireless traffic using both standard and modified MAC parameters. To imitate more pessimistic scenarios, P_{loss}^g of the bursty channel model is swept from 0% to 45% at 5% increments. b-WNCS with standard MAC parameters fails to operate under ambient wireless traffic with a percentage $eRMS$ exceeding 160% even at 0% $\overline{PLR_m}$, where $\overline{PLR_m}$ is the mean percentage loss rate of the channel, i.e., the weighted average of P_{loss}^g and P_{loss}^b with respect to steady state probabilities of the model being in a given state. When we repeat the experiments using modified MAC parameters, b-WNCS is again inoperative under bursty packet losses even though it is unaffected by the ambient wireless traffic.

Then, we consider the above set of experiments using WMBPNCS [31]. As given in Figure 10.7, percentage $eRMS$ of WMBPNCS is 53% at 7% $\overline{PLR_m}$ and never exceeds 85% when modified MAC parameters are used. Performance of WMBPNCS with standard MAC parameters degrades by at least 15% under ambient wireless traffic, nevertheless WMBPNCS remains stable under these conditions and clearly outperforms b-WNCS.

Finally, we present a time plot of motor position in response to a sawtooth reference signal under bursts of packet loss with $\overline{PLR_m} = 7\%$ without any ambient wireless traffic [31]. As shown in Figure 10.8, b-WNCS is completely unstable under these conditions with a percentage $eRMS$ of 726%, whereas WMBPNCS remains stable with a percentage $eRMS$ of 77% even though there are intervals during which the system is insensitive to changing reference signal due to bursty packet losses.

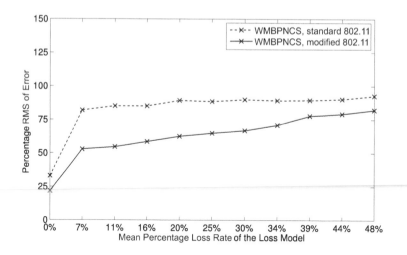

FIGURE 10.7
Controller performance of WMBPNCS over IEEE 802.11 under bursts of packet loss and ambient wireless traffic, using standard and modified MAC.

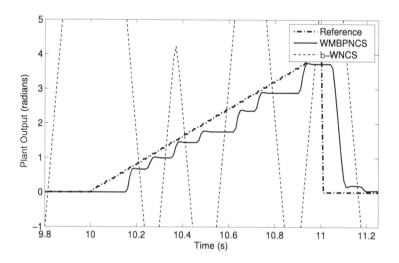

FIGURE 10.8
Sawtooth reference vs. plant output using IEEE 802.11 MAC under bursts of packet loss [31].

10.5.2   Performance Using COMAC

In this section, we evaluate the performance of the WMBPNCS over COMAC protocol under various levels of Rayleigh fading for various node distributions along a straight line and compare it to the performance of WMBPNCS over IEEE 802.11, as initially presented in [31]. In the experiments discussed next, we use the Rayleigh fading model to emulate severe wireless channel conditions and consider different node placements to observe the effects of path loss. The transmission power P_t and path loss exponent α parameters of the channel model are set to 1 mW and 4, respectively.

As illustrated in Figure 10.3, the nodes of WMBPNCS communicate with each other cooperatively in two stages when COMAC is utilized: In *stage 1*, the source node sends a packet to the destination node which is also overheard by the relay. In *stage 2*, the relay cooperates with the source for transmission of the packet to its destination. Diversity receiver at the destination node combines these two copies of the packet, significantly increasing chances of successful reception due to improved signal to noise ratio (SNR). In WMBP-NCS, the source can be either the controller or the sensor and the destination can be either the controller or the actuator. The relay node can be any neighboring node that can overhear and be heard by WMBPNCS nodes. In our experiments, we use a faithful implementation of the COMAC protocol whose details can be found in [31]. We use the same test-bed described in the previous subsection and utilize the third test-bed node, which is used to generate wireless traffic in the previous set of experiments, as the relay node. In the following, we use d to denote the distance between the controller and the sensor/actuator, whereas d_R stands for the ratio of the distance between the relay and the controller with respect to the distance between the controller and the sensor/actuator.

Mean and maximum packet loss burst lengths are of critical importance for the performance of an NCS. As the length of the packet loss burst, i.e., the number of consecutive packet losses, increases, controller performance degrades, risking the stability of the overall system. When the nodes use IEEE 802.11, mean packet loss burst length at the controller is 5 when $d = 70$ m and exceeds 20 when d reaches 85 m [31]. COMAC significantly reduces both mean and maximum packet loss burst lengths. When the relay is in the middle, COMAC reduces the mean packet loss burst length to 2 and keeps it below 6 for all $d \leq 85$ [31]. Even more disturbingly, when the nodes use IEEE 802.11, maximum packet loss burst length at the controller increases exponentially with d and exceeds 40 when d reaches 60 m [31]. For a 100 Hz control system such as the one we consider in this chapter, this corresponds to 0.4 seconds of reference insensitivity which renders the system unusable for most cases. COMAC, on the other hand, significantly reduces the variance in packet loss burst length and keeps the maximum packet loss burst length at the controller below 8 for all $d \leq 85$ [31].

Figure 10.9 shows the effects of both MAC protocols on the performance

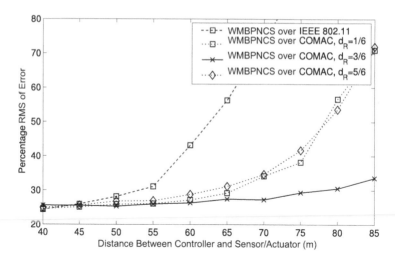

FIGURE 10.9
Controller performance of WMBPNCS over COMAC under Rayleigh fading vs. d [31].

of WMBPNCS under various scenarios [31]. When the nodes use IEEE 802.11, controller performance degrades with increasing d and percentage $eRMS$ exceeds 70% for d greater than 70 m. When COMAC is utilized and the relay is in the middle, percentage $eRMS$ always remains below 35% for all d values that we consider. However, WMBPNCS's performance depends heavily on the position of the relay and degrades when the relay is not in the middle. The intuition behind this is that, when the relay is closer to the source, it is more probable that the source and the relay will initiate a cooperative exchange; but there is a lower chance of successful cooperation as the destination node is away from both the source and the relay and SNRs of the C-DATA-I and C-DATA-II packets received at the destination will be lower. When the relay is closer to the destination, on the other hand, there is a lower chance of initiating a cooperative exchange since SNRs of C-RTS and ACO frames exchanged between the relay and the source will be lower. Since WMBPNCS is a closed loop system where nodes switch the roles of source and destination, both cases cause a degradation in controller performance due to increased packet loss and the best controller performance is achieved when the relay is in the middle.

Finally, Figure 10.10 illustrates the time plot of motor position obtained when WMBPNCS operates over both IEEE 802.11 and COMAC with the relay in the middle, $d = 70$ m and a sawtooth reference signal is applied to the controller [31]. When IEEE 802.11 is used, the system is insensitive to the changes in the reference signal during bursts of packet losses. On the other hand, the plant output follows the reference closely when COMAC is utilized.

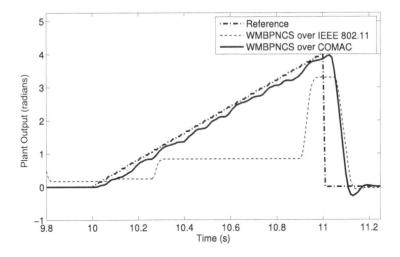

FIGURE 10.10
Sawtooth reference vs. plant output using COMAC under Rayleigh Fading
[31].

10.5.3 Performance Using IEEE 802.15.4

In this section, we compare the performance of WMBPNCS with that of
b-WNCS when both systems operate over IEEE 802.15.4. The results that
we present next are obtained through simulations using the TrueTime [4]
Matlab toolbox, which is designed to simulate real-time embedded computer
systems at the instruction execution level and communication network at the
data transport level. The IEEE 802.15.4 network model in TrueTime follows
the official standard operating in CSMA mode with the following parameter
values: $CW_{min} = 3$, $CW_{max} = 5$, data transmission rate = 250kbps and
minimum frame size = 248 bits. Number of predictions calculated by the
controller node of WMBPNCS is set to 20.

First, we compare the performance of WMBPNCS with that of b-WNCS
under bursts of packet loss generated by the Gilbert/Elliot model previously
discussed. P_{loss}^g of the bursty channel model is swept from 0% to 45% at
5% increments to imitate non-ideal channel characteristics in the good state.
As shown in Figure 10.11, b-WNCS is severely affected by bursty losses and
its percentage $eRMS$ exceeds 70% under bursty packet loss at 25% $\overline{PLR_m}$,
where $\overline{PLR_m}$ is the mean percentage loss rate of the channel model. On the
other hand, percentage $eRMS$ of WMBPNCS is 30% at 7% $\overline{PLR_m}$ and never
exceeds 45%.

Next, we evaluate the performance of two WMBPNCSs operating in close
proximity under bursty packet losses: the nodes of the first WMBPNCS are
located on the corners of a $1m \times 1m$ square and the nodes of the second
WMBPNCS are located on the corners of an $0.8m \times 0.8m$ square inset. Both of

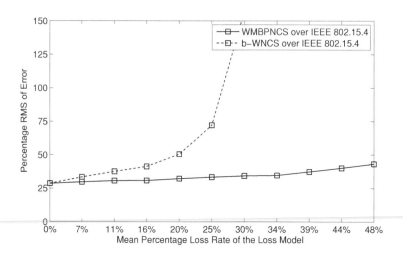

FIGURE 10.11
Comparison of the performances of WMBPNCS and b-WNCS over IEEE
802.15.4 under bursts of packet loss.

the WMBPNCSs perform satisfactorily under these conditions and percentage
$eRMS$ values do not exceed 60% even under bursts of packet loss with 48%
$\overline{PLR}_m$. We repeat the same experiment using two b-WNCSs, however they are
both unstable even under ideal channel conditions, with a percentage $eRMS$
value exceeding 150%.

Finally Figure 10.12 shows a time plot of the responses of both WMBP-
NCS and b-WNCS under bursts of packet loss with 30%$\overline{PLR}_m$. b-WNCS is
completely unstable under these conditions with a percentage $eRMS$ of 127%
whereas WMBPNCS follows the reference closely with a percentage $eRMS$ of
30%.

10.6   Conclusions

In this chapter we focus on the problem of designing and operating a system
with real-time constraints over a wireless network. We investigate the effects
of the wireless network on the real-time system and look at ways to improve
the overall performance of the system.

The results of our experiments strongly suggest that one can obtain very
satisfactory performance by considering both aspects of the wireless control
problem, i.e., the feedback control system and the underlying network, instead
of focusing on only either one of them. The Wireless Model Based Predictive
Networked Control System (WMBPNCS) that we consider in our experiments

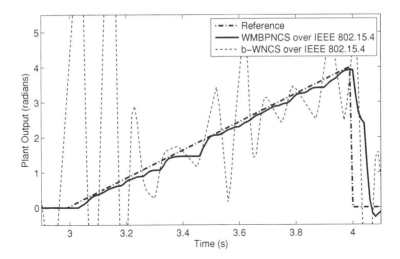

FIGURE 10.12

Sawtooth reference vs. plant output using IEEE 802.15.4 under $30\%\overline{PLR}_m$.

is a time-triggered wireless networked control system, which acknowledges the fact that the underlying network is imperfect and employs a model based predictive controller and an actuator state machine to tolerate random bursts of packet loss. Thanks to these mechanisms, WMBPNCS can operate satisfactorily in conditions where basic wireless networked control systems fail. Nevertheless, the performance of WMBPNCS is still limited by the quality of the wireless channel. In the presence of ambient wireless traffic, we show that the performance of WMBPNCS can be significantly improved by just a straightforward modification of MAC parameters of WMBPNCS nodes. This modification increases medium access priority of WMBPNCS nodes, and minimizes the effect of the ambient wireless traffic generated by neighboring nodes. We also observe that WMBPNCS suffers from severe multi-path (Rayleigh) fading which typically occurs in an industrial setting with numerous obstacles and no direct line of sight between the transmitting and receiving nodes. In this case, we show that we can mitigate the effects of multi-path fading by using a cooperative medium access control protocol such as COMAC which leverages neighboring nodes to improve the quality of the wireless link. We also evaluate the performance of WMBPNCS over IEEE 802.15.4 through simulations and show that reliable wireless networked control can be achieved using low-cost components with modest capabilities. Our experiments show that such systems can even share the same network with little or no performance penalty. Significant performance gains achieved by the integration of WMBPNCS with various considerations in the network level point out that challenges of the wireless networked control problem can be well addressed with such multi-disciplinary approaches.

References

[1] K. J. Astrom and B. Wittenmark, *Computer controlled systems*, 3rd ed., Prentice Hall, 1997.

[2] J. Baillieul and P. J. Antsaklis, *Control and communication challenges in networked real-time systems*, Proceedings of the IEEE **95** (2007), no. 1, 9–28.

[3] A. Bemporad, M. Heemels, and M. Johansson, *Networked control systems*, Lecture Notes in Control and Information Sciences, Springer, 2010.

[4] A. Cervin, D. Henriksson, B. Lincoln, J. Eker, and K. E. Arzen, *How does control timing affect performance? Analysis and simulation of timing using jitterbug and truetime*, IEEE Control Systems Magazine **23** (2003), no. 3, 16–30.

[5] A. Chamaken and L. Litz, *Joint design of control and communication in wireless networked control systems: A case study*, American Control Conference (ACC), 2010, July 2010, pp. 1835 –1840.

[6] E. O. Elliot, *Estimates of error rates for codes on burst-noise channels*, Bell Systems Technical Journal **42** (1963), 1977–1997.

[7] M. Garcia-Rivera and A. Barreiro, *Analysis of networked control systems with drops and variable delays*, Automatica **43** (2007), no. 12, 2054–2059.

[8] P. Gil, A. Paulo, L. Palma, A. Amancio, and A. Cardoso, *Model based predictive control over wireless sensor and actuator networks*, IECON 2011 - 37th Annual Conference on IEEE Industrial Electronics Society, Nov. 2011, pp. 2600–2605.

[9] M. S. Gokturk and O. Gurbuz, *Cooperation in wireless sensor networks: Design and performance analysis of a mac protocol*, IEEE International Conference on Communications, 2008.

[10] ———, *Cooperative mac protocol with distributed relay actuation*, IEEE Wireless Communications and Networking Conference, 2009.

[11] A. Goldsmith, *Wireless communications*, Cambridge University Press, 2005.

[12] V. C. Gungor and G. P. Hancke, *Industrial wireless sensor networks: Challenges, design principles and technical approaches*, IEEE Transactions on Industrial Electronics **56** (2009), no. 10, 4258–4265.

[13] R. A. Gupta and M. Y. Chow, *Networked control system: Overview and research trends*, IEEE Transactions on Industrial Electronics **57** (2010), no. 7, 2527–2535.

[14] V. Gupta, B. Sinopoli, S. Adlakha, A. Goldsmith, and R. Murray, *Receding horizon networked control*, September 2006.

[15] J. P. Hespanha, P. Naghshtabrizi, and Y. Xu, *A survey of recent results in networked control systems*, Proceedings of the IEEE **95** (2007), no. 1, 138–162.

[16] IEEE, Inc., IEEE Press, NY., *IEEE std. 802.15.4-2003: Wireless medium access control (mac) and physical layer (phy) specifications for low rate wireless personal area networks (lr-wpans)*, 2003.

[17] IEEE, Inc., IEEE Press, NY., *IEEE std. 802.11-2007: Wireless lan medium access control (mac) and physical layer (phy) specifications*, 2007.

[18] International Society of Automation (ISA), *ISA-100.11a-2011 wireless systems for industrial automation: Process control and related applications*, 2011).

[19] H. Jianqiang, H. Guowen, and X. Shuenqing, *Study on message optimization scheduling scheme of wireless networked control systems*, Electric Information and Control Engineering (ICEICE), 2011 International Conference on, April 2011, pp. 2679–2682.

[20] Z. Ju and Z. Haihua, *Explicit model predictive control of networked control systems with time-delay*, Control and Decision Conference (CCDC), 2011 Chinese, May 2011, pp. 907–911.

[21] H. Li, M. Chow, and Z. Sun, *Eda-based speed control of a networked dc motor system with time delays and packet losses*, IEEE Trans. Ind. Electron. **56** (2009), no. 5, 1727–1735.

[22] X. Liu and A. Goldsmith, *Wireless network design for distributed control*, IEEE Conf. Decision Control, vol. 3, 2004, pp. 2823–2829.

[23] G. Montenegro, N. Kushalnagar, J. Hui, and D. Culler, *Transmission of ipv6 packets over ieee 802.15.4 networks*, Internet Engineering Task Force RFC-4944, 2007.

[24] L. A. Montestruque and P. Antsaklis, *Stability of model-based networked control systems with time-varying transmission times*, IEEE Transactions on Automatic Control **49** (2004), no. 9, 1562–1572.

[25] A. Onat, T. Naskali, E. Parlakay, and O. Mutluer, *Control over imperfect networks: Model based predictive networked control systems*, IEEE Transactions on Industrial Electronics **58** (2011), no. 3, 905–913.

[26] P. Otanez, J. Moyne, and D. Tilbury, *Using deadbands to reduce communication in networked control systems*, Proc. Amer. Control Conf., 2002, pp. 3015–3020.

[27] K. Pahlavan and P. Krishnamurthy, *Principles of wireless networks: A unified approach*, Prentice Hall, 2002.

[28] IML Workgroup Plc, *WirelessHART specification released*, Control Engineering Europe **8** (2007), no. 3, 8–9.

[29] L. Schenato, B. Sinopoli, M. Franceschetti, K. Poolla, and S. S. Sastry, *Foundations of control and estimation over lossy networks*, Proceedings of the IEEE **95** (2007), no. 1, 163 –187.

[30] J. Song, S. Han, A. K. Mok, D. Chen, M. Lucas, M. Nixon, and W. Pratt, *Wireless hart: Applying wireless technology in real-time industrial process control*, Real-Time and Embedded Technology and Applications Symposium, 2008, pp. 377–386.

[31] A. Ulusoy, O. Gurbuz, and A. Onat, *Wireless model-based predictive networked control system over cooperative wireless network*, IEEE Transactions on Industrial Informatics **7** (2010), no. 1, 41–51.

[32] A. Willig, M. Kubisch, C. Hoene, and A. Wolisz, *Measurements of a wireless link in an industrial environment using an IEEE 802.11-compliant physical layer*, IEEE Transactions on Industrial Electronics **49** (2002), no. 6, 1265–1282.

[33] ZigBee Alliance, *ZigBee specifications, version 1.0*, 2005.

11

Medium Access Control and Routing in Industrial Wireless Sensor Networks

Aysegul Tuysuz Erman

Department of Computer Engineering, Isik University, Istanbul, Turkey.

Ozlem Durmaz Incel

NETLAB, Department of Computer Engineering, Bogazici University, Istanbul, Turkey.

CONTENTS

11.1   Introduction

Wireless Sensor Networks (WSNs) appear as a promising solution for industrial applications, especially for monitoring purposes, due to the advantages that they provide [19]. First of all, they overcome the wiring constraints present in wired industrial monitoring and control systems. Other advantages can be listed as ease of installation and maintenance, reduced cost, and better performance [50]. On the other hand, lack of standardization, strict real-timeliness, and reliability requirements of some industrial applications have limited their use in industrial domains [6]. Today, however, initial examples of industrial wireless sensor networks (IWSNs) do appear especially on process monitoring and control [28, 19], supported with standardization efforts such as IEEE 802.15.4 [2], WirelessHART [40], and ISA100.11a [25].

As illustrated by the current industrial wireless sensor networking specification [25] of ISA (International Society of Automation), the deployment of wireless sensors significantly improves the productivity and safety of industrial plants while increasing the efficiency of the plant workers. Wireless low-power field sensors in industrial environments enable industrial users to significantly increase the amount of information collected and the number of control points that can be remotely managed.

Industrial automation is segmented into two distinct application spaces, known as "process control" and "factory automation." While there are some overlapping requirements of wireless sensor network deployments for these two segments, they may differ according to different networking needs. For instance, an IWSN designed for an industrial process control tends to be more tolerant to network delay than what is required for factory automation. However, both process control and factory automation require highly reliable communication. The list below shows the six classes of industrial applications defined by the SP100 working group of ISA, starting with the applications that require the highest priority, i.e., highest reliability and minimum latency.

- Class 0: Emergency action

- Class 1: Closed-loop, regulatory control

- Class 2: Closed-loop, supervisory control

- Class 3: Open-loop control

- Class 4: Monitoring with short-term operational consequences

- Class 5: Monitoring without immediate operational consequences

For instance, in safety and mission-critical applications where sensor nodes are deployed to detect events, such as oil/gas leak, the actuators or controllers need to receive data from "all" the sensors "within certain deadlines" [11]. If the reception of data even from a single sensor fails, this may lead to unpredictable and catastrophic failures.

In this chapter, we focus on the MAC and routing protocol solutions proposed for IWSNs. The reason why we focus on MAC and routing layers is that the operation of the upper layers are dependent on these layers of the networking stack and the protocols proposed for these layers impact the overall performance of the network. We investigate the requirements of the protocols for industrial applications and then provide a survey on the existing solutions either explicitly proposed for IWSNs or proposed for traditional WSNs but can meet the requirements of the application areas. We also explore cross-layer solutions that both focus on the MAC and routing layers.

As we mentioned, the applications from different classes vary according to their requirements. However, *reliability* and *latency* requirements are the two common concerns for all industrial applications. Hence, the major requirement is the quality of service, which requires the correct data at the right time [50] in IWSNs.

Energy efficiency may not be as severe as in traditional WSNs since wired energy source or energy harvesting might be possible in some applications, for instance energy might be harvested from vibrations in a factory environment. However, considering the challenge of providing wired energy source for all sensors and the lack of not always having a source for energy harvesting still make energy efficiency as one of the requirements in IWSNs in order to have an extended network lifetime.

The network is required to *adapt* to the changing application requisites, dynamic wireless links and topology changes. Especially considering the harsh, such as metal dominating, operating environments for IWSNs, wireless link quality may exhibit fluctuations both spatially and temporally due to fading. *Adaptivity* is also required for different types of traffic flowing in the network. For instance, network control packets may require best effort delivery whereas packets carrying a detected event information, such as equipment failure, should be delivered to the sink, or to the controllers in real-time. The network should be adapted according to the different requirements of different traffic types.

Scalability is required considering the large number of sensor nodes densely deployed in the environment. Depending on the application requirements, the number of sensors may be in the order of hundreds, even thousands. The network is required to operate with these number of nodes and should be able to adapt to the size changes due to the addition/deletion of nodes [18].

WSNs mostly operate on unlicensed bands like the $868 - 915MHz$ or $2.4GHz$ industrial, scientific, and medical (ISM) bands. Co-existence of different networks operating on the shared medium may cause performance degradations due to interference and contention on the shared medium. Therefore, co-existence with other networks and electronic devices, such as microwave ovens, that share the same parts of the spectrum is another requirement for IWSNs.

The requirements that we have discussed so far are the metrics that can be fulfilled at the MAC and routing layers. Other requirements such as maximizing throughput and goodput, minimizing end-to-end delay from sources to the sink should be explored with the whole protocol stack in mind.

The rest of the chapter is organized as follows: Section 11.2 includes the requirements of MAC protocols for IWSNs, standardization efforts at the MAC layer and a taxonomy of MAC protocols. In Section 11.3, we present a taxonomy of routing protocols, again together with the requirements of IWSNs at the routing layer and standardization activities. Section 11.4 includes the cross-layer protocol examples that involve both MAC and routing solutions. In Section 11.5, we provide a list of identified open problems to be investigated for future research and finally Section 11.6 concludes the chapter.

11.2   Taxonomy of MAC Protocols

In this section we explore the MAC design space for IWSNs. There are various WSN MAC protocols in the literature [21]. The main motivation for all the MAC protocols proposed for traditional WSNs is energy-awareness considering the battery limitation of the sensor devices. However, expected properties of networking protocols for WSNs heavily depend on the application specifications and only a few of the MAC protocols are designed to meet the requirements of the IWSNs.

In the following, we explain how the requirements of IWSNs can be met at the MAC layer. Then, we discuss the standardization efforts related to MAC layer design in IWSNs. Finally, we give a taxonomy of the MAC protocols that are either explicitly proposed for IWSNs or proposed for traditional WSNs but have the objectives that can meet the requirements of IWSNs, such as latency.

11.2.1   Requirements of IWSNs at the MAC Layer

In this section we focus on how MAC layer can help to achieve the requirements of IWSNs discussed in Section 15.1. The MAC layer is basically responsible for controlling access to the shared wireless medium and hence controls the avoidance of interference between transmissions and mitigates the effects of collision, such as by the retransmission of packets. Besides, it controls the duty cycling and sleep scheduling of the sensor radios in order to conserve energy. Moreover, the MAC layer helps for transmission power control to mitigate the effects of varying quality wireless links and unpredictable environmental conditions.

In order to meet the requirements of IWSNs, the MAC layer plays a key role. For instance to meet the high reliability requirements of the applications, the MAC layer can provide reliable communication by controlling the retransmission of lost packets, using acknowledgements with methods such as automatic repeat request (ARQ), and by supporting transmission power control, the MAC layer can mitigate the effects of bad-quality wireless links by turning them into good-quality, reliable links.

To minimize latency, for instance, contention based MAC protocols can support dynamic backoff schemes or adaptive contention windows, according to the traffic types. For real-time traffic, contention window size and backoff interval can be minimized to reduce the medium access latency. In case of MAC protocols supporting scheduled access, such as TDMA, transmission of schedules can be organized to minimize latency as well. Moreover, the MAC layer can maximize concurrency, by efficient controlling of the medium, while limiting the impact of interference on parallel transmissions and hence minimize latency.

Since wireless transmission is the most energy consuming operation for WSNs, the MAC layer can contribute to energy efficiency by duty cycling the sensor nodes' transceivers and by minimizing collisions and hence retransmissions. Additionally, the MAC layer can control the transmission power of nodes' radios to minimize energy consumption instead of transmitting at maximum power level.

The MAC layer can also adapt its operation according to changing application requirements, varying wireless link qualities, topologies and traffic requirements and can fine tune its operation parameters such as contention window size and duty cycle according to its observations like traffic pattern, collisions, channel conditions, and topology.

Contention-based random access MAC protocols are known to be more scalable than schedule-based MAC protocols since sensor nodes do not require synchronization or coordination among themselves and hence the addition and deletion of sensor nodes can easily be maintained. However, schedule-based MAC protocols can also be adapted to meet the scalability issues by providing less strict synchronization requirements, such as in LMAC [45].

Interference and performance degradation due to co-existing networks that share the same parts of the spectrum can be mitigated at the MAC layer by selecting different operating frequencies or using transmission power control. There exist different examples of MAC protocols that utilize multi-channel communication in WSNs [22].

11.2.2 Outline of Standardization Activities at the MAC Layer

Standardization efforts for WSNs is still in progress yet there is no widely accepted complete protocol stack for IWSNs. The IEEE 802.15.4 protocol that provides a framework for low data rate communications systems is used as a basis for the Zigbee, WirelessHART, and ISA100.11a specifications for industrial environments. The standard proposes both a physical layer and MAC layer solution. It has been originally designed for low-rate wireless personal area networks. The standard is then adopted by WSNs, interactive toys, smart badges, remote controls, and home automation, operating on license-free ISM bands.

ZigBee, WirelessHART, and ISA100.11a specifications for IWSNs use IEEE 802.15.4 at the physical layer. Zigbee uses the IEEE 802.15.4 protocol also at the MAC layer. IEEE 802.15.4 is the mostly common protocol at the MAC layer for IWSNs in the literature. Most of the studies, as we discuss in Section 11.2.3, either focus on the performance of IEEE 802.15.4 for industrial environments or propose modifications to the protocol. It uses a combination of both random access and scheduled access. In the beacon-enable mode, the IEEE 802.15.4 MAC layer uses a superframe structure, which is defined by a beacon message transmitted by the coordinator. The superframe consists of an active period, where data transmissions take place and inactive period for energy conservation. The active period is composed of contention-access period (CAP) and contention-free period (CFP) where guaranteed time slots (GTS) are allocated to the requesting transmitters by the coordinator. Communication during contention access period is based on slotted CSMA/CA, whereas in the contention-free period timeslots are reserved for communication without contention. In the beaconless mode, the MAC layer is based on unslotted CSMA/CA.

WirelessHART is an emerging solution for the replacement of the wired HART protocol in industrial environments. It uses a MAC protocol based on time division multiple access (TDMA) and requires time synchronization and pre-scheduling of time slots by a centralized manager. The manager needs to update the schedule in case of changes in the networks, such as dynamic topologies due to spatio-temporal variations in the wireless link quality. The protocol uses channel hopping and channel blacklisting to reduce the effects of interference due to static channel usage.

At the MAC layer, ISA100.11a only shares the basic MAC frame with IEEE 802.15.4. Similar to WirelessHART, ISA100.11a specification also uses TDMA for medium access and supports channel hopping either by per-timeslot hopping or slow frequency hopping.

11.2.3   MAC Protocols Proposed for IWSNs

In this section, we explore the MAC protocols that are designed explicitly for industrial applications. As we elaborate in the rest of the section, most of the protocols revolve around the IEEE 8021.5.4 protocol, either modifying the protocol to provide real-time support or to enable multi-channel communication. Moreover, most of the protocols are proposed quite recently.

One of the first studies that evaluates the performance of the IEEE 802.15.4 protocol for industrial applications is presented in [5]. Experiments were carried out in a testbed with ten sensor nodes (T-mote sky platform) organized in a single-hop star topology, each at $0.5m$ distance from the master node. Sensors were set to generate alarms according to a Poisson process time distribution. The performance of the IEEE 802.15.4 protocol was tested in non-beacon mode in the presence of interference. As a result of the experiments, requirements and efficient parameter selection for polling time and alarm latency were presented to help network designers.

Compliance of the IEEE 802.15.4 protocol with the requirements of industrial applications, such as real-time support and reliability, is evaluated analytically in [9]. In this regard, specific protocol limitations are highlighted and modifications to the protocol to enable real-time operation have been proposed. The highlighted limitations are that the maximum seven GTSs limits the number of devices, minimum CAP length of 440 symbols limits the available period for CFP for the allocation of GTSs and one allocated GTS can only consist of an integer number of superframe slots which causes the length of superframe to grow exponentially and hence result in inefficient bandwidth use. In order to overcome these limitations, the superframe structure is modified such that the CAP section is removed since nodes do not put request on GTSs but instead GTSs are preallocated to each device. Additionally, the packet format is modified to reduce the large overhead of physical and MAC layers. With the modification, the new packet format contains only a payload of one byte and a Frame Checksum field with 2 octets in length at the MAC layer. According to the theoretical analysis of the proposals, the required guaranteed latency bounds can be satisfied for an example case study with 20 nodes in a star topology.

In [30] Kunert *et al.* propose a simple master-slave protocol with polling to be used on top of IEEE 802.15.4 in order to provide predictable latency and improve reliability. In their framework, sensor nodes periodically generate packets and traffic is transmitted in both directions between a central master node and its surrounding slaves. Earliest deadline first (EDF) scheduling is assumed and the master node polls the slaves according to an EDF schedule. In order to provide reliability, a transport layer with realtime ARQ scheme is proposed on top of IEEE 802.15.4. The performance of the solution is evaluated both with theoretical analysis and simulations on Matlab and the results show that message error rate can be improved several orders of magnitude with the retransmission scheme and real-time guarantee. However, performance of the proposal was not analyzed in a real testbed of an industrial application.

In a similar study [49], Yoo *et al.* propose modifications to the IEEE 802.15.4 protocol in order to provide real-time guarantees. They introduce a distance-constrained offline real-time message-scheduling algorithm that allows the inclusion of GTS allocation information in every beacon frame. Besides providing a schedule, the proposal defines a beacon interval, superframe duration. First, the algorithm checks the pre-schedulability of the given message set with their specific deadlines and sets an upper bound for the beacon interval. In the next step, periods of messages are harmonized such that message set is transformed into superframe slots. Then the algorithm checks the maximum GTS constraint such that it tries to assign the messages into the minor superframe considering the maximum number of seven GTS in a superframe. The performance of the proposal is first evaluated with simulations and the results reveal that the algorithm is capable of finding a feasible schedule for a given periodic real-time message set. Effect of message length on schedulability is also analyzed. Additionally, the proposal is implemented on T-Sink/Sensor node platform with IEEE 802.15.4-compliant CC2420 transceiver. Using 1 sink and 5 sensor nodes, the performance was tested in a small testbed. The schedule was computed offline on a PC and downloaded to the coordinator. The schedule was broadcast by the coordinator in beacon messages and associated nodes register for message exchange if any GTS belongs to that. The proposal mainly focused on real-time aspect of IWSNs and reliability was not considered.

In [10], i-MAC protocol (a MAC that learns) is proposed for the use of WSNs in manufacturing systems. The motivation is that, due to the repetitive operation of manufacturing machines, such as an assembly line, traffic patterns show strong temporal correlations and a MAC protocol can learn these patterns and use them for efficient scheduling of transmissions. In i-MAC, time is organized into frames similar to TDMA and each frame consists of slots where sensors transmit messages to the controller. The protocol assumes a controller with multiple radios, each operating on different frequencies, in order to receive packets from different sensors simultaneously and supports channel hopping. It differs from TDMA such that multiple sensors may be assigned the same slot but nodes that have a high probability of transmitting together (called burst set) are assigned different slots. The base station assigns the time slots and periodically updates if burst sets change in time. The performance of i-MAC was evaluated in a simulator environment which was built based on inputs provided by several machine users like *"machines have product arrival rates in the range of 1-4sec at full load while larger machines with small robotic parts have arrival rates of 5-10 sec."* The performance of the protocol was compared to T-MALOHA, multi-channel exponential back-off and frequency-time division multiple access (FTDMA) algorithms [11] in terms of meeting deadlines, network lifetime.

Another recent MAC protocol proposed for an industrial process automation system is GinMAC [42]. The protocol aims for both timely and reliable data collection. It consists of three phases: off-line network dimensioning, TDMA schedule and delay conform reliability control. In the first phase, the protocol identifies the channel conditions and constructs a tree topology. According to the results of the first phase, in the second phase a TDMA schedule is created in which each node is assigned one timeslot. According to the channel characteristics, redundant slots are also added in the schedule. In the delay conform reliability control the aim is to support temporal transmission diversity.

Generic Multi-channel MAC protocol (G-McMAC) is proposed in [34] to mitigate the effects of interference and enable co-existence for IWSN applications. The protocol uses a dedicated control channel for channel negotiation and operates in two segments of beacon period and contention plus data period (CDP). Beacons contain information about preferred channel list, channel schedules, time stamp information, hierarchy level, and beacon interval length. The sink node starts synchronization by initiating beacons. The nodes receiving the beacon message synchronize to the network and transmit their beacon messages. When a node intends to transmit a packet, first it senses the preferred channel and if the channel is free it sends a request message to the intended receiver with the channel information on the control channel. The receiver also senses the preferred channel and responds with an acknowledgement if it is free on the control channel. Data transmission takes place on the agreed channel. The performance of the protocol is evaluated both analytically and via simulations and the results reveal that the dedicated control channel of G-McMAC outperforms split-phase and common hooping approaches in terms of average delay and throughput, and under W-LAN interference G-McMAC supports coexistence of multiple sensor applications by efficiently avoiding interference.

A multi-channel extension to the IEEE 802.15.4 protocol is proposed in [44] for IWSNs. The paper introduces a multi-channel superframe scheduling (MSS) algorithm which avoids collisions by scheduling the IEEE 802.15.4 superframes on different channels while maintaining network connectivity. The protocol is introduced for cluster tree networks with bounded delay capabilities. The most severe problem that can occur in cluster-tree networks is identified as the not properly synchronized beacon messages. If beacon messages collide with other beacon or data messages of different clusters, nodes not receiving the beacons might lose synchronization and get disconnected from the network. Although a TDMA based approach can solve the beacon collision problem, it may limit the scalability since parallel transmissions can take place. MSS algorithm partitions the network by scheduling beacon frames in two different timeslices. The coordinators which are two hops away are assigned the same timeslice. The first partition includes the PAN Coordinator and all the clusters that can reach it in an even number of hops and their superframe takes place on time slice 1, and the second partition includes all the other clusters, i.e., those featuring an odd tree depth whose transmissions take place on time slice 2. Then, the MSS algorithm assigns channels to the clusters that share the same time slice in order to prevent interference and collisions with spatial reuse of the channels. Channel assignment is performed in a centralized manner with full knowledge of the network topology and physical location of the nodes. The location information is used to calculate interference between the clusters. The performance of the algorithm is evaluated both analytically and via simulations and in a small testbed of 6 TelosB motes. The performance is also compared with time division superframe scheduling and results show that the number of schedulable clusters increases with the proposed technique and MSS outperforms standard time division approaches.

In [11], Chintalapudi *et al.* propose MAC protocols for low-latency, hard real-time control applications. The first proposed protocol is called MALOHA and it is an extension of the ALOHA protocol into the multi-channel domain. The sink, i.e., the controller, is assumed to have multiple radios and sensors can reach the sink in one hop. Similar to slotted ALOHA, at each slot a node decides whether to transmit or not with a probability and next task is to select a channel uniformly. Another protocol proposed in the paper is called T-MALOHA where time is divided into frames in which a set of consecutive transmission pipelined time slots are followed by an acknowledgement slot. The performance of the protocols are compared with a FTDMA-based protocol in simulations and it was shown that they perform better than FTDMA since it does not scale well for larger networks.

11.2.4   WSN MAC Protocols with Latency Bound and to Support Real-Time Operation

As we discussed in Section 11.2.3, although the protocols were proposed for industrial applications, most of them were not evaluated in a real application setting. Instead, the focus was on meeting the requirements of industrial applications, such as providing bounded delay, real-time guarantee, reliability, and co-existence with other networks. In this section, we give an outline of the MAC protocols with the objective of bounded latency (guaranteed delay) and with real-time support proposed for traditional WSNs which can also be used to meet the latency requirement of IWSNs. Reliability and other issues will be discussed in Section 11.2.5.

In [41] a detailed survey of MAC protocols for mission-critical applications for WSNs is presented. The authors define mission-critical applications as *the applications that demand data delivery bounds in terms of time and reliability.* Applications of IWSNs can also be included in the category of mission-critical applications since they also require the correct data at the right time. In the survey, authors classify the existing MAC protocols for WSNs into delay-aware protocols, reliability-aware protocols, and both delay and reliability-aware protocols. For a more comprehensive explanation, the reader can refer to this survey.

When bounds on latency are desired, TDMA is perhaps the obvious choice by preventing conflicts, eliminating collisions, overhearing, and idle listening, which are the main sources of energy consumption in wireless communications. More importantly, it can provide provable guarantee on the delay bound. As we elaborate in this section, all the protocols with the objective of guaranteeing end-to-end latency use a TDMA based approach. Despite the benefits of TDMA approach, the main critic is that it is difficult to compute a conflict-free schedule without centralized algorithms and the schedule needs to be changed with the addition of new nodes. However, TDMA can work very well with static topologies and there exist protocols that can perform the scheduling in a distributed manner, like LMAC [45]. For a detailed analysis of protocols that minimize latency instead of guaranteeing end-to-end delay, the reader can refer to [41, 23].

In [12], the PEDAMACS protocol was proposed as a delay-aware MAC protocol for WSNs. The protocol assumes a powerful sink node that can reach all sensor nodes in one hop. The sink is responsible for topology discovery and TDMA scheduling. The protocol gives end-to-end delay guarantees in the computed schedule. The performance of the protocol was evaluated only with simulations.

RT-link protocol [39] is another TDMA based MAC protocol that aims end-to-end delay guarantee. For synchronization, the protocol assumes special hardware for achieving out-of-band time synchronization. RT-Link proposes two types of slots, namely scheduled slots and contention slots. Scheduled slots provide contention-free transmission whereas slotted Aloha algorithm is used for contention slots and these slots are used by the new nodes joining the network. The performance of the protocol was evaluated with a real WSN deployment in a coal mine facility.

In [24], Incel *et al.* explore and evaluate a number of different techniques using realistic simulation models to study the schedule length for data collection, i.e., convergecast, in WSNs with tree topologies. Initially, a simple spatial-reuse TDMA scheme is used to minimize the schedule length. Then the method is extended with multiple frequency channels and transmission power control to mitigate the effects of interference and minimize the schedule length. A receiver-based channel assignment (RBCA) scheme is proposed where the receivers (i.e., parents) of the tree are statically assigned a channel, and the children of a common receiver transmit on that channel. They show that, once multiple frequencies are used to completely mitigate the effect of interference, then the lower bound on the schedule length, i.e., total delay to receive data from all sources, is $max(2n_k - 1, N)$, where n_k is the maximum number of nodes in any branch of the tree and N is the number of source nodes in the network. The authors proposed a slot assignment scheme to achieve this bound.

HyMAC [33] is another protocol that combines TDMA with frequency division multiple access (FDMA). Time slots and frequencies are assigned according to the Breadth First Search (BFS) algorithm on a tree topology with the objective of guaranteed delay.

In [31], Li *et al.* focus on scheduling messages with deadlines in multi-hop, real-time WSNs. Providing timeliness guarantees for messages with specific deadlines in real-time robotic sensor applications has been studied and the problem is shown to be NP-hard. They propose a heuristic where a central scheduler schedules messages based on their per-hop timeliness constraints and associated routes, transmission ranges, and the location of the nodes. Performance of the heuristic is compared with CSMA-CA where nodes make local scheduling decisions independent of others and it was shown that the proposal outperforms CSMA-CA based scheduling in terms of the deadline miss ratio. For more examples of real-time MAC protocols, a detailed survey can be found in [43].

11.2.5 WSN MAC Protocols with Other Objectives Related to the Requirements of IWSNs

Reliability requirement of IWSNs is usually supported at the transport layer but also the MAC layer can contribute to reliability using methods such as ARQ. Energy efficiency is supported by almost all MAC protocols via duty cycling and sleep scheduling of sensor nodes [21]. Adaptivity is also usually addressed by the protocols to deal with topology changes, varying link qualities, etc. Moreover, adaptivity according to different types of traffic flowing in the network is usually addressed by QoS-aware protocols [48]. Co-existence with other networks sharing the same spectrum is usually addressed by multi-channel communication and the reader can refer to [22] for a detailed survey on multi-channel communication in WSNs.

11.2.6 Classification

In Table 11.1, we present a classification of the protocols that we have discussed according to their objectives, whether they provide latency guarantees or not, whether they guarantee reliability, whether energy awareness is taken into account, medium access scheme and the environment for the performance evaluations of the protocols. As we have already mentioned, most of the protocols that are explicitly designed for industrial applications either propose modifications to the IEEE 802.15.4 protocol or evaluate its performance. Algorithms that aim to provide delay bound mostly use a TDMA scheme and also the protocols that modify IEEE 802.15.4 for delay bound guarantee use the contention-free GTSs. Multi-channel communication is often used for co-existence issues. Most of the protocols have been implemented in a simulation environment and the protocols tested with real deployments is very limited. Real testing on industrial environments with harsh conditions should be performed since an industrial environment may challenge the operation of the protocols which may not be encountered during simulations.

11.3 Taxonomy of Routing Protocols

Many routing protocols have been designed for wireless sensor networks over the past decade [4, 8]. Most of them consider traditional WSN requirements such as energy efficiency in the first place. Only a small number of the routing protocols are proposed to meet the diverse IWSN application requirements.

In the first part of this section, we elaborate on specific requirements of IWSNs that can be satisfied by a routing protocol. Secondly, we discuss the routing protocols that are either particularly proposed for IWSNs or designed

TABLE 11.1
Comparisons of MAC Protocols for IWSNs

Protocol	Objective	Latency	Reliability	Energy-awareness	Medium Access	Evaluation
[9]	RT support for IEEE 802.15.4	yes	no	yes	IEEE 802.15.4	Theoretical Analysis
[30]	Reliability and latency guarantee	yes	yes	yes	IEEE 802.15.4	Simulations
[49]	Provide RT guarantee for IEEE 802.15.4	yes	no	yes	IEEE 802.15.4	small testbed
i-MAC [10]	A MAC learning the patterns of manufacturing systems	yes	no	yes	scheduled	simulations
GinMAC [42]	A MAC for industrial process automation systems	yes	yes	yes	TDMA	testbed
G-McMAC [34]	Enable co-existence	no	no	yes	contention-based	simulations
[44]	Multi-channel scheduling over IEEE 802.15.4	implicit	no	yes	IEEE 802.15.4	small testbed
[11]	providing hard real-time guarantees	yes	no	yes	TDMA, CSMA	simulations
PEDAMACS [12]	delay bounded TDMA Scheduling	yes	no	yes	TDMA	simulations
RT-link [39]	end-to-end delay guarantee	yes	no	yes	TDMA	Deployment in a coal mine facility
[24]	minimize schedule length in TDMA	yes	no	yes	TDMA	simulations
[33]	guaranteed delay	yes	no	yes	TDMA	simulations
[31]	RT scheduling	yes	no	yes	scheduled	simulations

for traditional WSNs by considering reliability and latency requirements of IWSNs.

11.3.1   Routing Requirements of IWSNs

Although it is hard to generalize the network topology, the majority of the existing industrial applications can be met by networks of 10 to 200 sensor nodes [37]. In an oil refinery, for instance, the total number of sensor nodes might reach one million; however, to achieve efficient networking, the nodes will be clustered into smaller networks that are interconnected with some common sensors or cluster heads. The network traffic of many industrial applications is mostly composed of real-time publish/subscribe sensor data from the field sensor nodes over a WSN towards one or multiple sinks. Sink nodes can be a backbone router, a controller/manager, or can be connected to a controller. The sensor data generally makes its way through the backbone router (sink) to the centralized controller where it is processed. In open-loop industrial control applications, an operator sees the information coming from field sensors and takes action; the control information is sent out to the actuator node in the network, or alerts are sent to workers as a results of data analysis. Figure 11.1(a) shows a typical industrial WSN.

As it is mentioned in Section 15.1, for WSNs to be adapted in the industrial environment, besides energy efficiency, the network needs to have two main qualities: *high reliability*, and *low latency*. These application specific performance requirements can be addressed both in MAC and routing layers. In this section, we focus on how routing layer can fulfil these requirements. Figure 11.1(b), adapted from [13], shows the mutual interaction between the network layers. As explained in [13], there are strong interactions between MAC and Routing layers. The topological information (i.e., connection graph) and the application requirements are combined in the routing layer. Based on the specific metric, the routing protocol plans forwarding of messages to distribute the end-to-end traffic flow in the network and determines a distribution of the actual traffic on each link of the network, where the communication is regulated by the MAC layer. As an output of the MAC layer link performances such as reliability, delay, or energy consumption are obtained. Moreover, link performance indicators may influence directly the routing metric, so the loop between MAC and routing layers is closed.

Traffic pattern related requirements have a great influence on routing. As we mentioned in Section 11.2.1, the industrial market classifies process applications into three board categories: (i) Safety (Class 0), (ii) Control (Class 1-2-3), and (iii) Monitoring (Class 4-5). Each application class initiates a different traffic pattern, i.e., periodic data, event data, or query/response model [37]. Because different traffic patterns have different service requirements, it is often better to have different routes for different data flows between the same two endpoints. The *adaptivity* of a routing protocol to different service requirements of different traffic patterns can be achieved by supporting different

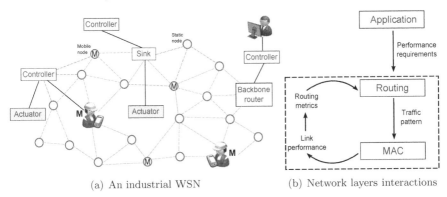

(a) An industrial WSN   (b) Network layers interactions

FIGURE 11.1
Networking in IWSNs.

metric types for each link, which is used to compute the path according to some objective function (e.g., maximize reliability or minimize latency) depending on the nature of the traffic.

To optimize an objective function such as minimizing latency, *prioritization of transmission* should be also considered. For packet transmissions, a sensor node has to decide which packet in its queue will be sent at the next transmission opportunity. Packet priority is used as one criterion for selecting the next packet. In addition, due to the lossy nature of the IWSN, the routing in plants should attempt to propose multiple paths towards the destination sink to satisfy reliability requirement. The availability of each path in a multipath route can change over time because of link failures and mobility. Therefore, it is important to measure the availability on a per-path basis and select a path (or paths for different traffic patterns) according to the *availability* requirements. Application may require availability of source-to-sink connectivity when the application needs it or when the application might need it.

Some peer-to-peer control applications may require transmission of information towards *multiple sinks*. Publication of sensor readings to more than one subscriber sink can be very useful in such applications. For instance, industrial automation systems are generally operated under a single authority as contrasted with the situation of Smart Grid system or building automation where some authorities share systems. For example, a power supplier and clients share monitoring systems in Smart Grid and the manager of the building and residents share permission to access lights in building automation. This variation brings some differences also in networking structure. In industrial automation, an IWSN typically has only one sink that has rights to control the whole system; therefore, all data collected from the network has to be forwarded towards this single authority (i.e., sink) and only the authority has the right to send commands to actuators in the network. On the other hand, in a Smart Grid system or building automation, there are multiple entities (i.e., sinks) that are interested in data collected from the network and have some rights to control the whole system or some parts of the system based on their roles in the network.

Many industrial applications involve mobile sensors that move in and out of plants, such as active sensor nodes attached on containers or vehicles. In addition, the workers are wirelessly connected to the sensor network on the plant. To be a part of the control/actuator system, a worker needs to be connected directly to the sensors or control points on the equipments near which he is working. In such a dynamic environment, the routing protocol designed for IWSNs should also satisfy *mobility* requirements. The routing protocol should support walking speeds for maintaining network connectivity as the handheld devices change position in the wireless sensor network. When we take into account the field sensors attached to moving parts of machineries, and on vehicles such as cranes or fork lifts, the routing protocol should support vehicular speeds of up to 35 km/h.

Although the standardization process for WSNs is ongoing especially on MAC Layer, there is not any widely accepted complete protocol stack for WSNs for control applications. The lack of standard routing protocol solutions is due to fact that the protocols for control applications face complex control and communication requirements changing from class to class [18].

11.3.2   Standardization Efforts for Routing

The IETF routing over low power and lossy networks (ROLL) working group [37] was created in 2007 with the aim of analyzing and eventually developing solutions for IP-based low power and lossy networks (LLNs[1]). The working group elaborated a list of requirements for a routing protocol on top of LLNs in industrial automation domain in [37]. The ROLL working group has worked on the specification of a new routing protocol, namely RPL [47], which is designed based on the needs of LLNs.

RPL can be considered as the first standardization effort for routing in IWSNs. RPL constructs destination-oriented directed acyclic graphs (DODAGs) over the network, according to optimization objectives based on metrics along the paths. In the RPL context, all edges are contained in paths oriented toward and ending at one or more root nodes (i.e., sink nodes). A network may run multiple DODAGs concurrently to fulfil multiple requirements. There are various metrics and constraints that can be used for path calculation and packet forwarding in RPL. The protocol supports both static and dynamic metrics. Link reliability, packet delay, and node energy consumption are measurements that can be used both as metric and constraints. Expected

---

[1]LLNs are a class of network in which both the routers and their interconnect are constrained. LLN routers typically operate with constraints on processing power, memory, and energy (battery power). Hence, a WSN is a typical LLN.

transmissions count (ETX) is a reliability metric that provides the number of transmissions a node expects to make to a destination sink in order to successfully deliver a packet. In addition, more metrics can be embedded in the same DODAG metric container.

Indeed, RPL is a typical gradient-based routing in which multicasting of DODAG information object (DIO) messages to the neighbors is performed periodically to achieve gradient setup from each node. In order to join a DODAG, a node listens to the DIO messages sent by its neighboring nodes and selects a subset of neighbor nodes as their parents in the DODAG. This tree-based routing approach can guarantee end-to-end delay performance; however, it needs periodic maintenance of DODAGs to handle dynamics of the network, such as mobility of nodes and link failures. The effect of mobility rate and speed on the reliability of tree-based routing is evaluated in [15]. If the number of mobile nodes and the mobility speed is high in the network, a tree-based routing protocol may not be able to guarantee reliable communication due to inaccurate DODAGs.

ISA100 is an ISA[2] committee whose charter includes defining a family of standards for industrial automation. ISA100.11a [25] is a working group within ISA100 that is working on a standard for monitoring and non-critical process control applications. In addition to IETF ROLL working group, ISA100.11a also promotes the use of graph based routing for industrial scenarios. However, ISA100.11a proposes the use of multipath graph routing in a centralized manner with the specifics of the implementation being left open. Also, no recommendations have been provided by ISA100.11a on how to make routing adaptable to the different network conditions and industrial application needs, which are basic requirements for industrial scenarios [46].

11.3.3 Routing Protocols Proposed for IWSNs

In this section, we explore the routing protocols that are designed explicitly for industrial applications. In [38], the authors propose a routing protocol, namely THVRG, which combines two-hop velocity based routing (THVR) [32] with gradient-based routing to reduce both energy consumption and end-to-end delay in IWSNs. THVRG uses the gradient setup scheme proposed by IETF ROLL in [47]. Simulation based evaluations show that THVRG achieves data delivery of 99% of the packets in a given deadline (i.e., 1350 ms) when there is only one source in the network. However, when the number of sources increases to 6, data delivery ratio within a deadline of 1200 ms drops around 85%. The protocol is not evaluated to see the effects of mobility on delay and reliability performance.

Energy Aware Routing for Real-Time and Reliable Communication in Wireless Industrial Sensor Networks (EARQ) has been proposed in [20]. In EARQ, a node estimates the energy cost, delay and reliability of a path to

---

[2]ISA is an ANSI-accredited standards-making society.

the sink node, based only on information from neighboring nodes. Then, it calculates the probability of selecting a path, using the estimates. A path with lower energy cost is likely to be selected, because the probability is inversely proportional to the energy cost to the sink node. To achieve real-time delivery, only paths that may deliver a packet in time are selected. To achieve reliability it may send a redundant packet via an alternate path, but only if it is a source of a packet. Simulation results show that EARQ decreases the ratio of deadline-missed packets around 1-2% and outperforms other real-time (RT) protocols (e.g., MMSPEED [17]) especially in energy consumption. However, in all simulation scenarios there is only one sink in the network. Multi-sink networks and mobility of nodes are not evaluated in the simulations.

In [46] Villaverde *et al.* propose a Q-learning based QoS aware route selection algorithm for IWSNs. The InRout algorithm uses a reward function based on two different QoS parameters: Packet Error Rate (PER) and energy. The reward is sent by any sink after receiving a data packet and the process is repeated through the nodes in the route until the source node is reached. This means that whenever the action of choosing a route to send a packet has been performed by a node, its Q-value is updated with the reward, that is, the node learns the goodness of the action (i.e., the route). InRout chooses the path with the lowest number of hops to deliver delay bounded packets on time. Simulation results show that InRout outperforms EARQ in terms of packet delivery ratio while consuming more energy than EARQ. InRout is proposed for static networks and does not consider multi-path routing.

In [16] Barac *et al.* evaluate the potentials of flooding as a data dissemination technique in IWSNs. Starting from the generic form of flooding, authors introduce some modifications such as handling duplicate and outdated packets and avoiding unnecessary retransmissions. Flooding delivers packets via multiple paths, which enhance redundancy and reliability. Flooding brings simplicity since there is no need for exchanging control messages between the nodes and in case of link failures or mobility of nodes it is not needed to recalculate the routing path. The simulation study shows that the flooding-based routing is capable of delivering most of the packets (between 80-100%) within a shorter deadline (around 200 ms). However, energy consumption of the protocol, which is expected to be high, is not evaluated in this work.

Energy-aware location-based routing, called e-GPSR, is proposed for industrial and logistics scenarios in [36]. The Greedy Perimeter Stateless Routing (GPSR) [27] is the core routing strategy of e-GPSR. To increase the reliability and robustness, and to decrease the power consumption and latency, e-GPSR makes a locally optimal choice for the next hop (greedy mode) taking into account the residual energy level of sensor nodes. All packets are marked by their originator with the locations of their destinations. e-GPSR makes a locally optimal choice for the next hop, that is the neighbor geographically closest to the packet's destination. The main drawback of greedy algorithms is that a packet may fall into a local maximum, namely the forwarding node is closer to the destination than all its neighbors. To increase reliability, e-GPSR enters in perimeter mode (i.e., face routing) to handle encountered routing voids (i.e., local maximum) along the route towards sink. Obviously, in geographical approach a high density of nodes improves the network connectivity, with consequent increase in the reliability of delivery. However, geographic routing may results in a non-optimal forwarding path between a source and a sink, which may cause higher latency.

11.3.4 WSN Routing Protocols with QoS Guarantee for Reliability and Timeliness

The protocol called MMSPEED [17] is designed to provide QoS differentiation in both reliability and timeliness domains so that packets can choose the most proper combination of service options depending on their timeliness and reliability requirements. The protocol defines multiple speed layer based on the end-to-end deadline and the geographic distance to the destination and meets a high reliability requirement by transmitting duplicated packets over multi-path routing. MMSPEED also achieves end-to-end QoS provisioning with local decisions at each intermediate node without end-to-end path discovery and maintenance; hence it provides self-adaptability to network dynamics. These property of MMSPEED protocol make it suitable for IWSNs. However, since the target applications of MMSPEED protocol are short-living sensor network applications such as emergency response, energy efficiency is not considered as a design issue. Because MMSPEED uses multipath forwarding by transmitting duplicated copies of the same packets and also uses a larger number of hops rather than only shortest path, the overall power consumption of the protocol is high.

Krogmann *et al.* present a novel Reliable, Real-time Routing protocol (3R) based on multipath routing for highly time-constrained Wireless Sensor and Actuator Networks in [29]. 3R partially reuses the idea of expected transmissions. Multi-path transmissions enhance the reliability of a transmission if necessary. The routing metric requires choosing non-interfering disjointed routes that satisfy the timing and reliability requirements of a packet. Similar to MMSPEED, 3R also uses PRR (Packet Reception Rate) estimations for calculating the necessary number of forwarding paths to ensure a certain reaching probability. The difference is that 3R does not consider the reaching probability and transmission latencies apart from each other. Instead, it correlates these factors to yield more accurate estimations. The results show that the chosen routing metric can decrease latencies and increase the reliability at the cost of a higher energy consumption. Adaptability of 3R to dynamic topology changes is not evaluated in the paper.

11.3.5 Classification

In Table 11.2, we present a classification of the routing protocols that we have summarized according to their objectives, whether they provide latency and/or reliability guarantees, whether energy awareness is taken into account, whether multi-path routing is used, whether mobility is considered or evaluated and the environment for the performance evaluations of the protocols. In the literature, there are some other routing protocols proposed for IWSN or time-constrained reliable WSNs. We have chosen a set of protocols considering the requirements discussed above to give an overview of the routing protocols in industrial applications domain. As we have already mentioned, most of the protocols that target reliability use multi-path routing between sources and sinks. The energy-aware protocols try to eliminate unnecessary multi-path transmissions. In most of the protocols, latency constraints are met generally by choosing the shortest possible path. Although mobile nodes (e.g., nodes attached to works or vehicles in the plants) are essential parts of an IWSN, only a small number of the protocols takes into account the mobility issues in their protocol design. The performance of all the protocols discussed above have been evaluated by simulations. Field testing of the protocols on harsh and dynamic industrial environments is really needed to see whether these protocols can handle challenging conditions which are not possible to be generated by simulations.

The last three protocols in the table are cross-layer protocols, which are discussed in the following section.

11.4   Cross Layer Protocols

To improve the performance of the routing protocol, network layer needs MAC-layer specific information, such as link performance, as illustrated in Figure 11.1(b). In what follows, some cross-layer protocols that address both network and MAC layers are mentioned. In these protocols, MAC layer is tightly coupled with the routing metrics to meet the requirements of IWSNs.

In the SERAN protocol [7], the packets are forwarded to a randomly chosen node within the next-hop cluster in the minimum spanning tree rooted at the sink. The MAC layer regulates the communication between the nodes of the transmitting cluster and the nodes of the receiving cluster within a single TDMA-slot. The random selection of the receiving node is achieved by multi-casting the packet over all the nodes of the receiving cluster, and by having the receiving nodes implement a random acknowledgement contention scheme to prevent duplication of the packets. However, this acknowledgement contention scheme has some disadvantages such as consuming energy to transmit acknowledgement messages and increasing the listening time both in the transmitting and receiving cluster. Although SERAN allows the network to operate with low energy consumption subject to delay requirements, it does not consider neither tunable reliability requirements nor duty-cycling

TABLE 11.2
Comparisons of Routing Protocols for IWSNs

Protocol	Objective	Latency	Reliability	Energy-awareness	Multi-path	Mobility	Evaluation
RPL [47]	metric-based gradient routing	yes	yes	yes	yes	yes	RFC - Proposed Standard No evaluations
THVRG [38]	metric-based gradient routing	yes	yes	yes	yes	no	simulations
EARQ [20]	energy-aware RT proactive routing	yes	yes	yes	yes	no	simulations
InRout [46]	Q-learning based reactive routing	implicit	yes	yes	no	no	simulations
Flooding [16]	Eliminating control messages	implicit	implicit	no	yes	implicit	simulations
e-GPSR [36]	energy-aware and low-delay geographic routing	yes	no	yes	no	yes	simulations
MMSPEED [17]	QoS differentiation in reliability and timeliness	yes	yes	no	yes	yes	simulations
3R [29]	Reliability, real-time guarantee	yes	yes	no	yes	no	simulations
Cross-Layer							
SERAN [7]	low delay and energy consumption	yes	no	yes	no	no	simulations
Breath [35]	energy-aware RT routing to a single sink	yes	yes	yes	no	no	small testbed
TREnD [14]	energy-aware RT routing between clusters	yes	yes	yes	no	no	small testbed

policies, which are essential to reduce energy consumption. Furthermore, SERAN focuses only on low traffic networks.

The Breath [35] protocol is designed for scenarios where a plant must be controlled over a multi-hop network. Breath uses a randomized routing, a hybrid TDMA at the MAC, radio power control at the physical layer. Similar to SERAN [7], nodes route data packets to next-hop nodes randomly selected in a forwarding region. Each node, which is either a transmitter or a receiver, does not stay in an active state all time, but sleeps for a random amount of time depending on the traffic and channel conditions. The MAC protocol of Breath is based on a CSMA/CA mechanism similar to IEEE 802.15.4. Breath assumes that all nodes are location aware. Location information is needed for adjusting the transmission power and to change the number of hops. Experimental results show that the protocol achieves the reliability and delay requirements, while minimizing the energy consumption. Breath outperforms the standard IEEE 802.15.4 in terms of both energy efficiency and reliability.

The TREnD [14] protocol is designed for environments where multiple industrial plants have to be controlled by a multi-hop network. Similarly to SERAN [7], the routing algorithm of TREnD is hierarchically sub-divided into two parts: (i) a static route at inter clusters level, and (ii) a dynamical routing algorithm at node level. This is supported at the MAC layer by an hybrid time division multiple access and carrier sensing multiple access (TDMA/CSMA) solution. To offer flexibility for the addition of new nodes, robustness to node failures, and support for the random selection of next-hop node, the communication stage between nodes during a TDMA-slot is managed at MAC layer by a p-persistent CSMA/CA scheme. In hybrid TDMA/CSMA solutions the p-persistent MAC gives better performance than the binary exponential backoff mechanism used by IEEE 802.15.4.

11.5   Future Research Directions / Open Problems

In this section we list the open problems as future research directions for MAC and routing protocol design in IWSNs:

- As we mentioned, most of the protocols are either tested with simulations or in limited testbed settings. Real testing on industrial environments with harsh conditions that may challenge the performance of the proposed protocols is required for a realistic evaluation of IWSNs.

- Mobile sinks and mobile sensor nodes, for instance worn by the workers in a factory environment, should be supported by the designed protocols.

- Although some examples of cross-layer solutions exist, as we discussed in Section 11.4, application-specific cross-layer solutions should be proposed to meet the requirements of IWSNs and optimize the network operation.

- Supporting real-time services is of paramount concern for IWSNs [49]. More research on real-time scheduling for IWSNs is required.

- Energy harvesting for extending the network lifetime is a viable solution for IWSNs, for instance considering harvesting from vibrations. In an energy-harvesting IWSN, the level of remaining energy may change according to the availability of the harvesting source whose availability is often highly variable. Therefore, harvesting-aware power management and protocols, that can predict when the energy source will be available, should be designed [26].

- As we mentioned, co-existence of different networks operating on the same parts of the spectrum may cause performance degradations due to interference and contention on the shared medium. It is highly likely that industrial environments may contain different wireless technologies that share the same spectrum. Therefore, efficient methods to mitigate the effects of interference should be studied besides frequency hopping.

- As reported in [3], error control techniques, such as forward error correcting (FEC) codes are usually neglected in order to skip power the consumption of decoding the packets in IEEE 802.15.4 based IWSNs. However, by using error control techniques the energy consumption may be reduced since less retransmissions should take place. This tradeoff should be studied in detail and energy-efficient error correction mechanisms should be proposed.

11.6 Conclusions

In this chapter, we have surveyed the MAC and routing protocol solutions proposed for IWSNs. We listed the networking requirements of IWSNs and classified the protocols according to these requirements for comparisons. Besides the protocols that are explicitly designed for IWSNs, protocols designed for traditional WSNs which may meet the requirements of IWSNs are also covered in our survey. The survey also includes examples of cross-layer solutions that involve both MAC and routing. As we discussed, since standardization activities revolve around the IEEE 802.15.4 protocol, especially using the physical layer, most of the protocols either propose modifications to the IEEE 802.15.4 protocol or evaluate its performance. In the network layer, standardization efforts are based on tree-based multipath routing; therefore, many other routing protocols focus on how to construct efficient metric-based trees for multipath routing. Most of the protocols have been implemented in a simulation environment and only a few protocols have been tested with real deployments. Real testing on industrial environments with harsh conditions are required for the acceptability of the protocols. Cross-layer solutions that involve interactions between MAC and routing layers, protocols supporting mobile nodes and sinks, supporting real-time services, energy harvesting-aware power management and protocols, co-existence issues and error control techniques have been identified as some of the important directions for future research.

References

[1] Routing Over Low power and Lossy networks (ROLL). http://www.ietf.org/dyn/wg/charter/roll-charter.html; accessed April 19, 2012.

[2] IEEE Standard for Information Technology Part 15.4: Wireless Medium Access Control (MAC) and Physical Layer (PHY) Specifications for Low- Rate Wireless Personal Area Networks (LR-WPANs), IEEE Std. 802.15.4), October 2003. http://standards.ieee.org/getieee802/download/802.15.4-2003.pdf; accessed April 21, 2012.

[3] J. Åkerberg, and Gidlund, M. Future research challenges in wireless sensor and actuator networks targeting industrial automation. In *9th IEEE International Conference on Industrial Informatics (INDIN), 2011*, pages 410–415, 2011.

[4] K. Akkaya and Younis, M. A survey on routing protocols for wireless sensor networks. *Ad Hoc Networks*, 3(3):325–349, 2005. Survey.

[5] M. Bertocco, G. Gamba, and A. Sona. Experimental characterization of wireless sensor networks for industrial applications. *IEEE Transactions On Instrumentation And Measurement*, 57(8):1537–1546, 2008.

[6] M. Bjorkbom, L. M. Eriksson, and J. Silvo. Technologies and methodologies enabling reliable real-time wireless automation. In Meng Joo Er, editor, *New Trends in Technologies: Control, Management, Computational Intelligence and Network Systems*, pages 85–112. Sciyo, 2011.

[7] A. Bonivento and C. Fischione. SERAN: a Protocol for Clustered WSNs in Industrial Control and Automation. *Distributed Computing*, 00(c):3–5, 2009.

[8] A. Boukerche, M.Z. Ahmad, B. Turgut, and D. Turgut. A taxonomy of routing protocols in sensor networks. In A. Boukerche, editor, *Algorithms and Protocols for Wireless Sensor Networks*, chapter 6, pages 129–160. Wiley, 2008.

[9] F. Chen, T. Talanis, R. German, and F. Dressler. Real-time enabled ieee 802.15.4 sensor networks in industrial automation. In *2009 IEEE International Symposium on Industrial Embedded Systems*, pages 136–139. IEEE, July 2009.

[10] K. K. Chintalapudi. i-Mac - a Mac that learns. In *Proceedings of the 9th ACM/IEEE International Conference on Information Processing in Sensor Networks*, IPSN '10, pages 315–326, New York, NY, USA, 2010. ACM.

[11] K. K. Chintalapudi and L. Venkatraman. On the design of mac protocols for low-latency hard real-time discrete control applications over 802.15.4 hardware. In *Proceedings of the 7th International Conference on Information Processing in Sensor Networks*, IPSN '08, pages 356–367, Washington, DC, USA, 2008. IEEE Computer Society.

[12] S. Coleri-Ergen and P. Varaiya. Pedamacs: Power efficient and delay aware medium access protocol for sensor networks. *IEEE Trans. on Mobile Computing*, 5(7):920–930, July 2006.

[13] P. Di Marco. *Modeling and Design of Multi-hop Energy Efficient Wireless Networks for Control Applications.* Licentiate Thesis, KTH School of Electrical Engineering, Stockholm, Sweden, 2010.

[14] P. Di Marco, P. Park, C. Fischione, and K. H. Johansson. TREnD: A Timely, Reliable, Energy-Efficient and Dynamic WSN Protocol for Control Applications. *2010 IEEE International Conference on Communications*, pages 1–6, May 2010.

[15] A. T. Erman, A. Dilo, L. van Hoesel, and P. Havinga. On mobility management in multi-sink sensor networks for geocasting of queries. *Sensors*, 11(12):11415–11446, December 2011.

[16] Barac F., J. Akerberg, and M. Gidlund. A lightweight routing protocol for industrial wireless sensor and actuator networks. In *Proceedings of the 37th Annual Conference on IEEE Industrial Electronics Society, IECON*, pages 2980–2985. IEEE Press, November 2011.

[17] E. Felemban, C.G. Lee, and E. Ekici. MMSPEED: multipath Multi-SPEED protocol for QoS guarantee of reliability and. Timeliness in wireless sensor networks. *Mobile Computing, IEEE*, 5(6):738–754, 2006.

[18] C. Fischione, P. Park, P. Di Marco, and K. Henrik Johansson. Design Principles of Wireless Sensor Networks Protocols for Control Applications. In *Wireless Network Based Control.* Springer New York, New York, NY, 2011.

[19] V. C. Gungor and G. P. Hancke. Industrial Wireless Sensor Networks : Challenges, Design Principles, and Technical Approaches. *IEEE Transactions on Industrial Electronics*, 56(10):4258–4265, October 2009.

[20] J. Heo, J. Hong, and Y. Cho. EARQ: Energy Aware Routing for Real-Time and Reliable Communication in Wireless Industrial Sensor Networks. *IEEE Transactions on Industrial Informatics*, 5(1):3–11, 2009.

[21] I. Demirkol, C. Ersoy, and F. Alagoz. Mac protocols for wireless sensor networks: a survey. *IEEE Communications Magazine*, 44(4):115–121, April 2006.

[22] O. D. Incel. A survey on multi-channel communication in wireless sensor networks. *Comput. Netw.*, 55(13):3081–3099, September 2011.

[23] O. D. Incel, A. Ghosh, and B. Krishnamachari. Scheduling Algorithms for Tree-Based Data Collection in Wireless Sensor Networks. In *Theoretical Aspects of Distributed Computing in Sensor Networks*. Springer, 2011.

[24] Ozlem Durmaz Incel, Amitabha Ghosh, Bhaskar Krishnamachari, and Krishna Chintalapudi. Fast data collection in tree-based wireless sensor networks. *IEEE Trans. Mob. Comput.*, 11(1):86–99, 2012.

[25] Wireless systems for industrial automation: Process control and related applications), May 2008. http://www.isa.org/Community/SP100WirelessSystemsforAutomation; accessed April 19, 2012.

[26] A. Kansal, J. Hsu, M. Srivastava, and V. Raghunathan. Harvesting aware power management for sensor networks. In *Proceedings of the 43rd Annual Design Automation Conference*, DAC '06, pages 651–656, New York, NY, USA, 2006. ACM.

[27] B. Karp and H. T. Kung. GPSR: greedy perimeter stateless routing for wireless networks. In *Proceedings of the 6th Annual International Conference on Mobile Computing and Networking*, MobiCom'00, pages 243–254, Boston, MA, USA, 2000.

[28] L. Krishnamurthy, R. Adler, P. Buonadonna, J. Chhabra, M. Flanigan, N. Kushalnagar, L. Nachman, and M. Yarvis. Design and deployment of industrial sensor networks: experiences from a semiconductor plant and the north sea. In *Proceedings of the 3rd International Conference on Embedded Networked Sensor Systems*, SenSys '05, pages 64–75, New York, NY, USA, 2005. ACM.

[29] M. Krogmann, M. Heidrich, D. Bichler, D. Barisic, and G. Stromberg. Reliable, real-time routing in wireless sensor and actuator networks. *ISRN Communications and Networking*, 2011:10:1–10:8, January 2011.

[30] K. Kunert, E. Uhlemann, M. Jonsson, E. Uhlemann, and M. Jonsson. Predictable real-time communications with improved reliability for IEEE 802.15.4 based industrial networks. In *8th IEEE International Workshop on Factory Communication Systems (WFCS), 2010*, pages 13–22, 2010.

[31] H. Li, P. Shenoy, and K. Ramamritham. Scheduling messages with deadlines in multi-hop real-time sensor networks. In *RTAS 2005: Proceedings of the 11th IEEE Real Time and Embedded Technology and Applications Symposium*, pages 415–425, March 2005.

[32] Y. Li, C. S. Chen, Y.-Q. Song, Z. Wang, and Y. Sun. Enhancing Real-Time Delivery in Wireless Sensor Networks With Two-Hop Information. *IEEE Transactions on Industrial Informatics*, 5(2):113–122, 2009.

[33] S. Mastooreh, S. Hamed, and K. Antonis. Hymac: Hybrid tdma/fdma medium access control protocol for wireless sensor networks. In *PIMRC 2007: The proceedings of the 18th IEEE Personal, Indoor and Mobile Radio Communications Symposium*, pages 1–5, September 2007.

[34] S. Nethi, Nieminen, J.; Jantti, R. Nethi, S. and J Riku. Exploitation of multi-channel communications in industrial wireless sensor applications: Avoiding interference and enabling coexistence. In *IEEE Wireless Communications and Networking Conference (WCNC), 2011*, pages 345–350, 2011.

[35] P. Park, C. Fischione, and A. Bonivento. Breath: an adaptive protocol for industrial control applications using wireless sensor networks. *Mobile Computing*, 10(6):821–838, 2011.

[36] S. Persia and D. Cassioli. Routing Design for UWB Sensor Networks in Industrial and Logistics Scenarios. In *Proceedings of the 16th IST Mobile and Wireless Communications Summit*, pages 1–5, Budapest, Hungary, July 2007.

[37] K. Pister, P. Thubert, S. Dwars, and T. Phinney. Industrial routing requirements in low-power and lossy networks. RFC 5673, Internet Engineering Task Force, October 2009. http://tools.ietf.org/html/rfc5673; accessed April 19, 2012.

[38] P.T.A. Quang and D.-S. Kim. Enhancing Real-Time Delivery of Gradient Routing for Industrial Wireless Sensor Networks. *IEEE Transactions on Industrial Informatics*, 8(1):61–68, 2012.

[39] A. Rowe, R. Mangharam, and R. Rajkumar. Rt-link: A global time-synchronized link protocol for sensor networks. *Ad Hoc Netw.*, 6(8):1201–1220, November 2008.

[40] J. Song, S. Han, A.K. Mok, D. Chen, M. Lucas, and M. Nixon. Wirelesshart: Applying wireless technology in real-time industrial process control. In *RTAS '08: Proceedings of the IEEE Real-Time and Embedded Technology and Applications Symposium, 2008*, pages 377–386, April 2008.

[41] P. Suriyachai, U. Roedig, and A. Scott. A survey of mac protocols for mission-critical applications in wireless sensor networks. *Communications Surveys & Tutorials, IEEE*, PP:1–25, 2011.

[42] P. Suriyachai, J. Brown, and U. Roedig. Time-critical data delivery in wireless sensor networks. In *Proceedings of the 6th IEEE International*

Conference on Distributed Computing in Sensor Systems, DCOSS'10, pages 216–229, Berlin, Heidelberg, 2010. Springer-Verlag.

[43] Z. Teng. A Survey on Real-Time MAC Protocols in Wireless Sensor Networks. *Communications and Network*, 02(02):104–112, 2010.

[44] E. Toscano. Multichannel Superframe Scheduling for IEEE 802.15. 4 Industrial Wireless Sensor Networks. *Industrial Informatics, IEEE*, 8(2):337–350, 2011.

[45] L. van Hoesel and P. Havinga. A lightweight medium access protocol (LMAC) for wireless sensor networks. In *INSS' 04: Proceedings of the 1st International Conference on Networked Sensing Systems*, Tokyo, Japan, June 2004.

[46] B. C. Villaverde, S. Rea, and D. Pesch. Inrout a QoS aware route selection algorithm for industrial wireless sensor networks. *Ad Hoc Networks*, 10(3):458–478, 2011.

[47] T. Winter, P. Thubert, A. Brandt, J. Hui, R. Kelsey, P. Levis, K. Pister, R. Struik, JP. Vasseur, and R. Alexander. RPL: IPv6 Routing Protocol for Low-Power and Lossy Networks. RFC 6550, Internet Engineering Task Force, March 2012. http://datatracker.ietf.org/doc/rfc6550/; accessed April 19, 2012.

[48] M. A. Yigitel, O. D. Incel, and C. Ersoy. QoS-aware MAC protocols for wireless sensor networks: A survey. *Comput. Netw.*, 55(8):1982–2004, June 2011.

[49] S.-e. Yoo, P. K. Chong, D. Kim, Y. Doh, M.-l. Pham, E. Choi, and J. Huh. Guaranteeing Real-Time Services for Industrial Wireless Sensor Networks With IEEE 802.15.4. *IEEE Transactions on Industrial Electronics*, 57(11):3868–3876, 2010.

[50] G. Zhao. Wireless Sensor Networks for Industrial Process Monitoring and Control: A Survey. *Network Protocols and Algorithms*, 3(1):46–63, April 2011.

12

QoS-Aware Routing for Industrial Wireless Sensor Networks

Berta Carballido Villaverde, Susan Rea, and Dirk Pesch

Nimbus Centre for Embedded Systems Research, Cork Institute of Technology, Cork, Ireland

CONTENTS

12.1   Introduction

Machine to machine communication systems in industrial environments have traditionally relied on cable technologies. The most recent of these use IP based systems such as Foundation Fieldbus High Speed Ethernet [20], PROFInet [24], etc. The advantages of cable include high reliability, safety, and stability. For these reasons, it is likely that these wired systems will not be easily replaced by wireless systems in the near future, but rather that wireless systems will perform complementary tasks. Although wireless technologies suffer from the reliability problems of the wireless transmission medium, they have a number of advantages that should not be underestimated. Wireless low power devices allow industries to significantly increase the number of systems and parameters that can be measured and controlled. They can be placed in positions where cabling would be too expensive to make the installation of sensors viable or where cabling would not be possible or too difficult such as in moving machinery.

When designing and deploying a wireless sensor network for industrial environments, the designer must address a number of questions such as which communication technologies to employ, what type of requirements those technologies should meet, etc. Within the space of industrial monitoring two standards have recently been published that attempt to support the designer in their choice of technology, e.g., ISA100.11a [26] and WirelessHART [17]. Nevertheless, these standards only provide very basic recommendations with regards to a number of critical design considerations such as routing functionality. This lack of specification at the routing layer has led another standardization body, the Internet Engineering Task Force (IETF), to work on the definition of a standard protocol for routing in low power lossy networks, where industrial networks are included, referred to as RPL (Routing for IPv6 Low Power Lossy Networks) [49]. While this routing protocol defines how the routes (graphs) should be built, it leaves the specification of the objective functions that should be employed to build the routes open. The reason for this is that different industrial sectors will have their own range of monitoring and control applications, each with distinct requirements, that they need to implement. Each application will typically have its own related parameters and objective functions to satisfy application Quality of Service (QoS) demands, such as the rate of successful packet delivery, network lifetime, or end-to-end delays within required bounds.

This chapter provides an overview of industrial application classes and requirements as well as the design requirements and challenges that must be considered when developing any routing mechanism in wireless embedded monitoring and control networks for industrial environments. A survey is provided on existing approaches for routing in industrial environments as well as mechanisms that may be employed to provide QoS at the routing level. Routing is a key process in WSNs when dealing with QoS requirements as routing decisions impact on network lifetime, packet delivery ratios, and end-to-end delays.

12.2 Industrial Applications: QoS Requirements and Key Performance Indicators

Wireless sensor networks have an immense potential to improve and simplify the way in which industrial factories are managed, monitored, and controlled. Deploying WSNs in industrial environments has significant advantages such as high adaptivity and low cost in terms of installation and retrofitting [11, 34, 50]. Specific applications of wireless sensor networks in industrial scenarios include condition monitoring systems for small electric motors [3], agent-based steady-state motor analysis [43], temperature measurement for end-mill inserts [51], vibration-based monitoring for tool breakage [41], sensing of current, voltage, and acoustic emission signals [14], process manufacturing [28], discrete manufacturing [27], and many more [50].

Crucial to the success of this technology is the ability to satisfy monitoring and control application QoS requirements such as successful monitoring data delivery, prolonged network lifetime, or bounded end-to-end data delivery delays. Communication protocols play a key role in satisfying those QoS requirements and influence the overall system performance. For instance, at the routing layer, a routing mechanism that aims at prolonging network lifetime by distributing the load may in fact select as appropriate routes those that offer poor delivery ratios if the delivery requirement is not included in the routing decision process. Therefore, routing protocols must be carefully designed and, with this purpose, traffic classes, application types, and requirements relating to industrial environments must be identified. Moreover, metrics must be defined so that the overall network performance can be quantified in terms of QoS.

Following from the need to identify the types of industrial monitoring and control applications and their associated data traffic requirements, several classifications and categories have been proposed. For example, [8, 7] proposes a classification for factory control applications depending on their data traffic requirements (periodic, aperiodic, critical, non critical ...). In [46] applications are classified depending on their relevance to the automation pyramid [1]. Several efforts have been made recently by standardization bodies and working groups to identify the different requirements associated with industrial applications. These include the ISA100 [13] standard and the IETF Routing over Low Power Lossy Networks Working Group [37]. The ISA100 standard focuses on the establishment of standards, recommended practices, and related information for implementing wireless systems in factory environments.

On the other hand, the ROLL group focuses on analyzing the functional requirements for routing protocols in industrial Low-power and Lossy Networks (LLNs), which include industrial wireless sensor networks. Industrial application classifications provided by ISA100 and ROLL are surveyed in Sections 12.2.1 and 12.2.2. These classifications may be used as reference when designing appropriate QoS aware communication mechanisms for industrial environments. For instance, the objective functions and metrics utilized to build routes in industrial environments, such as those proposed in Section 12.4.5, may be carefully selected considering these classifications and the corresponding application requirements. Finally, several performance indicators that are suitable for evaluating the overall network performance and communication algorithms operation in terms of QoS are presented in Section 12.2.3.

12.2.1   Classification Based on Type of Application Data

The ROLL working group classifies applications depending on the type of traffic they generate as follows:

- Periodic data: In periodic applications such as temperature monitoring, data is generated periodically and has deterministic and predictable bandwidth requirements. The main requirement for this application is the timely delivery of data and, as such, permanent resources must be allocated to assure availability of the required bandwidth. Moreover, newer readings for this type of application data usually obsolete previous ones. Therefore, the periodic update of data to the plant application is considered more relevant than the end-to-end delivery latency. One example of this type of application is temperature monitoring for plastic machinery [13]. In this type of application, the most important requirement is to assure the constant availability/update of data.

- Event data: Applications that generate event based data include alarms and aperiodic data reports with bursty data bandwidth requirements. Moreover, some alarms may be critical requiring service differentiation from the network. Events take place continually in everyday factory operations. Production commences and ends, stock arrives at the loading dock, machines report failures, monitoring and control systems notify out-of-bounds conditions, etc. One example of an event based application can be an application controlling water and fuel tanks fullness in a thermic power plant [32]. This type of application is event based and it would have a tight timely delivery requirement.

- Client/Server data: Many applications in industrial environments are based on client/server interactions following a command response protocol. For these applications, the required data bandwidth is often bursty and the round-trip latencies would be in the order of hundreds of milliseconds. The transmission of this data would generally be based on cost-based fair-share best-effort service. An example application here could involve a technician sending a request to obtain information on which valves are opened in a thermic power plant [32]. This example application would have a timely delivery requirement under human tolerated delay.

• Bulk transfer data: Applications that require the transmission of blocks of data in multiple packets can be classified under bulk transfer. Temporary resources are assigned for these applications depending on the data rates and file sizes to meet the specific application service requirements. One example of an event based application is machinery condition monitoring [11]. Machinery condition monitoring applications typically have requirements on reliable data transfer that must be initiated when necessary (i.e., following the detection of an "interesting" event or periodically). This type of application would not have strict timely delivery requirements; however, it would have bandwidth requirements (typical applications will generate tens of Kbytes of data [37]).

12.2.2 Classification Based on Application Data Criticality

In addition to the detailed classification provided by the ROLL group, the ISA 100.11a standard [26] also classifies industrial applications. In this case the classification is performed depending on the timely delivery requirements of the applications (see below). The ISA-SP100.11a's main focus is on the non-critical application space with support for delays down to 100ms with optional support for lower delays (classes 1 to 5).

• *Class 5 - Monitoring without immediate operational consequences:* Applications belonging to this class usually do not present strong timeliness requirements. Some may require high reliability (i.e., sequence-of-events reporting) while others may support the loss of some samples (i.e., slowly changing information report).

• *Class 4 - Monitoring with short-term operational consequences:* Example applications belonging to this class could be those that require sending a technician to perform maintenance after detecting some fault or abnormal behavior (i.e., alerting).

• *Class 3 - Open loop control:* In this type of application an operator "closes the loop" between output and input. An example application could be an operator controlling the factory machinery operation.

• *Class 2 - Closed loop supervisory control:* This class of closed-loop control is usually non-critical with latency requirements ranging from 100ms to above 1s [37].

• *Class 1 - Closed loop regulatory control:* This class is usually critical. Examples could be pressure and flow control.

TABLE 12.1

Example Applications of Industrial Environments [40].

SP100.11a App. Class	5	4	4
Industry	Automotive	Wastewater Treatment	Natural Gas
Application	Wireless Torque Wrench Monitoring	Monitor the ammonia and oxygen levels in waste water	Pipeline Monitoring and Control
Reporting Interval [s] MUST	2	600	60
Min. Reliability for Single Message [%] MUST	99.990	80	90
Latency [s] MUST	2	60	120

- *Class 0 - Emergency action:* This application class is always critical and includes safety-related actions. Most of this class functions utilize dedicated wired networks to reduce the chances of failure or even attacks. Examples are fire control or emergency shutdown.

Among all the possible applications the primary focus across industry is on the monitoring related classes, 4 and 5 as well as the non critical portions of classes 2 and 3, mainly due to the reliability concerns imposed by the wireless channel [37]. This interest towards less critical applications was also showcased in a market survey [40] carried out by the International Society of Automation where, when considering deploying a wireless sensor network in a factory installation, 88.8% of interested companies (users and vendors) would use the network for monitoring purposes (condition and process monitoring). More critical uses, such as high-speed control applications were only considered by 13% of those companies surveyed. Table 12.1 provides a brief selection of representative sample applications typical in industrial environments and their associated requirements.

In general, it is possible to conclude that industrial applications have a wide range of requirements ranging from soft latency to tight latency, constant or variable bandwidth needs, soft to strict reliability, etc. Therefore, in this environment, it is important that the communication protocols can adapt and satisfy the requirements imposed by the diverse applications.

12.2.3 Key Performance Indicators

It is well known that the concept of Quality of Service (QoS) may be perceived and interpreted in different ways depending on the targeted technology. For example, in the communication networks domain, from the application perspective, QoS usually refers to the quality as perceived by the user or application. From the network perspective, QoS reflects a measure of the service quality that the network offers to the applications or users. The network's main purpose is to satisfy the QoS required by the applications while maximizing network resource utilization.

QoS requirements and parameters that influence QoS in WSN differ from those of traditional networks since sensor nodes have many specific constraints that make them a unique. As a result, additional QoS parameters to measure the network performance need to be identified for this type of network. The performance should be expressed by parameters that:

1. Consider all aspects of the service from the user's point of view,

2. Center on user-perceivable effects, instead of their causes within the network,

3. Can be easily related to network performance parameters,

4. Can be objectively or subjectively measured.

Considering this and the application requirements detailed in Sections 12.2.1, and 12.2.2, the following key indicators can be outlined as having an important impact from the user and application perspective in industrial wireless sensor networks:

1. Network Lifetime: this is also a performance indicator to take into account in WSNs as the sensor nodes are battery operated. Depending on the lifetime, the network will be able to serve the user or application for a smaller or longer period of time.

2. Information Loss: this has a very direct consequence on the quality of the information finally presented to the user. In the wireless sensor networks domain information loss is not limited to the effects of bit errors or packet loss during transmission, but it also includes the packet loss due to buffer overflows that for instance is also related to the bandwidth allocation or medium contention. Thus, any metric used to measure information loss should consider both sources of information loss.

3. Bandwidth Availability: this is also an important concern for some of the industrial wireless sensor applications outlined in the previous sections and the lack of it will have consequences on the quality of the information delivered to the user. For example, lack of bandwidth may translate into lost packets due to memory restrictions of the sensor nodes. Bandwidth should be used and allocated wisely to avoid packet loss and maximize network throughput.

4. Delay Requirements: Although the targeted monitoring applications have soft-delay requirements, delay still needs to be taken into account as it has a very direct impact on user satisfaction depending

on the application, and includes delays introduced by the network and at the sensor node.

5. Application Data Relevance: this is another parameter to consider for QoS provision. The data generated by the different sensor nodes can have several degrees of relevance, therefore not all data packets may be treated in the same way - i.e., it is more important to transmit a fuel pipe leak packet than a normal pipe pressure reading packet. Therefore, another parameter to consider is how well the application needs are satisfied depending on the importance of the generated data.

12.3 General Considerations for Routing in Industrial Environments

Several considerations must be taken into account when designing QoS aware routing mechanisms in industrial wireless sensor networks. Those include the expected topologies and connectivity among sensors, the challenges imposed by sensor network technology as well as the limitations and requirements imposed by the wireless channel.

12.3.1 WSN Topologies in Industrial Scenarios

Being aware of the expected working topology is crucial to the design of any QoS aware routing protocol in wireless sensor networks. The need for topology awareness is driven by the following: end-to-end delays in low power wireless embedded networks are greatly influenced based on the number of hops [4] (i.e., the higher the number of hops the greater the delay); the higher the number of hops a packet has to traverse in a wireless network the more likely it is that packet will be lost due to bad channel conditions or even dropped in an overloaded buffer; finally, the higher the number of wireless nodes a device has as leaf nodes or children, will impact severely on its battery usage as it will be required to forwarding information for leaf nodes. So even though a QoS aware routing protocol designed for industrial wireless sensor networks should be able to provide optimal performance under any conditions, in reality this is a very challenging task. Thus expected topologies should be considered in analyzing how well or how realistically the application requirements described in Sections 12.2.1 and 12.2.2 can be met by the communication protocols.

In a typical industrial environment, according to the Networking Working Group (NWG) of the Internet Engineering Task Force (IETF) [37], there may be multiple sinks with the number of sinks being far smaller than the total number of nodes. Networks may be composed of between 10 to 200 field devices and usually the maximum number of hops is 20. An example of a typical industrial topology is presented by ISA SP100.11 in [22] (see Figure 12.1). Field devices themselves would act as routers and would be stationary. Some moving devices without routing capabilities may be placed on moving machinery or similar. One example of physical topology could be a multi-square-kilometer refinery with isolated tanks distributed across the plant [37]. In this scenario, a few hundred sensor node devices would be deployed providing total coverage with a self-forming self-healing wireless mesh network of 5 to 10 hops in length (note that in mesh networks every device can communicate directly and forward data from any device in range). An example of an extreme opposite case is where a backbone network is deployed along a factory plant and most nodes are in direct sight of one or more backbone routers. In this case, the majority of communications between field devices takes place across the backbone.

Finally, with regards to logical topologies, a basic requirement for routing protocol design in industrial networks is the provision of multiple paths towards the destination with potentially different costs [37]. The protocol must route over paths that are capable of supporting different QoS application requirements, such as those referred to in Sections 12.2.1 and 12.2.2 and have the ability to recompute paths based on underlying link attributes/metrics that may change dynamically. Moreover, path redundancy is vital to assure reliability due to the unpredictability and harshness of the wireless medium since a single link failure could compromise the communication flow if only one route is available. In summary, a routing protocol design for wireless communications in industrial scenarios must provide path redundancy and be capable of adapting to different application requirements and dynamic network conditions. Possible metrics that can be utilized at the routing level to satisfy the QoS application requirements referred to previously will be analyzed in Section 12.4.

12.3.2 Challenges

Although the limited size of the sensor nodes makes them attractive for use in industrial environments, their size imposes restrictions on the available resources such as the energy, computational power, and memory. This can make satisfying the application requirements a challenging task. Moreover, the lossy connectivity found in industrial environments posses further demands to the communication protocols when trying to satisfy the application requirements. The factors and tradeoffs that must be considered when designing routing protocols in industrial environments are discussed next. These factors should be considered in conjunction with the QoS application requirements outlined previously when designing or selecting appropriate objective functions or metrics, such as those described next in Section 12.4, for the provision of QoS aware routing.

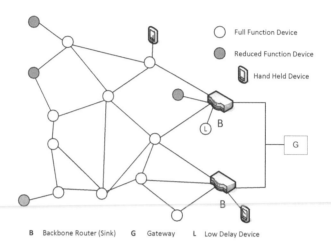

FIGURE 12.1
ISA SP100.11a Basic Network [22].

12.3.2.1   Low Power Operation & Delay

Satisfying the delay requirements of any application in wireless sensor networks can be a challenging task and this is particularly true in multihop topologies. This is due to the fact that in order to reduce the energy consumption, the sensor node must sleep as much as possible, that is, it must have as small a duty cycle as possible. This implies that at each hop a node has to wait for some time frame depending on the duty cycle of the destinations and the scheduling in order to send a packet. Generally, the bigger the distance in hops the bigger the incurred delay will be. Thus delay and low energy consumption are tradeoff requirements and any algorithm designed for low power wireless sensor networks must consider this. This can have implications on applications with very tight delay requirements such as emergency event activation or some types of automation applications where delay constraints are in the order of tens of milliseconds [9]. For these fast action applications, star topologies are more suited since smaller delays can be guaranteed. Note as well that the ISA 100.11a Standard [26], which has been developed specifically for industrial environments, does not consider hard delays. It considers applications that require delays down to 100ms -not time critical, with the support for lower delays being optional. For less critical applications such as monitoring, where higher latencies can be tolerated, a mesh network topology or a mix of a star and mesh network are suggested as appropriate network topologies.

12.3.2.2 Lossy Connectivity & Reliability

Industrial environments usually involve harsh wireless channels. This is typically due to the abundance of metallic structures which obstruct the line of sight among communicating devices causing refraction, reflection, and scattering of the signals [33, 44]. Thus, satisfying reliability requirements in these environments can be challenging. Redundancy of routes and routing metrics that reflect the status of the channel can be utilized by routing protocols to mitigate the effects of those adverse conditions. Note that the ROLL group identifies path redundancy as a necessary requirement for routing in industrial low power lossy networks [37].

12.3.2.3 Memory Footprint & Control Overhead

In order to satisfy QoS requirements for applications, intelligent algorithms must be developed so that the information available from the network and the sensor nodes can be successfully interpreted to provide the best possible performance. However, any algorithm design for wireless sensor networks must consider the fact that sensor nodes have very restricted memory capacity and energy budget as well as limited computational capabilities. Therefore another tradeoff requirement in wireless sensor networks is the design of intelligent algorithms, capable of analyzing nodes and network status, that require little memory, have low computational needs and use limited energy (for example in terms of transmitting/receiving control packets).

12.3.2.4 Conflicting QoS Requirements

In multihop topologies satisfying reliability and delay requirements conflicts with the demand for long network lifetimes. This is due to the fact that in order to deliver the sensed data from any node to the sink, the most reliable path or the quickest path may not be the most energy efficient. Therefore, finding a balance between these tradeoff requirements when performing the routing tasks is a necessary condition in order to have good network performance over a longer period of time. Moreover, the routing protocol may choose different paths for different applications depending on their requirements. Again, objective functions or metrics for QoS aware routing such as those described next in Section 12.4 must be carefully designed.

12.3.2.5 Resource Allocation & Priority

Application data relevance is also a parameter to consider for QoS provision. The data generated by the different sensor nodes can have several degrees of relevance, therefore not all data packets can be treated in the same way - i.e., it is more important to transmit say a fuel pipe leak packet than a normal pipe pressure reading packet. Due to the fact that the bandwidth availability and the storage capacity are reduced, application data relevance must be taken into account when allocating the available resources.

12.4   Current Approaches for Routing in Industrial Environments

To conclude this chapter, this section presents current approaches for routing in industrial wireless sensor networks. More specifically, routing mechanisms proposed by different standardisation bodies are surveyed as well as different existing metrics and objective functions that may be utilized to provide QoS aware routing based on the requirements identified before in Sections 12.2.1 and 12.2.2. Note that when designing any QoS aware routing mechanism the application requirements should be evaluated together with the general considerations identified before in Section 12.3 as any parameter from topologies to selected objective functions can have multiple implications in the overall system performance.

12.4.1   WirelessHART & ISA100

Looking at industrial standardization groups, both WirelessHART [17] and ISA100.11a [26] employ Graph Routing and Source Routing algorithms, with Graph Routing being used in most cases. For the Graph Routing, both standards use a central manager that, after obtaining information relating to the nodes connectivity and status, computes multiple routes (routing graphs) for every node and downloads them to the individual nodes. To send a packet, the source device writes a specific graph ID, depending on the destination, in the network header. All network devices on the route to the destination must be pre-configured with graph information that specifies the neighbors to which the packets may be forwarded. Details on specific mechanisms to compute these routes or select among them are left open, thus every company has to develop its own proprietary solution. On the other hand, Source Routing is intended to supplement Graph Routing for network diagnostics. With Source Routing, the source device includes in the header an ordered list of devices through which the packet must travel to reach a destination. Each routing device utilizes the next network device address in the list to determine the next hop. The routes are found on demand by broadcasting requests. In summary, WirelessHART and ISA100 propose without the use of Graph Routing for most industrial WSN communications, but leave the selection of objective functions that are used to build the graphs or select among the available routes open and at the discretion of the designer. Thus, neither standard provides comprehensive recommendations for QoS provisioning where routing is considered.

12.4.2   ZigBee

The Zigbee specification [52] also provides recommendations for routing and has been adopted for some industrial scenarios [19, 29]. Routing mechanisms in Zigbee are based on a simplified version of AODV and Tree Routing that do not comply with the requirement for multipath routing. It is unclear, as it has not been formally evaluated, whether ZigBee can achieve a similar performance to WirelessHART or ISA100 in industrial environments for different QoS requirements relating to industrial applications given the fact that it does not offer path redundancy - which can also mean that it can use the same path for every application and so does not provide QoS at the routing layer.

12.4.3   Proprietary Wireless Sensing in IWSN

Some companies producing wireless sensor solutions for industrial environments, such as OneWireless from Honeywell [21], are currently adopting the ISA 100.11a/WirelessHART standards which allow freedom of placement of sensor devices across the industrial plant, thanks to mesh connectivity and routing, with low power consumption. In addition, other companies are developing proprietary products, with for instance, ProSoft [25] providing devices for industrial sensing based on cellular communications or IEEE 802.11 standard, both power hungry technologies that require the availability of power lines for supplying energy to each device. Additionally, WISA from ABB [18], provides wireless devices for industrial sensing logically connected in clusters based on the IEEE 802.15.1 standard. WISA devices rely on the existence of a magnetic field, created by cable loops surrounding the working area, for power supply. In summary, common among solutions that do not consider standards such as ISA 100.11a for low power communications in industrial environments is that they provide wireless sensing based on power hungry devices, logically connected in cells or clusters and as such routing capabilities are not required. The advantage of these approaches is that they can potentially satisfy the most demanding QoS requirements due to the fact that these devices can remain awake all the time and they can utilize higher data rates, etc. Nevertheless, these devices cannot be placed freely as they rely on cables for power supply, and they have the inconvenience of imposing high costs for deployment and maintenance which makes them prohibitively expensive compared to low power wireless sensor network technology. Generally, this type of solution would only make sense when deploying a wireless network in a confined space to satisfy very stringent QoS requirements such as those of fast automation (i.e., maximum tolerated delays in the order of 10ms) where multihop topologies would not be recommended. Finally note that proprietary solutions lead to vendor lock-in as well as the need for the design and maintenance of a proprietary set of interfaces.

12.4.4 RPL

The IPv6 Routing Protocol for Low power and lossy networks (RPL) [49] is a routing protocol currently being developed by the IETF for communications between low power wireless embedded devices in lossy environments such as industrial wireless sensor networks. This routing protocol, similar to the protocols proposed by WirelessHART and ISA100, is based on Graph Routing. However, in this case a central manager is not needed since the graphs are calculated in a distributed fashion. With the purpose of building a directed acyclic graph (DAG) for inward traffic (i.e., traffic destined to the sink), the network sink node or gateway (i.e., the graph root) issues a control message that travels through the nodes in the network. Similarly, another control message is issued by the end devices to build DAGs for outward traffic (i.e., traffic destined to the end devices). The DAG construction is performed based on an objective function whose specification is left open. Following on from this, the IETF ROLL group is currently working on the definition of several routing metrics [45] that could be applied, together with a user defined objective function, to build DAGs. These metrics may be used by the routing protocol, in this case RPL, to satisfy different application QoS requirements and to adapt to different environments. Again the selection of these metrics and how they should be used in meeting specific requirements (i.e., in the form of an objective function) and, in some cases, how they should be calculated is left open. The metrics in question are defined in [45] and are:

1. Node State and Attribute (NSA): this metric is defined to provide information on node characteristics and may be used by the routing algorithm to create device aware routes. This metric includes a flag to define whether the device is in an *overload* state, this can also be used by the routing algorithm to avoid routing through nodes with that flag triggered.

2. Node Energy: this metric allows for the identification of a node as being mains powered, battery powered or as a scavenger (energy harvesting). In addition, it provides an estimation on the expected lifetime of a device. The information provided by this metric may be used to create energy aware routes.

3. Hop-Count: this metric refers to the number of traversed nodes along a path, i.e., the hop count. This information can be used for instance by the routing algorithm to limit the nodes per path.

4. Throughput: In low power networks, throughput may be highly variable depending on duty cycles. To optimize the overall network throughput, it may be desirable in some cases that nodes advertize their available or supported throughput. The throughput metric can be used by the routing algorithm to maximize throughput across paths.

5. Latency: the latency metric can be used by the routing algorithm

to measure the latency along a path and thus decide which paths meet application requirements.

6. Link Quality Level (LQL): this metric quantifies the link reliability as a discrete value and can be used by the routing algorithm to identify the most reliable paths or the paths that meet the application requirements in terms of reliability.

7. Expected Transmission Count (ETX): another possible metric with regards to reliability is the ETX that defines the number of transmissions a node expects to make in order to successfully deliver a packet to a destination. Note that this metric not only reflects the reliability of a path but also represents a measure of delay as the ETX also considers retransmissions (i.e., the higher the number of transmissions to reach a destination the longer it takes to reach it).

8. Link Color: this metric can be used to avoid or attract specific links for specific traffic types. It uses a 10 bit field where each bit can represent a different user defined rule/condition. For instance, one condition could relate to the use of a specific encryption method say for sensitive traffic. In such a case, the routing algorithm would be able to select nodes that run this specific encryption mechanism.

These metrics are still being defined and as such very much a work in progress. Nevertheless, this initial specification shows the importance of introducing a means to provide QoS at the routing layer in low power lossy networks, as routing decisions can severely impact on network lifetimes, latencies and throughput, etc.

12.4.5   Metrics for QoS Aware Routing

In order to provide Quality of Service aware routing, the routing algorithm in order to compute paths or select among different routes, must use some objective function based on different QoS metrics such as those described previously. Different metrics may be used to satisfy different application QoS requirements such as those identified in Sections 12.2.1 and 12.2.2. Moreover, different objective functions combining those metrics may be used to balance the conflicting requirements outlined in Section 12.3.2. This section focuses on analysing metrics and objective functions proposed by the research community for QoS aware routing in wireless sensor networks and their suitability for industrial environments.

12.4.5.1   Single Metric Routing

While it is possible to satisfy a single performance requirement typically using a single metric the focus here is one using multiple metrics that can be considered in a single objective functions or routing algorithm.

Equalizing energy consumption has been one of the primary concerns for routing algorithm design in wireless sensor networks. Several metrics and objective functions have been employed with this purpose in mind. For instance, a simple way to equalize energy consumption can be by selecting different routes on a round robin basis as proposed in [48]. Another possibility can be choosing the route based on the devices with the most available energy as proposed in [31] and [35]. The residual energy can be computed based on the currently available energy divided by the initial energy of the device. These metrics and similar ones can be readily used for industrial wireless sensor networks. It would also be interesting to consider, for energy balancing and network lifetime optimisation, the device type and characteristics, as suggested in [45], as in an industrial environment there may be line or battery powered devices as well as scavengers.

Delay has also been addressed in several works. Some example metrics used to find the paths that introduce the lowest delay include the hop count (i.e., shortest path) or time to next available transmission interval [42]. The shortest path metric is used to reduce delay by reducing the number of nodes that need to be traversed. On the other hand, the time to the next possible transmission is used to reduce waiting times between transmissions and thus reduce delay. Another metric considered is the distance between nodes, obtained for instance using global positioning systems [12], where the shorter the path between two nodes based on real distance infers a lower delay in transmit time. This metric however would not be suitable for indoor industrial scenarios unless the position is pre-loaded on devices (i.e., not obtained online) since obtaining accurate positioning in indoor industrial environments is at the moment extremely challenging and costly [30]. It has to be noted that the MAC layer has a key influence on delays, that is, a node may know accurately when it can access the channel (i.e., waiting delay) when it has been assigned some time slot for its transmissions (i.e., TDMA) or it may have to contend for the medium (i.e., CSMA) and then cannot accurately estimate the delay. Thus combining MAC layer scheduling such as TDMA with the routing algorithm, as proposed by WirelessHART and ISA100, is an efficient way to satisfy stricter delay requirements for some industrial applications.

Reliability has also been a major concern for routing in wireless sensor networks due to the instability of the wireless medium. Metrics for reliability include the previously introduced ETX employed in [10], with the Link Quality Indicator (LQI) and Received Signal Strength Indicator (RSSI) having also been used as metrics to compute the most reliable paths within the network [15]. Both metrics represent the local status of a link among two devices in terms of quality. On the other hand, buffer occupancy is another metric that has also been considered for reliable routing as information routed through heavily loaded nodes has an increased likelihood of being dropped [4]. It is usually good practice for a routing protocol to avoid varying paths too often so that instability in the network is prevented. Thus, metrics and objective functions that smooth variability would be preferable when designing routing algorithms for industrial wireless sensor networks.

Routing algorithms with load balancing based objective functions have also been developed for wireless sensor networks with the purpose of optimizing throughput and balancing energy consumption. The number of links per node and the required transmission power to reach those links is used to modify routing paths and equalize load as referred to in [36]. Moreover, the available energy per node can also be utilized to balance load [38]. Energy and load balancing are related, in order to equalize energy consumption among nodes they must receive and transmit similar amounts of information with transmitting and receiving operations being the biggest energy consumers. Furthermore, the routing algorithm can be combined with a bandwidth scheduling algorithm to ensure the availability of bandwidth for those applications requiring it either in a centralized manner, such as defined by WirelessHART and ISA100, or in a distributed fashion as in [5].

12.4.5.2 Multiple Metric Routing

More elaborated routing algorithms that aim to simultaneously consider several QoS requirements have also been proposed within the existing literature. For instance, researchers in [6] propose a route selection algorithm that considers energy (residual energy) and reliability (lost packets at buffer and wireless channel) for industrial environments with the objective of satisfying reliability requirements first and then equalizing energy consumption as a second objective. This work showed how conflicting requirements such as prolonged network lifetime and reliability may be balanced with a single objective function that combines different QoS metrics. On the other hand, researchers in [16] propose an objective function for routing in industrial wireless sensor networks that considers delay (propagation, queuing, and retransmission time), energy (residual energy) and reliability (successfully delivered packets), again with the objective of satisfying several performance metrics simultaneously. For wireless sensor networks in general, several objective functions combining different routing metrics are available in the literature [47, 2, 39]. Nevertheless, it is worth highlighting that some of these works do not take into consideration the inherent properties of wireless sensor networks such as duty cycling or energy and buffer limitations which would make them unsuitable candidates to satisfy application QoS requirements in real environments.

12.5 Conclusions

With ever increasing competition among industries for greater market share, companies are faced with the need to increase efficiencies and to comply with regulations within a financially prudent scope. As an answer to this, wireless sensor networks have been viewed by the industrial sector as being a dynamic retrofit low-cost solution that can be used to greatly increase the productivity and efficiency of industrial processes. To further push the growth of wireless sensor networks in the industrial marketplace confidence in their ability to deliver data reliably, on time and on an energy efficient manner must be underpinned by robust QoS provisioning. As a consequence of the diverse industrial application requirements and challenges imposed by constrained wireless sensor network devices, technical challenges still remain where QoS based routing is concerned. Among the currently available standards none provide an open and comprehensive solution for QoS based routing in industrial environments. QoS provisioning has been left open by industrial standardization bodies such as WirelessHART or ISA100, thus every company must adopt its own proprietary solution. Moreover, even though IETF is proposing several metrics that can be used to measure and provide QoS in low power lossy networks, the manner in which the routing algorithm should employ those metrics to match the application requirements is left open. For QoS aware routing, some proprietary solutions such as the ABB WISA specification target QoS provisioning by deploying networks in a star based fashion. This has the advantage of providing potentially lower delays depending on the selected scheduling but limits the deployment of the sensor network to the availability of line powered sinks in the range of every sensor node, which does not fully exploit the freedom of placement that wireless sensor nodes can provide. Few research efforts have addressed QoS aware routing within the industrial environment although a promising approach is presented in [6]. This may be due to the fact that it is difficult to provide a general solution given the conflicting requirements that exist in this type of network as identified in Section 12.3.2 as well as the broad range of requirements of industrial applications as detailed in Section 12.2. Moreover, specific QoS requirements for different application or traffic classes have not yet been defined (only broad classifications are available as detailed in Section 12.2) which puts additional constraints to the design of QoS aware communication protocols for industrial environments. Nevertheless, with the increasing use of the technology, boosted by the recent development of standards, specific requirements for different industrial applications should become standard. In such a scenario, the availability of formal definitions for specific application QoS requirements will allow designers to target explicit application classes, or indeed a broad range of categories, by including adaptive strategies in the design of protocols.

12.6   Glossary

CSMA: Carrier Sense Multiple Access

DAG: Directed Acyclic Graph

ETX: Expected Transmissions Number

IETF: Internet Engineering Task Force

ISA: International Society of Automation

IWSN: Industrial Wireless Sensor Network

LQI: Link Quality Indicator

MAC: Media Access Control

QoS: Quality of Service

ROLL: Routing over Low Power Lossy Networks

RPL: IPv6 Routing Protocol for Low power and Lossy Networks

RSSI: Received Signal Strength Indicator

TDMA: Time Division Multiple Access

WISA: Wireless Interface for Sensors and Actuators

WSN: Wireless Sensor Network

References

[1] ANSI/ISA95.00.01-2000. Enterprise Control System Integration, The Instrumentation, Systems and Automation Society, 2000.

[2] A. B. Bagula and K. G. Mazandu. Energy Constrained Multipath Routing in Wireless Sensor Networks. In *Proceedings of the 5th international conference on Ubiquitous Intelligence and Computing*, UIC '08, pages 453–467, Berlin, Heidelberg, 2008. Springer-Verlag.

[3] L. Bin, T.G. Habetler, R.G. Harley, J.A. Gutierrez, and D.B. Durocher. Energy Evaluation Goes Wireless. *Industry Applications Magazine, IEEE*, 13(2):17–23, 2007.

[4] B. Carballido Villaverde, S. Rea, and D. Pesch. Multi-objective Cross-Layer Algorithm for Routing over Wireless Sensor Networks. In *Sensor Technologies and Applications, 2009. SENSORCOMM '09. Third International Conference on*, pages 568–574, June 2009.

[5] B. Carballido Villaverde, S. Rea, and D. Pesch. Guaranteeing Reliable Communications in Mesh Beacon-Enabled IEEE802.15.4 WSN for Industrial Monitoring Applications. In *ADHOCNETS*, pages 359–370, 2010.

[6] B. Carballido Villaverde, S. Rea, and D. Pesch. InRout A QoS Aware Route Selection Algorithm for Industrial Wireless Sensor Networks. *Ad Hoc Networks*, 10(3):458–478, 2012.

[7] G. Cena, I.C. Bertolotti, A. Valenzano, and C. Zunino. Evaluation of Response Times in Industrial WLANs. *Industrial Informatics, IEEE Transactions on*, 3(3):191–201, 2007.

[8] G. Cena, I.C. Bertolotti, A. Valenzano, and C. Zunino. Industrial applications of IEEE 802.11e WLANs. In *Factory Communication Systems, 2008. WFCS 2008. IEEE International Workshop on*, pages 129–138, May 2008.

[9] F. Chen. Improving IEEE 802.15.4 for Low-latency Energy-efficient Industrial Applications. *in Echtzeit 2008 - Fachtagung der GI-Fachgruppe Echtzeitsysteme. Boppard, Germany: Springer*, 2008.

[10] S. Dawson Haggerty, A. Tavakoli, and D. Culler. Hydro: A Hybrid Routing Protocol for Low-Power and Lossy Networks. In *Smart Grid Communications (SmartGridComm), 2010 First IEEE International Conference on*, pages 268–273, Oct. 2010.

[11] C. Emmanouilidis, S. Katsikas, and C. Giordamlis. Wireless Condition Monitoring and Maintenance Management: A Review and a Novel Application Development Platform. *Proceedings of the 3rd World Congress on Engineering Asset Management and Intelligent Maintenance Systems Conference (WCEAM-IMS 2008)[Springer]*, pages pp. 2030–2041, 2008.

[12] E. Felemban, C. Lee, and E. Ekici. MMSPEED: Multipath Multi-SPEED Protocol for QoS Guarantee of Reliability and Timeliness in Wireless Sensor Networks. *Mobile Computing, IEEE Transactions on*, 5(6):738–754, 2006.

[13] A. Flammini, D. Marioli, E. Sisinni, A. Taroni, and M. Pezzotti. A Wireless Thermocouples Network for Temperature Control in Plastic Machinery. In *Factory Communication Systems, 2006 IEEE International Workshop on*, pages 219–222, 0-0 2006.

[14] N. Ghosh, Y.B. Ravi, A. Patra, S. Mukhopadhyay, S. Paul, A.R. Mohanty, and A.B. Chattopadhyay. Estimation of Tool Wear During CNC Milling Using Neural Network-Based Sensor Fusion. *Mechanical Systems and Signal Processing (v21, n1), pp. 466-479.*, 2007.

[15] C. Gomez, A. Boix, and J. Paradells. Impact of LQI-Based Routing Metrics on the Performance of a One-to-One Routing Protocol for IEEE 802.15.4 Multihop Networks. *EURASIP J. Wirel. Commun. Netw.*, 2010:6:1–6:20, February 2010.

[16] J. Heo, J. Hong, and Y. Cho. EARQ: Energy Aware Routing for Real-Time and Reliable Communication in Wireless Industrial Sensor Networks. *IEEE Transactions on Industrial Informatics*, 5(1):3–11, 2009.

[17] *http://wirelesshart.hartcomm.org/*. WirelessHART, HART Communication Foundation, [Last Accessed May 2012].

[18] *http://www.abb.com*. ABB WISA. [Last Accessed May 2012].

[19] *http://www.an solutions.de/applications/industrial_automation.html*. Adaptive Network Solutions. [Last Accessed May 2012].

[20] *http://www.fieldbus.org*. Foundation Fieldbus High Speed Ethernet. [Last Accessed May 2012].

[21] *http://www.honeywellprocess.com*. OneWireless Honeywell. [Last Accessed May 2012].

[22] *http://www.isa.org/filestore/ISASP100_11_CFP_14Jul06_Final.pdf*. ISA-SP100.11 Call for Proposals. Wireless for Industrial Process Measurement and Control., July 2006.

[23] *http://www.isa.org/isa100*. International Society of Automation SP100 Standards Committee, 2009.

[24] *http://www.profibus.com*. PROFINET System Description - Technology and Application. [Last Accessed May 2012].

[25] *http://www.prosoft technology.com*. RadioLinx ProSoft Technology. [Last Accessed May 2012].

[26] ISA100.11a 2009. Wireless Systems for Industrial Automation: Process Control and Related Applications, 2009.

[27] H.-J. Korber, H. Wattar, and G. Scholl. Modular Wireless Real-Time Sensor/Actuator Network for Factory Automation Applications. *Industrial Informatics, IEEE Transactions on*, 3(2):111–119, May 2007.

[28] K. Koumpis, L. Hanna, M. Andersson, and M. Johansson. Wireless Industrial Control and Monitoring Beyond Cable Replacement. *Profibus International Conference*, 2005.

[29] J. Lee, C. Chuang, and C. Shen. Applications of Short-Range Wireless Technologies to Industrial Automation: A ZigBee Approach. In *Telecommunications, 2009. AICT '09. Fifth Advanced International Conference on*, pages 15–20, May 2009.

[30] H. Liu, H. Darabi, P. Banerjee, and J. Liu. Survey of Wireless Indoor Positioning Techniques and Systems. *Systems, Man, and Cybernetics, Part C: Applications and Reviews, IEEE Transactions on*, 37(6):1067–1080, 2007.

[31] Y. Liu, L. Guo, H. Ma, and T. Jiang. Energy Efficient On-Demand Multipath Routing Protocol for Multi-Hop Ad Hoc Networks. In *Spread Spectrum Techniques and Applications, 2008. ISSSTA '08. IEEE 10th International Symposium on*, pages 592–597, 2008.

[32] C. Lizzi, L. Bacon, E. Becquet, and E. Gressier Soudan. Prototyping QoS based architecture for power plant control applications. In *Factory Communication Systems, 2000. Proceedings. 2000 IEEE International Workshop on*, pages 119–126, 2000.

[33] S. Luo, N. Polu, Z. Chen, and J. Slipp. RF channel modeling of a WSN testbed for industrial environment. In *Radio and Wireless Symposium (RWS), 2011 IEEE*, pages 375–378, Jan. 2011.

[34] A. Muller, A. Crespo Marquez, and B. Lung. On the Concept of e-Maintenance: Review and Current Research. volume 93, pages 1165–1187, 2008.

[35] C. Ni, H. Zin, B. Lee, D. Hwang, and C. Kim. Dynamic Packet Balancing Agent in MANETs Based on AOMDV. In *Proc. First Asian Conf. Intelligent Information and Database Systems ACIIDS 2009*, pages 362–367, 2009.

[36] P. H. Pathak and R. Dutta. Impact of Power Control on Relay Load Balancing in Wireless Sensor Networks. In *Wireless Communications and Networking Conference (WCNC), 2010 IEEE*, pages 1–6, April 2010.

[37] K. Pister, P. Thubert, S. Dwars, and T. Phinney. Industrial Routing Requirements in Low Power and Lossy Networks, RFC 5673, October 2009.

[38] A. Ranganathan and K.A. Berman. Dynamic State-Based Routing for Load Balancing and Efficient Data Gathering in Wireless Sensor Networks. In *Collaborative Technologies and Systems (CTS), 2010 International Symposium on*, pages 103–112, May 2010.

[39] A. Razzaque, M. M. Alam, M. Or Rashid, and C. Hong. Multi-Constrained QoS Geographic Routing for Heterogeneous Traffic in Sensor Networks. *IEICE Transactions*, 91-B(8):2589–2601, 2008.

[40] P. Sereiko and D. Caro. What Process Control Professionals Think About Wireless, Analysis of an ISA100 and Control Magazine Survey. ISA Expo. Houston, Texas, October 2007.

[41] V. Sudararajan, A. Redfern, M. Schneider, and P. Wright. Wireless Sensor Networks for Machinery Monitoring. *ASME International Mechanical Engineering Congress and Exposition*, 2005.

[42] G. Sun and B. Xu. Delay-Aware Routing in Low Duty-Cycle Wireless Sensor Networks. In *Wireless Communications Networking and Mobile Computing (WiCOM), 2010 6th International Conference on*, pages 1–4, Sept. 2010.

[43] V. Sundararajan, A. Redfern, W. Watts, and P. Wright. Distributed Monitoring of Steady-State System Performance Using Wireless Sensor Networks. *ASME Conference Proceedings*, 2004(47136):23–29, 2004.

[44] E. Tanghe, W. Joseph, L. Verloock, L. Martens, H. Capoen, K. Van Herwegen, and W. Vantomme. The industrial indoor channel: large-scale and temporal fading at 900, 2400, and 5200 MHz. *Wireless Communications, IEEE Transactions on*, 7(7):2740–2751, July 2008.

[45] JP. Vasseur, M. Kim, K. Pister, N. Dejean, and D. Barthel. Routing Metrics used for Path Calculation in Low Power and Lossy Networks. RFC 6551. 2012.

[46] K. Walzer, J. Rode, D. Wunsch, and M. Groch. Event-Driven Manufacturing: Unified Management of Primitive and Complex Events for Manufacturing Monitoring and Control. In *Factory Communication Systems, 2008. WFCS 2008. IEEE International Workshop on*, pages 383–391, May 2008.

[47] P. Wang and T. Wang. Adaptive Routing for Sensor Networks using Reinforcement Learning. In *Proc. Sixth IEEE Int. Conf. Computer and Information Technology CIT '06*, 2006.

[48] Y. Wang, H. Mao, C. Tsai, and C. Chuang. HMRP: Hierarchy-Based Multipath Routing Protocol for Wireless Sensor Networks. In Tomoya Enokido, Lu Yan, Bin Xiao, Daeyoung Kim, Yuanshun Dai, and Laurence Yang, editors, *Embedded and Ubiquitous Computing*, volume 3823 of *Lecture Notes in Computer Science*, pages 452–459. Springer Berlin / Heidelberg, 2005.

[49] T. Winter, P. Thubert, A. Brandt, T. Clausen, J. Hui, R. Kelsey, P. Levis, K. Pister, R. Struik, and JP. Vasseur. RPL: IPv6 Routing Protocol for Low Power and Lossy Networks. RFC 6550. 2012.

[50] P. Wright, D. Dornfeld, and N. Ota. Condition Monitoring in End-Milling Using Wireless Sensor Networks (WSNs). *Transactions of NAMRI/SME*, 36:177183, 2008.

[51] P. K. Wright, D. A. Dornfeld, R. G. Hillaire, and N. K. Ota. A Wireless Sensor for Tool Temperature Measurement and its Integration within a Manufacturing System. *Trans. North American Manufacturing Research Institute*, 34, 2006.

[52] ZigBee 053474r17. ZibBee Alliance Specification-Document. 2008.

13

Reliable and Robust Communications in Industrial Wireless Sensor Networks

Sasan Khoshroo

University of Western Ontario, Canada

Honggang Wang

University of Massachusetts, USA

Yalin Wang

PCOC, China

CONTENTS

13.1 Introduction

Wireless Sensor Networks (WSNs) have been utilized in various types of Industrial applications recently. However, reliability and data quality (distortion reduction) of the sensing data transmissions are still among the most important research challenges in WSNs. In addition, energy consumption is one of the most important constraints in such networks. Cross-layer design has been shown to be effective solution to address these challenges. Therefore, in this chapter, a cross-layer design of Resource Allocation and Channel Coding is introduced to protect Distributed Source Coding (DSC) based data transmission in a wireless sensor network environment. Resource allocations consider the rate adaptation and Automatic Repeat-reQuest (ARQ) retransmissions. The joint design of resource allocation, channel coding and DSC not only improves the reliability of the network, but also can improve the network energy efficiency and information quality under strict latency requirements. Simulation studies show that the described joint design significantly improves the DSC based data transmission quality, reliability, and the network energy efficiency. Finally, selective channel coding is introduced. It is demonstrated that, applying this technique can improve the network performance, even more.

One of the recent applications of wireless sensor networks (WSNs) is the enhanced perception of the surrounding environment in the form of audio feeds, still images, and streaming video. The nodes used in a WSNs environment are equipped with capacity-limited batteries and recharging these batteries is difficult or even impossible due to the inaccessibility to the device. Also, the functionality of sensor devices and thus, the entire network typically depends on the battery life. Thus, optimizing the energy efficiency is critical for improving the lifetime of sensor nodes, and thus assuring the reliability of wireless systems [22]. Algorithms and protocols such as MAC protocols, routing protocols and signal processing algorithms have been hugely under studies to be optimized for such resource-limited applications. See [32, 17, 8, 20, 14, 4, 18] for recent studies on energy efficiency improvement in wireless sensor networks. Also see [2, 15, 6, 1, 19, 3] for reliability analysis in such networks.

The high correlation among the data collected by the adjacent sensor nodes is one of the unique features of application-driven WSNs [22]. There are two types of correlations among the data being sensed: Spatial correlation and Temporal correlation, which mean that the data under studies does not change drastically over a small region or over a small period of time, respectively. Based on this characteristic of WSNs, Distributed Source Coding (DSC) technique has been proposed in both space and time domain in order to improve the energy efficiency of the network [31, 7].

Compared with traditional, one-to-many source coding schemes in which the transmitting nodes must undertake a complicated encoding task, DSC's encoders can be designed as simply and efficiently as possible while the decoder is at the base station, which is connected to an unlimited power source, performs the joint decoding [31]. DSC works in a way that, the information bits are collected independently, at different data rates and under a certain delay requirement, from source sensors to reach the sink node for information decoding. Therefore, one of the most important characteristics of DSC in sensor networks is the requirement of *Unequal Error Protection (UEP)* and multirate transmission over the network [27]. DSC can compress multiple correlated signals from sensors that do not exchange data with each other (hence distributed coding).

Using DSC can greatly improve the energy efficiency of the network because the burden of a complicated encoding task will be shifted from the encoder side to the decoder side. Slepian and Wolf [26] proved that separate encoding of correlated sources in DSC is as efficient as joint encoding in traditional source coding for lossless compression. The joint entropy of two correlated discrete random variables X and Y, $R = H(X, Y)$, is proved to be sufficient even for the case of separate encoding of correlated sources. Similar results were obtained later by Wyner and Ziv [28, 29] with regards to lossy coding of joint Gaussian sources.

Although using DSC has the potential of reducing the network energy consumption by reducing the network data redundancies and avoiding unnecessary data transmission from the source, for DSC to be widely utilized in many wireless sensor applications, one should carefully develop and design a DSC-based approach that can work both reliably and efficiently in the resource-limited wireless networks. In such networks, the data transmitted over wireless channels could get corrupted due to the noise, fading, and other impairments. Techniques such as adaptive power and modulation control, Automatic Repeat reQuest (ARQ) control, and channel coding, have been used to improve the integrity and reliability of the transmitted data. Applying these techniques alone cannot improve the network performance and efficiency significantly. Thus, many studies have been carried out on the joint design of DSC and channel coding. But such joint design has limitations in dealing with the power issues because it lacks direct power control and efficient bandwidth allocation, which are addressed using Resource allocation [9, 11]. An appropriate resource allocation will have a significant impact on both performance and the energy consumption. In [27], the authors propose a joint design of DSC and resource allocation and develop an algorithm to optimize the information efficiency, defined as the information quality per unit of energy consumption at the sink node. The continuation of this work is [12] in which a complete joint design of DSC, resource allocation, and channel coding is proposed and shown to have a great effect on the network performance and efficiency.

In this chapter, we first analyze the DSC technique and study the relationship between source coding rates and the source nodes correlation. Then, we cover resource allocation techniques and show how two important control variables, transmission rate and retry limit, trade off packet error rate and energy-latency performance to provide unequal error protection (UEP) among sensors. After that, the channel coding technique is investigated as a promising technique to improve the overall network reliability and robustness. We also show how to formulate a cross-layer information quality optimization problem with the energy efficiency and delay constraints, and describe how it can be simplified and solved efficiently. Based on the simulation results in [12], we introduce selective channel coding already proposed in [13] to further improve the information quality and the network reliability.

13.2   DSC Information Quality and Resource Allocation

13.2.1   Definition of Information Quality with Multirate DSC Compression Dependency

Let A and B be two correlated sources which transmit their data as two equiprobable binary streams of length 3, A_0 and B_0, i.e., $A_0, B_0 \in \{0,1\}^3$, respectively. Then the entropy of both sources is 3 bits, i.e., $H(A_0) = H(B_0) = 3b$. We assume that the correlation between the data being collected by these two sources is such that A_0 and B_0 are different at most by one bit. Therefore, the Hamming distance between A_0 and B_0 is less than or equal to 1 (based on the correlation information). Thus, for a given B_0, there are four equiprobable choices of A_0. For instance, given $B_0 = 001$, $A_0 \in \{000, 011, 101, 001\}$, the conditional entropy of A_0 given B_0 is 2 bits, i.e., $H(A_0 \mid B_0) = 2b$. Using traditional source coding scheme in which the encoding task is done on the encoder side jointly, 3 bits are needed to transmit B_0 (which is considered as side information) and 2 bits are essential to index the four possible choices of A_0 associated with B_0. Therefore, instead of sending 6 bits, the joint encoder transmits only 5 bits without losing accuracy at the decoder. On the other hand, using DSC scheme, if the set of all possible outcomes of A_0 is first partitioned into four cosets W_{00}, W_{01}, W_{10}, and W_{11} each containing two possible outcome with Hamming distance 3, i.e., $W_{00} = \{000, 111\}$, $W_{01} = \{001, 110\}$, $W_{10} = \{010, 101\}$, and $W_{11} = \{011, 100\}$, 2 bits will still suffice to convey A_0. For example, if $A_0 = 000$ and $B_0 = 001$, the encoder will transmit 00, conveying that A_0 belongs to the coset W_{00}. At the receiver, knowing the value of B_0, the decoder will easily decode A_0 as 000, because 000 is the one closer to the value of Y, i.e., $d(000, 001) = 1$, $d(111, 001) = 2$. Thus, we observe that, using distributed source coding scheme in which the side information B_0 is known only at the decoder and not at the encoder, it is still possible to send only 2 bits instead of 3 bits to represent A_0. Thus, the Slepian-Wolf limit of $H(A_0, B_0) = H(B_0) + H(A_0 \mid B_0) = 5$ bits can be achieved with the lossless decoding.

In [27, 12] the above modulo coset based DSC compression scheme proposed in [5] is used. Equation (1) in [12] shows the relationship between the absolute value of the distance between samples from two sensor nodes and the source coding rate. It can be seen, as explained earlier, that the larger the correlation between two correlated sources, the shorter the distance between sample values from them, thus fewer bits need to be used to represent the source samples at the encoder. These source bits can be successfully decoded with the side information available at the decoder [27]. Since the compressed DSC packets are decoded using their side information, the performance of DSC relies deeply on the successful decoding of the side information packets, thus introducing a tradeoff between compression and data quality over error prone wireless channels. Unequal error protection addresses this tradeoff by *protecting* the important side information packets against noises and other impairments on the one hand, and *not protecting* or *protecting partially* the less important packets, on the other hand.

For a decodable DSC data packet sent to the sink node, all the packets from its ancestors must also be successfully delivered and decoded. Thus, the expected total information quality is defined as [27, 12]

$$E[IQ] = \sum_{i=1}^{N} IQ_{s,i} (1 - p_i) \prod_{k \in S_i} (1 - p_k) \qquad (13.1)$$

where N represents the total number of source coding nodes, $IQ_{s,i}$ indicates the information quality of the DSC sample from node i, and p_i denotes the packet loss probability of each packet which is directly related to the source coding rate of the transmitting sensor node and the Bit Error Rate, P_b, in the form of $p_i = 1 - (1 - P_b)^{L_i}$. P_b is directly related to the transmission rate r and the modulation scheme. On the other hand, the source coding rate, L_i, is directly related to the DSC parent selection. The parent (side information) node of sensor node i is represented by ψ_i. Also, S_i is defined as the set of all the parent and ancestor nodes of node i. For example, if node X selects node Y as its parent, i.e., $\psi_X = Y$, and node Y selects node Z as its parent, i.e., $\psi_Y = Z$, then we have $S_X = \{Y, Z\}$. The defined information quality provides a quantitative metric to evaluate the system performance and will be used to form the quality optimization problem.

13.2.2 Resource Allocation for Multirate Wireless Transmissions

The resource allocation technique is utilized to optimize the energy consumption while maintaining data integrity and meeting total delay constraints by implementing an optimization of the transmission rates, ARQ and other transmission properties. Usually high transmission rate leads to a denser modulation encoding, resulting in a higher bit error probability. Thus, there is a tradeoff between data rates and bit error rate (BER). The desirable BER can be presented as a function of SNR (Signal to Noise Ratio) per bit [30, 24, 10]. For example, the BER for M-PSK modulation scheme over an AWGN channel is expressed as follows [21]

$$P_b = \frac{2}{log_2 M} Q \left(\sqrt{2 log_2 M \frac{E_b}{N_0}} \, sin \frac{\pi}{M} \right) \qquad (13.2)$$

where (E_b/N_0) is the energy per bit to noise power density ratio which represents the channel SNR, and $log_2 M = b$ is the number of bits per modulation symbol. The above equation is a good approximation of BER for large SNR and for all values of M. In addition to physical layer transmission rate, the link layer retry limit is another important resource adaptation factor. The average packet error rate with consideration of retransmissions, assuming that

Automatic Repeat reQuest (ARQ) technique is applied to the link layer transmission for packet loss recovery, can be simply approximated as [23]

$$\bar{p} = p^m \tag{13.3}$$

where m is the maximum number of retransmissions.

We can see that different values of the number of bits per modulation symbol (b) and the maximum number of retransmissions (m) result in different packet error rates. Under certain channel conditions, there is an optimal combination of b and m that can achieve the minimum packet error rate. It is also worth noting that, the lower transmission rate and higher retry limit result in the highest packet delivery ratio. On the other hand, considerable energy saving can be achieved by dynamically adjusting transmission rate and retry limit. In fact, a higher transmission rate reduces the active time of the radio transceivers, resulting in better energy efficiency and lower latency performance. Furthermore, with the help of retransmissions at the link layer, the total energy consumption and latency can be approximated as [23]

$$\bar{E}_i = \bar{m} \times E_i \ , \ \bar{T}_i = \bar{m} \times T_i \tag{13.4}$$

where $\bar{m}$ is the average number of retransmissions and is given in [27] and E_i and T_i are the energy consumption and latency of the source node i, calculated in [27].

In summary, the resource allocation provides two major important control variables, transmission rate and retry limit, to trade off packet error rate and energy-latency performance and to achieve UEP among DSC sensors. In fact, the lowest transmission rate and highest number of retransmissions can be assigned to the node which transmits the side information and a set of other control parameters can be assigned to other nodes to optimize the information efficiency (quality) while meeting the transmission latency requirements.

13.3 Channel Coding

The term *reliability* in transmission systems and computer networks usually refers to the probability that a transmitted packet will be received at the destination with no error, either at any required particular instant or for a certain length of time. Channel coding is a practical technique to protect the message against channel noises, by adding redundancy to the *information sequence*, thus increasing the reliability and information rate transmitted through the channel. In fact, channel coding gives the receiver the ability of error detection and error correction. There are two types of channel codes [21], *block codes* and *convolutional codes*. The information sequence is represented by a binary sequence of length k, and is mapped to another binary sequence of length n, called the *codeword* $(n > k)$, resulting in a (n, k) code with rate $R_c = k/n$. The generated codeword in block codes depends only on the current k-bit information sequence, therefore they are memoryless codes. Examples of block codes are Repetition codes, Hamming codes, Maximum-length codes, Reed-Muller codes, and Hadamard codes. One important class of block codes, in which the decoding computational complexity is reduced, is the class of cyclic codes. The important coding schemes of BCH and Reed-Solomon belong to this class of block codes. Depending on the redundancy levels and Hamming distance between the codewords, these codes have different error correction and error detection properties. Let d_{min} denote the Hamming distance between the codewords. The error correcting capability, t, defined as the maximum number of guaranteed correctable errors per codeword is given by [25]

$$t = \left\lfloor \frac{d_{min} - 1}{2} \right\rfloor \qquad (13.5)$$

where $\lfloor a \rfloor$ gives the largest integer that does not exceed a. Error detecting capability, on the other hand is given by

$$e = d_{min} - 1 \qquad (13.6)$$

In this chapter we focus only on Hamming and Reed-Solomon codes.

13.3.1 Hamming Codes

A Hamming code is a kind of linear block codes with parameters $n = 2^m - 1$ (codeword length) and $k = 2^m - m - 1$ (information sequence length), for $m \geq 3$. The Hamming distance of these codes is 3, independent of the value of m. Therefore, the error correcting and detecting capability of Hamming codes are $t = 1$ and $e = 2$, respectively, meaning that they can detect two or fewer errors and correct single errors within a block. Assuming hard decision decoding, the bit error probability can be calculated as [25]

$$P_b \approx \left(\frac{1}{n} \right) \sum_{j=2}^{n} j \binom{n}{j} p^j (1-p)^{(n-j)} \qquad (13.7)$$

where n is the codeword length and p is the channel symbol error probability which depends on the modulation scheme and channel properties, given by (13.2) for M-PSK modulation over an AWGN channel.

13.3.2 Reed-Solomon (RS) Codes

Reed-Solomon (RS) is a kind of cyclic block codes which has found a wide variety of applications such as communication and data storage systems. RS codes can achieve a very large Hamming distance, which enables the decoder to correct the entire symbol even if all the bits within a symbol are corrupted.

This is why they are commonly used in wireless communications where burst noise is a common phenomenon because of multi-path fading. Reed-Solomon codes are 2^m-ary (n, k) BCH codes with the Hamming distance of $d_{min} = n - k + 1 = 2t + 1$, where $n = 2^m - 1$, $k = 2^m - 2t - 1$, and $1 \leq t \leq 2^{m-1} - 1$ is the maximum number of symbol errors that can be corrected. For example, a 31-error correcting RS (127,65) is a Reed-Solomon code in which $n = 127$, $k = 65$, $m = 7$, $t = 31$, and $d_{min} = 63$. Assuming hard decision decoding, the RS bit error probability in terms of the channel symbol error probability, p, is approximated as [12]

$$P_b \approx \left(\frac{1}{2^m - 1} \right) \sum_{j=t+1}^{2^m - 1} j \binom{2^m - 1}{j} p^j (1 - p)^{(2^m - 1 - j)} \qquad (13.8)$$

Figure 2 in [12] shows the bit error probability of M-PSK modulation over an AWGN channel with respect to SNR per bit, for different values of M and for three different scenarios: 1. without using channel coding, 2. using Hamming (7,4), and 3. using RS (31,16) with $t = 7$. It is shown that for a specific channel condition (SNR per bit), the bit error performance gets worse as M increases. This is due to the fact that for a larger M, the Euclidean distance between the modulation symbols decreases and the probability that the receiver estimates a wrong codeword increases. Also, it can be seen that, utilizing channel coding improves the bit error performance significantly, as predicted. Finally, we note that RS (31,16) demonstrates a better performance than Hamming (7,4). This result is also quite predictable because the hamming distance of the Hamming code (7,4) is 3, but that of the RS(31,16) is 15, giving more power to the decoder to detect and correct the errors.

13.3.3 Cross-Layer Design

As mentioned earlier, using the channel coding technique can reduce the packet error rate significantly. This technique, along with already discussed variables (the transmission rate and link layer retransmissions) helps to control the packet error rate. On the other hand, in channel coding, the information redundancy increases the length of the source coding packets. As mentioned above, increasing the packets length results in more energy consumption and delay. Therefore, it is possible that using channel coding, the total energy consumption and delay increases. However, this is not true because the total energy consumption and delay depend also on the transmission rate. Because the packet error rate reduces significantly with the help of channel coding, higher transmission rates can be allocated to the transmitters, thus reducing the overall energy consumption and latency and increasing the information efficiency at the decoder end. Furthermore, the channel coding can improve the energy efficiency by reducing the number of retransmissions.

13.4 Information Efficiency Optimization Problem Formulation

The problem is to find the optimal parent node (DSC data dependency), ψ_i, and transmission rate, r_i, as well as retransmission limit, m_i, for each DSC sensor node, which maximize the information efficiency which is defined as the correctly decoded DSC information bits (data quality gain) per unit of energy consumption, subject to a total delay constraint. The overall optimization problem can be formulated as follows [27]

$$\{\psi_i, r_i, m_i\} = argmax \left(\frac{E[IQ]}{\sum_i E_i} \right) \qquad (13.9)$$

subject to

$$\sum_i T_i \leq T_{max} \qquad (13.10)$$

where $E[IQ]$ is the expected total information quality and $(E[IQ]/\Sigma_i E_i)$ is defined as information efficiency at the receiver end. In this joint design, the resource allocation strategies work with the DSC and channel coding to improve the overall network performance, in terms of information efficiency. In fact, an ideal UEP scenario can be achieved by allocating lower rate and more retransmissions to side information packets and higher rate and less number of retransmissions to those less important packets. The variable source coding rates at each node due to DSC require the efficient support at the lower link and physical layers.

It is worth mentioning that, the full design of DSC, channel coding and resource allocation is a cross-layer design and like other cross-layer designs in wireless sensor networks, there is a tradeoff between the design complexity, system compatibility and adaptability, and performance enhancement [16]. The complexity related to the proposed resource optimization and control is not ignorable in resource-constrained devices. It is important to reduce the size of control variable space (transmission rate and maximum number of retransmissions) without greatly degrading the optimization performance. The approach described in [12] is basically initializing all the factors to the most robust ones, i.e., minimum transmission rate and maximum number of retry limits, calculating the information efficiency and total delay based on these values and recording the results, starting to increase the transmission rate and decrease the retry limits, recalculating the information efficiency and total delay and comparing the new results with older ones and finally, choosing the set of control variables which result in the maximum information efficiency while satisfying the delay constraints. The readers can be referred to [12] for further details on solving the optimization problem.

13.5   Cross-Layer Design Performance

Extensive simulation results are shown in figures 3-9 in [12] on the performance analysis of the described joint design of DSC, resource allocation and channel coding, for different scenarios, using two channel codes, Hamming (7,4) and RS(31,16). The summary of the simulation results includes: 1. When the channel condition is not so good ($SNR = 5\ dB$), applying the channel coding technique improves the information efficiency of the network significantly (For some values of the maximum latency budget this improvement reaches 300%). But, when the channel condition improves ($SNR = 15\ dB$), the information efficiency of both coded schemes is approximately equal to that of the uncoded scheme and the effect of applying the channel coding technique will become as low as 11%. This is because using channel coding does not improve the packet error rate significantly in good channel conditions, and on the other hand, it affects the energy consumption and delay by increasing the packet length. The information efficiency with respect to the channel condition (in SNR per bit) is shown in Fig. 7 in [12]. It is observed that, up to $SNR \approx 8\ dB$, using channel coding improves the information efficiency, but for SNR values more than $8\ dB$, fairly no improvement is achieved by utilizing channel coding. This introduces the idea of selective channel coding proposed in [13] and described in the next subsection. The information efficiency with respect to the number of nodes in the network is also shown in figures 8 and 9 in [12]. It is observed that, information efficiency decreases with the increase in the number of nodes. This is because of the increased traffic amount and less error resilient DSC coding dependency due to a longer data gathering chain.

13.5.1   Selective Channel Coding

The idea of selective channel coding is to encode the data packets only when the current channel condition is bad, i.e., low SNR. Therefore, when the channel SNR is high, the source node sends the packet without applying the channel coding technique. A threshold value for SNR should also be declared, based on which the sensor node can decide which value of SNR is low and which one is high. Note that, the SNR can be monitored at the receiver and the sender can estimate the SNR through acknowledgement packets from the receiver.

Using this technique, the optimization problem can be formulated as [13]

$$\{\psi_i, r_i, m_i, \alpha_i\} = argmax \left(\frac{E\,[IQ]}{\sum_i E_i} \right) \tag{13.11}$$

subject to

$$\sum_i T_i \leq T_{max} \tag{13.12}$$

where α_i is the number of selected data blocks for channel coding, transmitted by node i. The goal of this optimization problem is to find the number of encoded data blocks related to each node, along with other parameters discussed above, to maximize the information efficiency, while meeting the delay requirements. Solving this optimization problem is similar to the previous one. The performance of this scheme is also investigated in [13] through extensive simulations, using the SNR threshold value of 8 dB. In fact, the effect of selective channel coding is analyzed by comparing its performance with that of the *full* channel coding (Figures 3 and 4 in [13]). It is shown that, using selective channel coding further improves the network performance in terms of information efficiency.

13.6 Conclusions

Industrial WSNs require reliable and robust communications to support many applications. In this chapter, we introduced a cross-layer solution that jointly designs channel coding, resource allocation with DSC to improve the performance of industrial WSNs. The cross-layer solution not only improves the reliability of data transmissions, but also can improve the energy efficiency while meeting strict latency requirements. Compared with the existing solutions, the approach described in this chapter is optimal. Further, we discussed a novel selective channel coding scheme which can choose either appropriate channel coding or resource allocation based on the various channel conditions. The simulation studies show the effectiveness of the proposed cross-layer design.

References

[1] Al-Habashneh, A. Y., Ahmed, M. H., and Husain, T. Reliability analysis of Wireless Sensor Networks for forest fire detection. In *IWCMC*, pages 1630–1635, 2011.

[2] Balouchestani, M., Raahemifar, K., and Krishnan, S. Increasing the reliability of wireless sensor network with a new testing approach based on compressed sensing theory. In *Proceedings of the 8th International Conference on Wireless and Optical Communications Networks, WOCN 2011, Paris, France, 24-26 May 2011*, pages 1–4. IEEE, 2011.

[3] Campobello, G., Leonardi, A., and Palazzo, S. Improving energy saving and reliability in wireless sensor networks using a simple CRT-based

packet-forwarding solution. *IEEE/ACM Trans. Netw.*, 20(1):191–205, February 2012.

[4] Canthadai, A. M., Radhakrishnan, S., and Sarangan, V. Multi-Radio Wireless Sensor Networks: Energy Efficient Solutions for Radio Activation. In *GLOBECOM*, pages 1–5. IEEE, 2010.

[5] Chou, J., Petrovic, D., and Ramchandran, K. A distributed and adaptive signal processing approach to reducing energy consumption in sensor networks. In *Proceedings INFOCOM'03*, pages 1054–1062, San Francisco, CA, USA, 2003.

[6] Doohan, N. V., Mishra, D. K., and Tokekar, S. Reliability Analysis for Wireless Sensor Networks Considering Environmental Parameters Using MATLAB. In *CICSyN*, pages 99–102, 2011.

[7] Draper, S., and Wornell, G. Side information aware coding strategies for sensor networks. *IEEE Journal on Selected Areas in Communications*, 22(6):966–976, Aug. 2004.

[8] Fang, W., Liu, F., Yang, F., Shu, L., and Nishio, S. Energy-efficient cooperative communication for data transmission in wireless sensor networks. *IEEE Transactions on Consumer Electronics*, 56(4):2185–2192, 2010.

[9] Han, Z., and Liu, K. J. *Resource Allocation for Wireless Networks: Basics, Techniques, and Applications*. Cambridge University Press, 2008.

[10] Haykin, S. *Communication System*. Wiley, New York, USA, 3rd edition, 1994.

[11] Huang, J., Han, Z., Chiang, M., and Poor, H. V. Auction-Based Resource Allocation for Cooperative Communications. *IEEE Journal on Selected Areas on Communications, Special Issue on Game Theory*, 26(7):1226–1238, Sep. 2008.

[12] Khoshroo, S., Wang, H., Xing, L., and Kasilingam, D. A joint resource allocation-channel coding design based on distributed source coding. *Wireless communications and mobile computing*, Feb. 2012. Major Revision.

[13] Khoshroo, S., Wang, H., Xing, L., and Kasilingam, D. Selective Channel Coding for Resource-aware DSC Based Applications. In *The 4th International Conference on Communications, Mobility, and Computing*, Guilin, China, May. 2012.

[14] Kuntz, R., Gallais, A., and Nol, T. Auto-adaptive MAC for energy-efficient burst transmissions in wireless sensor networks. In *WCNC*, pages 233–238. IEEE, 2011.

[15] Lee, E., Park, S., Park, H., and Kim, S.-H. Quality-Based Event Reliability Protocol in Wireless Sensor Networks. *IEICE Transactions*, 94-B(1):293–296, 2011.

[16] Li, S., and Ramamoorthy, A. Rate and power allocation under the pairwise distributed source coding constraint. *IEEE Transactions on Communications*, 57(12):3771–3781, Dec. 2009.

[17] Li, W., Bandai, M., and Watanabe, T. Tradeoffs among Delay, Energy and Accuracy of Partial Data Aggregation in Wireless Sensor Networks. In *24th IEEE International Conference on Advanced Information Networking and Applications (AINA)*, pages 917–924, Japan, Apr. 2010.

[18] Padmanabhan, K., and Kamalakkannan, P. Energy-Efficient Dynamic Clustering Protocol for Wireless Sensor Networks. *International Journal of Computer Applications*, 38(11):35–39, January 2012. Published by Foundation of Computer Science, New York, USA.

[19] Pai, H.-T. Reliability-Based Adaptive Distributed Classification in Wireless Sensor Networks. *IEEE Transactions on Vehicular Technology*, 59(9):4543–4552, 2010.

[20] Pensas, H., Valtonen, M., and Vanhala, J. Wireless Sensor Networks Energy Optimization Using User Location Information in Smart Homes. In *BWCCA*, pages 351–356. IEEE, 2011.

[21] Proakis, J. G., and Salehi, M. *Digital Communications*. McGraw-Hill, 5th edition, 2007.

[22] Ramamoorthy, A. Minimum Cost Distributed Source Coding Over a Network. *IEEE Transactions on Information Theory*, 57(1):461–475, Jan. 2011.

[23] Schaar, M., and Turaga, D. Cross-layer packetization and retransmission strategies for delay-sensitive wireless multimedia transmission. *IEEE Transactions on Multimedia*, 9(1):185–197, Jan. 2007.

[24] Schurgers, C., and Aberthorne, O., and Srivastava, M. Modulation scaling for energy aware communication systems. In *International Symposium on Low Power Electronics and Design 2001*, pages 96–99, Aug. 2001.

[25] Sklar, B. *Digital Communications: Fundamentals and Applications*. Prentice Hall PTR, New Jersey, 2nd edition, 2001.

[26] Slepian, D., and Wolf, J. K. Noiseless coding of correlated information sources. *IEEE Transactions on Information Theory*, 19(4):471–480, Jul. 1973.

[27] Wang, W., Peng, D., Wang, H., Sharif, H., and Chen, H. H. Cross-layer multirate interaction with Distributed Source Coding in wireless sensor networks. *IEEE Transactions on Wireless Communications*, 8(2):787–795, Feb. 2009.

[28] Wyner, A. D. On source coding with side information at the decoder. *IEEE Transactions on Information Theory*, 21(3):294–300, May 1975.

[29] Wyner, A. D., and Ziv, J. The rate-distortion function for source coding with side information at the decoder. *IEEE Transactions on Information Theory*, 22(1):1–10, Jan. 1976.

[30] Xiao, J., Cui, S., Luo, Z., and Goldsmith, A. Power scheduling of universal decentralized estimation in sensor networks. *IEEE Transactions on Signal Processing*, 54(2):413–422, Feb. 2006.

[31] Xiong, Z., Liveris, A., and Cheng, S. Distributed source coding for sensor networks. *IEEE Signal Processing Magazine*, 21(5):80–94, Sep. 2004.

[32] Zhang, P., Xiao, G., and Tan, H.-P. A preliminary study on lifetime maximization in clustered wireless sensor networks with energy harvesting nodes. In *8th International Conference on Information, Communications and Signal Processing (ICICS)*, pages 1–5, Singapore, Dec. 2011.

14

Network Security in Industrial Wireless Sensor Networks

Nouha Oualha

CEA, LIST, Communications Systems Laboratory, France

CONTENTS

14.1   Introduction

Wireless sensor networks (WSNs) consist of a large number of autonomous devices that are spatially distributed to monitor physical or environmental conditions. WSNs can be deployed anywhere and do not require heavy infrastructure or wires. Such flexibility and inexpensive deployment motivate their wide adoption, notably by industrial applications. WSNs are then used to monitor and control industrial processes, assets, and physical environments, as well as to track mobile workers and monitor their health.

In industries, the vertical integration is in the path to be replaced by more open and flexible structures, in which integration with other networks (e.g., Internet) and interoperability between different federated domains (with the all-IP paradigm [15]) become standard practice in industrial applications. In this context, security is of greater concern, when integrating WSNs into industrial applications; however, performance and usability aspects are still to be taken with great attention when designing security solutions for WSNs.

From the industrial and research communities perspective, designing security techniques for WSNs has drawn strong attention. Efforts in the research literature and a number of standardization bodies (e.g., IETF) have focused on security techniques that aim at securing the operation of the WSN at the network level. With this respect, this chapter presents a comprehensive state of the art of the network security protocols and standards proposed for industrial WSNs that aim at securing the operation of the sensor node, as the central element of a WSN, at different points of its lifecycle. Specifically, it focuses on authentication, network access control, key management, identity management (privacy protection), and security maintenance mechanisms.

14.2   Industrial Wireless Sensor Networks

The introduction of WSNs in the industry has been motivated by the reduced cost, complexity, and time of their deployment and operation. Besides this financial incentive, other aspects like safety, worker health, and environment protection benefit from their introduction. The research efforts around industrial WSNs are still active and growing. Nevertheless, real deployment of WSNs within industrial applications has been already enabled in practice. The following section first describes the security operations required during the lifecycle of a sensor, and then it gives an example of an industrial WSN: the smart metering approach [45].

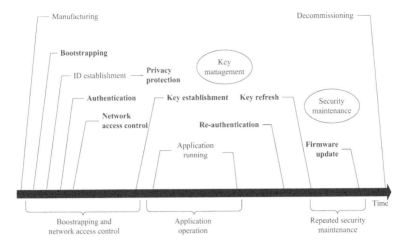

FIGURE 14.1
Main security operations (highlighted) during the lifecycle of a sensor.

14.2.1 Lifecycle of a Sensor

The network security operations that each sensor node in the WSN should accomplish are the set of mechanisms applied at the network layer to provide trusted operation of the WSN. Mechanisms range from routing and network protections to link-layer protections. The network security operations are not dependent on the sensor functionalities (e.g., temperature, humidity, position, pressure, vibration sensors), but on the features that these sensors have (e.g., mobility patterns, density, privacy protection). These security operations are required at different points during the lifetime of a sensor [13] (Figure 14.1).

14.2.2 Wireless Sensor Networks for Smart Energy Supply and Demand Optimization

The smart metering approach consists in deploying sensors (i.e., smart meters) within households that communicate to an energy suppliers control system (refer to Figure 14.2). Sensors report information about customer energy consumption to the control system. On the other hand, they also provide information to customers about the quantity of energy they are using, the overall electrical demand in their region, and the spot price of electricity that will impact their bill.

The approach calls for several security services. First of all, customer equipment (i.e., the smart meter) is remote-control enabled. Therefore, it is prone to remote attacks, in particular to DoS (Denial of Service) attacks. Therefore, authentication of the provenance of commands to customer equipment is another strong requirement.

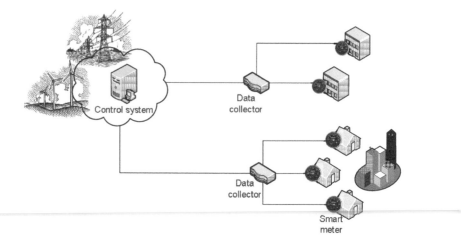

FIGURE 14.2
The smart metering approach.

In the electrical industry, lifetime of equipment is estimated to be 30 years. Customer devices lifetimes are meant to approach this threshold. But due to the rapid evolution of the ICT industry in recent years, it is not feasible. It presents major issues not only on the reliability of equipment but also for the long term security of the system. If a customer owns ten-year-old devices, the external computing industry will have progressed and attacks that were not envisaged at the time when the systems were initially designed will become available. Therefore another key challenge of this scenario is the remote management of the authentication key and security model of the customer devices.

The information about customer energy profile allows the energy supplier to act accordingly with respect to energy supply. However, in Europe the confidentiality of the consumer information is ensured by national regulators. A key challenge of this scenario is the consumers' privacy preservation, while enabling the electricity supplier to adapt the energy demand to the supplier.

14.3 Security Challenges in Industrial Wireless Sensor Networks

Generally, security issues are not handled independently, a tradeoff with usability and performance should be addressed (Figure 14.3). Indeed, sensor nodes are known to be resource-limited devices (CPU, memory, battery, communications, etc.). Some of them may be mobile, and may require then to

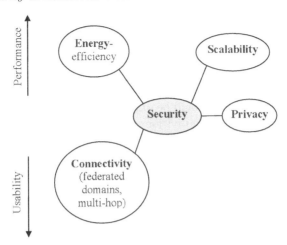

FIGURE 14.3
Security challenges.

re-authenticate to gain access to the WSN or to a new WSN at a different administrative domain. Because of their mobility or failure of some other network nodes, some sensor nodes may fail to reach the application infrastructure. Additionally, the industrial application may require that sensor nodes should be privacy protected. All these concerns inherent to the WSN or needed by industrial applications should be addressed by the proposed security solutions intended for the industrial WSN. These concerns are discussed in the following sub-sections and reflect the security challenges associated with industrial WSNs.

14.3.1 Resource Constraints

Sensor nodes are resource-constrained devices in terms of battery life, CPU capability, and memory capacity. Enabling WSNs with security may increase the overhead in sensor node resource consumption. For example, supplementing a node message containing a set command (e.g., 1 bit) with integrity protection (e.g., 128 bit-MAC) increases the message size by approximately 100 times. With respect to cryptographic primitives, low cost primitives (e.g., symmetric cryptographic algorithms, hash algorithms) are privileged to build security protocols. Also, since wireless communication is the more expensive operation for a sensor from energy point of view, security protocols should be built requiring limited bandwidth (e.g., reduced number of exchanged messages, small packet size).

14.3.2   Scalability

Since sensor nodes are manufactured as cheap and small devices, they are generally deployed in large scale and densely within the WSN. If security operations should be performed for each of these sensor nodes, these operations may become complex and impractical. In order to distribute the load of security overhead, sensors could be organized into clusters (e.g., using cluster-heads, cluster keys) and sub-divisions of clusters.

14.3.3   Mobility Support

Some industrial applications require sensor nodes to be mobile (e.g., sensors attached to workers). Nodes may move within the same WSN and connect to new neighbor nodes, or move to another WSN. In the first case, security operations may require to be quickly performed, depending on the sensor mobility pattern. In the second case, the visited WSN may be subscribed to an administrative domain different from the one to which subscribed the moving nodes. In this case, the security solutions should support cross-domain capability.

14.3.4   Intermittent Connectivity

WSNs have generally dynamic network topologies due to potential node break-down or mobility. Moreover, some nodes may fail to route communications because they entered a sleep mode or their battery depleted, which results in unreliable communications between sensor nodes and the application infrastructure. As another result of such intermittent connectivity, reliance of security operations on servers located in a remote infrastructure should be limited.

14.3.5   Privacy

Privacy protection becomes more and more an important issue in industrial applications, in particular if sensor nodes move across different administrative domains. Sensor nodes handle sensitive data and contextual information (e.g., worker/consumer identity, location) that should be kept secret to the outside world. Security protocols should be then built with privacy-preserving techniques.

14.4   Authentication and Network Access Control

Before the application starts running, the first operation performed by the sensor node is an initial authentication to the WSN through a bootstrapping operation. Whenever the sensor node reconnects to the WSN in a regular-basis or because of, for instance, of mobility or network node failure, it performs a network access authentication followed by a key-generating method to enforce network access control at lower layers.

14.4.1   Bootstrapping

When first deployed in the network, sensors should be bootstrapped prior to their normal operation in the network. The IETF Internet draft [41] defines bootstrapping as the transfer of settings (e.g., wireless channels, network addresses, link-layer keys, and application keys) to nodes at all their layers. The draft presents methods for bootstrapping that rely on initial authentication based on public keys that may be certified by a certification authority, before establishing symmetric keys used for subsequent authentications. The draft proposes also for authentication to use cryptographically generated addresses (CGAs) for address ownership verification. The use of CGAs requires executing the protocol SeND [4] (Secure Neighbor Discovery, RFC 3971). Based on the use of a CGA, a random interface identifier is bound to a public key, thus providing a proof of interface identifier ownership, but without a guarantee of network prefix ownership. CRYPTRON in [21] proposes to extend the concept of CGA towards network prefixes for routers using long cryptographically generated sub-prefixes. The public key is then bound to the prefix of the network address.

As discussed in [41], an asymmetric key-based authentication protocol like EAP or HIP-DEX can be used to bootstrap a WSN node. Subsequent authentications are performed based on the symmetric key generated following the bootstrapping operation. In the ETSI M2M specification [3], the M2M device derives connection keys with every authentication operation that serve themselves to derive application keys based on a key hierarchy structure. The specification provides examples of methods for M2M service bootstrapping which are either based on access network layer (e.g., GBA, EAP-SIM/AKA) or independent of the network layer (e.g., EAP-TLS, EAP-IBAKE).

The node may not reach the network directly; it may rely on other nodes in the path to the gateway (or the sink). Once bootstrapped, these intermediary nodes acquire the PANA Relay Element functionality, as described in [11]. Their immediate neighbors can then perform their own bootstrapping procedure through them. As per the bootstrapping itself, it occurs as multiple expanding rings centered on the PANA authentication Agent (PAA). The first nodes to be bootstrapped are those that are one IP hop away from the PAA.

14.4.2   Authentication and Network Access Control

In WSNs, it is critical to restrict the network access only to eligible sensors, while messages from unauthorized nodes will not be forwarded. Otherwise, the network may be vulnerable to denial-of-service (DoS) attacks that aim at resource depletion at sensor devices. To gain access to the network, nodes generally perform an authentication operation with a remote infrastructure. Due to the intermittent connectivity of nodes, network authentication may not be required, only network access control is reinforced within the WSN. Authentication takes place then between two neighboring sensor nodes.

Since authentication is performed more often than bootstrapping, authentication schemes are generally based on simple symmetric cryptographic primitives to be suitable to WSNs with their limited-computational resources. Authentication between the sensor node and the network generally relies on a centralized trusted authority (e.g., EAP-PSK [43], [48]). For instance, the authentication scheme in [48] allows the mutual authentication between a mobile user agent and a sensor node. The scheme adopts the Kerberos-like protocol; however, it does not require synchronization, and provides privacy protection.

For authentication between sensor nodes within the same WSN, the protocol in [8] allows nodes to authenticate to their neighbors based on challenge-response tuples that are pre-calculated beforehand with a key that is later erased. It uses a key hash chain for authentication of nodes similarly to TESLA (version adapted to WSNs [25]). The responsibility of network access control can be delegated to a coalition of nodes close to the authenticating node. In [27], a threshold number of them enable the node and its intermediary neighbor to authenticate to each other and at the same time compute a shared symmetric key. Network-wide keys can be also envisioned. Standards like ZigBee and Bluetooth use symmetric encryption for network authentication whereby two nodes can mutually authenticate each other by using random challenges with responses based on a secret key.

Network access schemes have been also proposed at lower layers (e.g., [51], [17]) providing hop-by-hop authentication. For instance, with LHAP [51], a node joining the network needs to perform first TESLA localized broadcast operations to bootstrap trust relationship with its neighbors. Then, the node switches to one-way key chains for traffic authentication.

14.4.3   Mobility-Supported Authentication

Because of their mobility, nodes generally move from a position where they have been authenticated to the network to another new position where they need to perform again an authentication procedure. The authentication procedure caused by node mobility is called re-authentication.

To address handover latency requirements, the IETF RFC 5836 [36] discusses quick re-authentication of a mobile device roaming from one attachment point to another using either security context transfer or early EAP authentication. Horizontal context transfer between attachment points is quickly reviewed, and discarded for security reasons (e.g., the domino effect). Vertical transfer of reusable keys from the EAP server to the new attachment point is considered interesting, and studied in detail. Within this latter model, the EAP Re-authentication Protocol (ERP) [33] has been proposed to allow fast device pre-authentication based only on two message exchanges. The IETF RFC 5873 [35] on PANA pre-authentication support defines extensions to the PANA protocol to carry EAP messages related to pre-authentication. It is supposed to work in the direct pre-authentication model as defined in RFC 5836, and therefore, describes exchanges between the PANA client and a candidate PANA authentication agent. As another early authentication model, the Authenticated Anticipatory Keying (AAK) scheme in [6] proposes to establish cryptographic keying material between a mobile device and one or more candidate attachment points (CAPs) before the device handovers to them. The serving attachment point (SAP) is involved in EAP authenticated anticipatory keying signaling communicated through the AAA server.

Secure schemes based on horizontal context transfer between attachment points (APs) without the involvement of the remote authentication server have been proposed in the literature (e.g., [24], [20], [46], [29]). For instance, the proactive key distribution (PKD) approach in [24] for fast re-authentication within IEEE 802.11i authentication framework, introduces a data structure, named Neighbor Graph, which dynamically tracks the potential APs to which a station may handoff. In [20], the PKD approach is security enhanced with a ticket-based approach whereby the station generates by itself the keys and securely sends them to the target AP through the serving AP. This enhancement does not introduce considerable signalling overhead compared to the PKD pre-authentication approach, while, ensuring the conformance with the IEEE 802.11i security requirements. In the scheme [20], the mobile device relies on transfer tickets generated by the serving AP that dispenses from re-authenticating to a new AP. In the Evolved Packet System for LTE, a similar procedure [5] of transfer of security parameters is used whereby the serving attachment point (i.e., eNodeB) derives a new key that is transferred to the target attachment point when handover occurs.

14.4.4 Discussion

The discussed related works draw the outline of bootstrapping and authentication procedures. Bootstrapping can be instantiated as a specific case of network entry authentication. Instead of letting authentication derives an M2M root key, as per ETSI M2M specification, bootstrapping allows deriving the short-term shared secrets used in subsequent authentications carried out by the sensor node. Other parameters such as the network-wide key or the cluster key (as in LEAP [50]), are derived and transported to the node. These parameters can be transported end-to-end or hop-by-hop through other network entities (e.g., PANA authentication Agent, PANA Enforcement Points). These parameters are used to enforce the network access control within the WSN.

With the urge to all-IP communications, authentications and bootstrapping could be performed at higher layers. For example, PANA over UDP could be chosen as the transport protocol for EAP instead of lower layer protocols, e.g., 802.1x.

14.5   Key Management

Keys are required to be established to protect security associations for each communicating node, each group of nodes, and each layer. At the initialization phase, nodes share an initial key using key pre-distribution scheme. These keys are either installed at nodes before their deployment in the WSN or derived following a bootstrapping operation. Whenever a new node joins or another one leaves the network, re-keying of shared keys is needed to provide backward and forward secrecy. Keys should be also refreshed periodically to mitigate their compromission.

14.5.1   Key Pre-distribution

A key management solution may rely on network-wide keys. As proposed in [22], the master-key is used in combination with random nonces exchanged by nodes, in order to establish a session key. With this solution, the compromission of one node triggers re-keying operation at all nodes in the network.

The use of pair-wise private keys allows each pair of nodes to share a secret; however, it requires the pre-distribution and storage of n -1 keys in each node (where n is the number of nodes in the WSN). Due to the large amount of memory required, pair-wise schemes are not viable when the network size is large. Probabilistic schemes (e.g., [9], [7], [14]) whereby each node receives only a subset of the keys are used instead. The probabilistic scheme in [9] allows nodes to share pair-wise keys. But, it guarantees that any two nodes with IDs at one unit Hamming distance from each other are able to share a common secret key. Perfect connectivity is not guaranteed but it can be achieved if the scheme uses for instance location information. A location-aware scheme is thus proposed in [9]. Perfect connectivity can be achieved with the scheme of Gupta and Kuri in [14] where the network is represented by a fully connected graph, which is static though. Standards like ZigBee 2007 [49] provides also perfect connectivity based on network-wide keys (Symmetric-Key Key Establishment-SKKE approach). The common secret key shared between two nodes is computed based on exchanged-randomly-generated challenges and using the master key that may be either pre-installed or transported from the remote trust center. Other approaches are provided in the ZigBee application profile, like the Public-Key Key Establishment (PKKE) and the Certificate-Based Key Establishment (CBKE) which allow establishing a key based on shared static and ephemeral public keys. Keys shared between nodes can be discovered based on the exchange of key IDs. The probabilistic scheme in [7] proposes to combine identity-based cryptography with probabilistic random key pre-distribution. In the proposed scheme, nodes do not exchange key IDs but instead these IDs are derived from the identity of their holders. Key IDs are discovered through routing information or other practical means.

14.5.2 Key Update

The shared key between two nodes should be continuously refreshed in order to mitigate key compromission. Keys could be refreshed by means of frequent fast re-authentication with the authentication authority (e.g., the key hierarchy for the ETSI M2M service layer [3]). For scalability considerations, the key management scheme proposed in [28] relies on a spanning tree for key update triggered by the base station, when this latter, for instance, detects malicious or captured nodes. However, the solution is not applicable for a mobile network, i.e., sensor nodes are assumed static; otherwise the spanning tree should be established often which incurs considerable bandwidth overhead in the network. Two neighboring nodes can refresh their shared keys on a regular-basis without the involvement of the base station. In the key evolution scheme in [26], nodes renew continuously their shared keys by changing keys based on the initial key values and also the exchanged messages. Moreover, nodes may use different keys for sending and receiving messages.

14.5.3 Discussion

From the discussed literature, different models for key management are proposed and applicable for different types of WSNs. If the WSN is composed of sparse nodes, key management could be handled by the remote key management authority. On the other hand, for dense WSNs, nodes may rather manage their keys and refresh them by themselves. In this case, probabilistic pre-distribution of keys could be envisioned to bootstrap communication between neighboring nodes. Not only pair-wise keys could be used, cluster and network-wide keys should be also considered. Cluster keys could serve for local broadcast communication and to refresh keys at the cluster level. The network key serves for network access control (e.g., ZigBee).

14.6   Security Maintenance

During the lifetime of a sensor, security mechanisms (e.g., keys, trusted identities addresses, firmware) may need to be upgraded. Most of the work in security mechanism update focuses on key management. Some efforts have been made to study firmware update, but aimed at providing firmware dissemination reliability or at minimizing update messages. Even tough, less effort is devoted to security of firmware update, some schemes, notably related to securing the Deluge protocol [18], have been proposed.

14.6.1   Software Update

In WSNs, Deluge [18] is the widely used protocol for network re-programming. The Deluge protocol divides the program into pages that are divided into packets. Then, program packets are propagated in the network in a multi-hop fashion. To efficiently authenticate the program packets, the scheme in [44] proposes to include in each propagated page an authentication segment and a key update segment. Based on multiple one-way hash chains, the authentication segment comprises a hash of the program page along with a chain key. The key update segment allows authenticating the used key based on the key used in the previous sent page. The first page will be associated with the commitment hash chain key stored at sensors. To prevent man-in-the-middle attacks due to multi-hop programming, nodes are divided into different groups according to their hop distance from the base station, and each node group uses a different hash chain.

In Deluge, since packets are propagated hop-by-hop and only a whole page is authenticated at once, a compromised node may send a bogus packet to the next hop neighbor nodes, thus making them fail in decoding a correct page. Cryptographic mechanisms to authenticate neighbors may not be sufficient, if nodes have been compromised and not detected yet. The Sreluge protocol [23] based on Rateless Deluge proposes mechanisms to prevent such attacks, known as pollution attacks, based on neighbor classification system and a time series forecasting technique to isolate polluters, and a combinatorial technique to decode data packets in the presence of polluters before the isolation is complete.

14.6.2   Discussion

For now, security mechanism update is performed hop-by-hop and protected based on link-layer security. The main security issues associated with mechanisms update are related to the local broadcast authentication of the received messages. Solutions have been proposed to provide end-to-end authentication at least at one-side (i.e., from the source node), even though such authentication is not packet-level granular since update propagation is realized at lower-layers.

14.7   Privacy Management

To ensure sensor privacy, several elements should be protected. Sensor identity (ID), along with the contextual information (e.g., location) that can be derived from sensor traffic, should be hidden to any attacker in the network e.g., an eavesdropper. Privacy protection can be provided with ID randomization (e.g., pseudonymity) or traffic randomization techniques.

14.7.1   Pseudonymity

To hide identities of sensors, it is not practical to encrypt node IDs using pair-wise keys shared between communicating nodes. The repeated encryption and decryption of IDs at each received packet are consuming operations for resource constrained sensor nodes. A more convenient solution is to rely on pseudonymous IDs. For each privacy protected sensor, pseudonyms are used to replace the common identifiers and addresses present in the fields of its transmitted packets' headers, such that packets appear unlinkable to the very sensor. Moreover, to achieve unlinkability between communications originating from the same sensor, pseudonyms are frequently changed at the different levels of the network stack.

The IETF RFC 4941 [34] on Privacy Extensions for Stateless Address Autoconfiguration in IPv6 proposes to use temporary addresses generated based on random interface identifiers. A temporary address has to be used for a limited short time and is associated with the connections being carried out. Whenever it expires, another freshly-generated temporary address is used for the new connections. CALM (Continuous Air interface for Long and Medium distance) in [2] provides a location privacy for anti-tracking protection system, based on MAC address randomization. It suggests relying on cross-layer synchronization for the anonymization system where security certificates are periodically changed.

The MOBIKE [12] extension to IKE allows a mobile or multi-homed host to update its IP addresses associated with ongoing IKE and IPsec security associations (SAs). Either, the host sends an informational request using the new address and containing the notification of the updated address, or it includes proactively one or more notifications of additional addresses in the IKE authentication exchange. When updating addresses, MOBIKE takes into account security considerations like using a "return routability" check performed by the responder to verify the addresses provided by the host. If the addresses are frequently changed to provide privacy, the mechanism requires at least two additional handshakes to check every used pseudonymous address. The mechanism is therefore not practical for frequently changing pseudonymous IP addresses.

Random identifiers provide anonymity, but they do not allow destination nodes to identify the sender at least at the application level, unless they have a trapdoor (i.e., a secret information) allowing them to link the random identifiers to the sender.

In the Simple Anonymity Scheme (SAS) [31], sensor nodes are associated with pseudonym ranges pre-assigned to them by an authority entity. If a sensor wants to send a message to its neighbor, it uses a random pseudonym from the appropriate range. Rather than storing a large number of pseudonyms, sensor nodes can generate them based on shared keys, as proposed in the Cryptographic Anonymity Scheme (CAS) [30]. Between both schemes, a hybrid approach could be considered whereby keys for pseudonym generation and pre-assigned pseudonyms may be both used, in order to achieve a certain tradeoff between CPU usage and memory efficiency. Similarly to CAS [30], in the first method of [37], sensor pseudonyms are dynamically changed by means of keyed one-way hash chains. The second method in [37] uses the chain in the reverse order. The advantage of this second method over CAS is that, if the keys are compromised, the attacker cannot impersonate the sensor.

In these latter schemes ([37], [31], and [30]), pseudonyms are secretly shared between sensor nodes from the same WSN. Credentials e.g., keys, pseudonym ranges, are either pre-distributed to nodes or securely provided to them through a remote authority entity. For instance, the multi-layered pseudonym architecture proposed in [39] is built with the aim to provide user privacy during initial EAP authentication and during fast handoff process. Three different types of pseudonyms are used depending on the type of the server to which the node authenticates. A bootstrapping pseudonym is used during the bootstrapping phase or during peer authentication with its home server. A home authentication pseudonym is used during node re-authentication to its home domain. And finally, a visited authentication pseudonym is used during node handoff to a visited domain. The distribution of pseudonyms is protected thanks to a tunneled method of the authentication protocol EAP. Each different server will control the generation of the pseudonyms associated with its node authentication process. It may be not have knowledge of the used pseudonyms related to other servers.

14.7.2   Anonymization Techniques

Even though, sensor ID might be sufficiently concealed, sensors may still leak sensitive contextual information associated with sensor current status (e.g., indication of battery level or CPU load) or the location and timing of events. Such information may be sufficient to deanonymize the sensor. To hide this type of information, some noise can be introduced into the sensor traffic to distort the leaked information.

Traffic sources can be protected by using fake sources and bogus messages introduced into the network. For instance, to prevent traffic analysis, the authors of [10] proposes to make each sensor send messages at a constant rate, and thus produce random messages. The authors of [16] propose to add random delays to message transmissions at each intermediate sensor. The technique does not need dummy messages; but, it is not practical for networks with minimal traffic. Message flooding techniques (e.g., [38]) have been also proposed, where messages are flooded into the network to reach the sink. Random walk and path confusion techniques (e.g., [47]) allow hiding traffic sources location. However, these two latter types of techniques produce high communication overhead.

Attacks against sensor privacy may target different layers. Anonymization at the network layer should be therefore envisioned, to prevent attackers from tracking sensors using the network identifier (e.g., IP address). Approaches for hiding network IDs like Crows [40] or Tor [1] have not yet adapted to the WSN context; even though, some efforts have been undertaken like AnonySense ([19], [42]) to anonymize sensor network IDs using Tor and Mixmaster [32]. Still, the proposed approach suffers from high latency, which depends on the population of MIX users and the data flow rate.

14.7.3 Discussion

When used for privacy protection, temporary pseudonyms should be mutually verifiable by the two end communicating entities. However, end-to-end anonymous communication is only possible at the expense of performance, since routing tables should be frequently changed and filled with the newly changed pseudonyms. Also, achieving security entails more performance consideration, since end-to-end anonymous nodes are required to authenticate again to gain access to the network. Pseudonyms should be then verifiable by at least some of the intermediary entities e.g., forwarding nodes, routers, gateways, access points, authentication server. Along with the ID anonymization system, a resolution counterpart should be then implemented. As for traffic anonymization techniques, they generally result in high latency and large communication overhead and are difficult to be adapted to the WSN context. In particular, in a large scale and dense network, communication capabilities are a limiting factor. Repeated collisions caused by traffic anonymization may result in resource exhaustion at sensors.

14.8 Conclusions

This book chapter surveys the security solutions proposed in the context of WSNs with the aim to study their integration into industrial applications.

A particular focus is put on network authentication and access control, key management, firmware update, and privacy protection solutions. The chapter discusses their applicability in the industrial domain based on a number of derived security challenges associated with this type of applications, which demonstrates that significant progress has been made toward acceptable and integrated solutions to secure industrial WSNs.

The research community has defined several security mechanisms for WSNs. Standardization efforts (e.g., 3GPP, ETSI, IETF) are ongoing to incorporate and interoperate security protocols and techniques. At the same time, investigations are still under way regarding other aspects of industrial networks:

- Mobile devices may move within a multi-domain setting; even initial authentication of devices may require, in some industrial scenarios, to be performed across a different domain network (e.g., cellular networks).

- Heterogeneity of nodes within the WSN is not deeply explored whereby more powerful nodes assist less powerful nodes. For example, a trusted module (e.g., Hardware Security Module) may be installed at some nodes and used to reinforce the security of the other nodes lacking such trusted module.

- The role of network entities (e.g., gateways, cluster-heads) with respect to security is not well defined. For example, in some scenarios where end-to-end security could not be provided because of protocol translation problem, a private gateway may share security credentials with nodes and the infrastructure and step between them to secure communications (i.e., decrypting and then encrypting with the appropriate keys).

- If the WSNs, like the Internet-of-Things and Machine type communications, would evolve toward the all-IP paradigm [15], the proposed security mechanisms should eventually support protocol translation in order to provide end-to-end security (e.g., tunnel mode in EAP, PANA).

- As for application layer security, the network security could be used to establish application keys (e.g., [3]) between nodes in the WSN, due to the limitation of resources within these networks. Sharing security mechanisms between layers could include, along the keying materials, common security mechanisms. For example, a bootstrapping or authentication network operation is performed once for all layers.

- Demonstrations of the feasibility of the security mechanisms are generally provided; but rare are implementations of the overall system operations through test beds. Such test beds are necessary for the extensive performance evaluation of the secure industrial WSN.

References

[1] Tor project anonymity. Available online. https://www.torproject.org/.

[2] CALM, Continuous Communications for Vehicle. SeVeCOM Workshop, February 2006.

[3] Draft ETSI TS 102 690 v0.14.2, Machine- to-Machine communications (M2M) - functional architecture, September 2011.

[4] J. Arkko, J. Kempf, B. Zill, and P. Nikander. SEcure Neighbor Discovery (SEND). IETF RFC 3971, March 2005.

[5] Rolf Blom, Karl Norr man, Mats Näslund, Stefan Rommer, and Bengt Sahlin. Security in the evolved packet system. Ericsson Review, February 2010.

[6] Z. Cao, H. Deng, Y. Wang, Q. Wu, and G. Zorn. EAP re-authentication protocol extensions for authenticated anticipatory keying (ERP/AAK). IETF Internet Draft, October 2011.

[7] D. W. Carman. New directions in sensor network key management. *International Journal of Distributed Sensor Networks*, 1, No. 1:3–15, 2005.

[8] O. Delgado, A. Fúster, and J. M. Sierra. A light-weight authentication scheme for wireless sensor networks. *Ad Hoc Networks*, 2011.

[9] F. Delgosha, E. Ayday, and F. Fekri. MKPS: a multivariate polynomial scheme for symmetric key-establishment in distributed sensor networks. In *The ACM Int'lWireless Communications and Mobile Computing Conference*, pages 236–241, Honolulu, Hawaii, USA, August 2007.

[10] J. Deng, R. Han, , and S. Mishra. Intrusion tolerance and anti-traffic analysis strategies for wireless sensor networks. In *The International Conference on Dependable Systems and Networks*, pages 637–646, Washington, DC, USA, 2004.

[11] P. Duffy, S. Chakrabarti, R. Cragie, Y. Ohba, and A. Yegin. Protocol for carrying authentication for network access (PANA) relay element. IETF RFC 6345, August 2011.

[12] P. Eronen. IKEv2 mobility and multihoming protocol (MOBIKE). IETF RFC 4555, June 2006.

[13] O. Garcia-Morchon, S. Keoh, S. Kumar, R. Hummen, and R. Struik. Security considerations in the IP-based internet of things. IETF Internet Draft, October 2011.

[14] A. Gupta and J. Kuri. Deterministic schemes for key distribution in wireless sensor networks. In IEEE Computer Society, editor, *The Third International Conference on Communication Systems Software and Middleware and Workshops (COMSWARE 08)*, pages 452–459, Washington, DC, USA, 2008.

[15] Tobias Heer, Oscar Garcia-Morchon, René Hummen, Sye Loong Keoh, Sandeep S. Kumar, and Klaus Wehrle. Security challenges in the IP-based internet of things. *Wireless Personal Communications*, 61, no. 3:527–542, December 2011.

[16] X. Hong, P. Wang, J. Kong, Q. Zheng, and J. Liu. Effective probabilistic approach protecting sensor traffic. In *The Military Communications Conference*, volume 1, pages 169–175, 2005.

[17] H. Hsu, S. Zhu, and A. R. Hurson. LIP: a lightweight interlayer protocol for preventing packet injection attacks in mobile ad hoc network. *International Journal of Security and Networks*, 2, Nos.3/4:202 – 215, 2007.

[18] J.W. Hui and D. Culler. The dynamic behaviour of a data dissemination protocol for network programming at scale. In ACM, editor, *The 2nd international conference on Embedded networked sensor systems (SenSys 04)*, pages 81–94, 2004.

[19] A. Kapadia, D. Peebles, C. Cornelius, N. Triandopoulos, and D. Kotz. AnonySense: Opportunistic and privacy-preserving context collection. In *The 6th Conference on Pervasive Computing*, pages 19–22, Sydney, Australia, May 2008.

[20] Mohamed Kassab, Jean Marie Bonnin, and Karine Guillouard. Securing fast handover in WLANs: a ticket based proactive authentication scheme. In *Globecom Workshops*, pages 1–6, November 2007.

[21] Ana Kukec, Marcelo Bagnulo, and Antonio de la Oliva. CRYPTRON: CRYptographic Prefixes for Route Optimization in NEMO. *Communications (ICC), 2010 IEEE Communications Society*, pages 1–5, May 2010.

[22] B. Lai, S. Kim, and I. Verbauwhede. Scalable session key construction protocol for wireless sensor networks. In *IEEE Workshop on Large Scale Real-Time and Embedded Systems*, 2002.

[23] Y.W. Law, Y. Zhang, M. Palaniswami J. Jin, and P. Havinga. Secure rateless deluge: Pollution-resistant reprogramming and data dissemination for wireless sensor networks. *EURASIP Journal on Wireless Communications and Networking, Special Issue on Security and Resilience for Smart Devices and Applications*, 2011. Article ID 685219.

[24] Celia Li and Uyen Trang Nguyen. Fast authentication for mobility support in wireless mesh networks. *Wireless Communications and Networking Conference, 2011*, pages 1185–1190, March 2011.

[25] D. Liu and P. Ning. Multi-level uTESLA: Broadcast authentication for distributed sensor networks. *ACM Trans. Embedded Computing Systems*, 3, no. 4:800–836, 2004.

[26] Zhihong Liu, Jianfeng Ma, Qingqi Pei, Liaojun Pang, and YoungHo Park. Key infection, secrecy transfer, and key evolution for sensor networks. *IEEE Transactions on Wireless Communications*, 9 n.8:2643–2653, August 2010.

[27] Leonardo Maccari, Lorenzo Mainardi, Maria Antonietta Marchitti, Neeli R. Prasad, and Romano Fantacci. Lightweight, distributed access control for wireless sensor networks supporting mobility. In *IEEE International Conference on Communications, ICC*, Beijing, China, May 2008.

[28] M.-L. Messai, M. Aliouat, and H. Seba. Tree based protocol for key management in wireless sensor networks. *EURASIP Journal on Wireless Communications and Networking*, 2010. Article ID 910695.

[29] Arunesh Mishra, Min Ho Shin, Jr. Nick L. Petroni, T. Charles Clancy, and William A. Arbaugh. Proactive key distribution using neighbor graphs. *Wireless Communications*, 11, no. 1:26–36, February 2004.

[30] S. Misra and G. Xue. Efficient anonymity schemes for clustered wireless sensor networks. *International Journal of Sensor Networks*, 1(1/2):50–63, 2006.

[31] S. Misra and G. Xue. SAS: A simple anonymity scheme for clustered wireless sensor networks. In *The IEEE International Conference on Communications 2006. ICC '06*, volume 8, pages 3414–3419, June 2006.

[32] U. Möller, L. Cottrell, P. Palfrader, and L. Sassaman. Mixmaster protocol, version 3. IETF Internet Draft.

[33] V. Narayanan and L. Dondeti. EAP extensions for EAP re-authentication protocol (ERP). IETF RFC 5296, August 2008.

[34] T. Narten, R. Draves, and S. Krishnan. Privacy extensions for stateless address autoconfiguration in IPv6. IETF RFC 4941, September 2007.

[35] Y. Ohba. Pre-authentication support for PANA. IETF RFC 5873.

[36] Y. Ohba and G. Zorn. Extensible authentication protocol (EAP) early authentication problem statement. IETF RFC 5836.

[37] Y. Ouyang, Z. Le, Y. Xu, N. Triandopoulos, S. Zhang, J. Ford, and F. Makedon. Providing anonymity in wireless sensor networks. In *The IEEE International Conference on Pervasive Services*, pages 145–148, Istanbul, Turkey, July 2007.

[38] C. Ozturk, Y. Zhang, and W. Trappe. Source-location privacy in energy-constrained sensor network routing. In ACM, editor, *The 2nd ACM workshop on Security of ad hoc and sensor networks*, pages 88–93, New York, NY, USA, 2004.

[39] F. Pereniguez, G. Kambourakis, R. Marin-Lopez, S. Gritzalis, and A. F. Gomez. Privacy-enhanced fast re-authentication for EAP-based next generation network. *Computing Communications*, 33, 14:1682–1694, 2010.

[40] M. Reiter and A. Rubin. Crowds: Anonymity for web transactions. *ACM Transactions on Information and System Security*, 1:66–92, 1998.

[41] B. Sarikaya, Y. Ohba, R. Moskowitz, Z. Cao, and R. Cragie. Security bootstrapping of resource-constrained devices. IETF Internet Draft, October 2011.

[42] M. Shin, C. Cornelius, D. Peebles, A. Kapadia, D. Kotz, and N. Triandopoulos. AnonySense: A system for anonymous opportunistic sensing. *Journal of Pervasive and Mobile Computing*, May 2010.

[43] D. Simon, B. Aboba, and R. Hurst. The EAP-TLS authentication protocol. IETF RFC 5216, March 2008.

[44] Hailun Tan, John Zic, Sanjay K. Jha, and Diethelm Ostry. Secure multihop network programming with multiple one-way key chains. *Mobile Computing, IEEE Transactions on Mobile Computing*, 10, no. 1:16–31, January 2011.

[45] Felix von Reischach, Nouha Oualha, Alexis Olivereau, David Bateman, and Emil Slusanschi. Deliverable D2.1: Scenario definitions and their threat assessment. Technical report, TWISNet: Trustworthy Wireless Industrial Sensor Networks, March 2011.

[46] Ralf Wienzek and Rajendra Persaud. Fast re-authentication for handovers in wireless communication networks. *Networking*, pages 556–567, 2006.

[47] Y. Xi, L. Schwiebert, and W. Shi. Preserving source location privacy in monitoring-based wireless sensor networks. In *The IEEE International Parallel and Distributed Processing Symposium*, Los Alamitos, CA, USA, 2006.

[48] Sungjune Yoon, Hyunrok Lee, Sungbae Ji, and Kwangjo Kim. A user authentication scheme with privacy protection for wireless sensor networks. In *The 2nd Joint Workshop on Information Security*, pages 233–244, Tokyo, Japan., 2007.

[49] E. Yüksel, H.R. Nielson, and F. Nielson. ZigBee-2007 security essentials. In *The 13th Nordic Workshop on Secure IT-systems (NordSec 2008)*, pages 65–82, Copenhagen, Denmark, 2008.

[50] S. Zhu, S. Setia, and S. Jajodia. LEAP: Efficient security mechanisms for large-scale sensor networks. In *The 10th ACM Conference on Computer and Communications Security (CCS 03)*, Washington D.C., October 2003.

[51] S. Zhu, S. Xu, S. Setia, , and S. Jajodia. LHAP: A lightweight network access control protocol for ad-hoc networks. *Elsevier Ad Hoc Networks Journal*, 4, Issue 5:567–585, September 2006.

15

Cognitive Radio Sensor Networks in Industrial Applications

A. Ozan Bicen and Ozgur B. Akan

Next-generation and Wireless Communications Laboratory (NWCL)
Department of Electrical and Electronics Engineering
Koc University, Istanbul, Turkey

CONTENTS

15.1 Introduction

The need for industrial monitoring and process control has arisen with the demand to overcome production capacity limitations, improve process efficiency, comply with the environmental regulations, predict machine failures, and precaution against natural accidents [14, 18]. Collected information from industrial equipment is used to diagnose efficiency decreases and faults in industrial applications. Therefore, reliable and timely information gathering from industrial equipment is extremely crucial to prevent possible inefficiencies and malfunctions in industrial applications.

Wireless monitoring and control of factory automation can provide improvement in productivity due to removed wires between control and sensing entities. To this end, wireless sensor networks (WSN) [4] have been employed to enable reliable wireless monitoring and control functionality in a wide variety of industrial applications, such as process control, precaution of industrial accidents, smart grid, monitoring of contaminated areas, healthcare applications, and intelligent transport [18, 19]. Furthermore, delay-bounded industrial process control, on-time diagnosis of machine faults, and machine vision are some of the vital real-time industrial applications which require timely transfer of sensed information to control devices, and these can be performed without installation of wires for communication via WSN. Recently, challenges, and design principles for deployment of WSN on the critical industrial equipment are investigated to realize reliable wireless monitoring and control in industrial applications [13]. The achievement of these benefits of WSN requires advanced communication skills in time-varying and complex industrial environments that impose significant challenges for connectivity, reliability, latency, and energy efficiency.

Furthermore, capabilities of cognitive radio can be utilized to mitigate the unique communication challenges of industrial applications for WSN, i.e., heterogeneous spectrum characteristics varying over time and space, application-dependent quality of service (QoS) requirements, dynamic connectivity, and node failures. With its adaptability to existing spectrum characteristics in the deployment field, cognitive radio sensor networks (CRSN) can enhance the overall energy and spectrum-efficient reliability performance for wireless monitoring and control [1, 7]. CRSN nodes can send the collected information over spectrum bands unused by licensed users in a multi-hop fashion to meet the industrial application reliability and latency, i.e., quality of service (QoS), requirements. Although the recent interest in WSN for industrial applications, the employment of CRSN in industrial applications is a non-investigated area. To the best of our knowledge, there is no extensive study of deployment of CRSN in industrial applications. The objective of this chapter is to point out the benefits of CRSN in industrial applications and discuss the open research directions for this promising research area.

The remainder of this chapter is organized as follows. In Section 15.2, the potentials of CRSN in industrial applications are introduced. In Section 15.3, the CRSN architecture configurations in industrial applications are discussed. Algorithm needs of CRSN for spectrum management in industrial applications are presented in Section 15.4. The open research issues on communication protocol development for CRSN in industrial applications are given in Section 15.5. Finally, chapter is concluded in Section 15.6.

15.2   Advantages of CRSN for Industrial Applications

WSN communications confined to unlicensed bands may cause significant challenges in reliability and energy efficiency for industrial applications [1]. Many sensor nodes are placed in close proximity, and actor nodes for automation may also be a part of the deployment. In CRSN, opportunistic spectrum access (OSA) provided by cognitive radio can assist to alleviate these challenges due to spectrum scarcity. Wireless monitoring and control networks, such as factory automation, intelligent transport, smart grid, and healthcare applications, can exploit the unused portions of the spectrum in the licensed bands via the cognitive radio capability. Advantages of CRSN in industrial applications can be outlined as

- *Minimization of environmental effects*: Interference from other wireless communication systems, equipment noise, topology changes, and fading in industrial environments yield challenging varying spectrum characteristics. Therefore, communication delay and rate in industrial applications are both time and location dependent for wireless links. Considering challenging wireless link conditions, sensor nodes in industrial applications must be able to adjust their communication parameters adaptively without requiring any modification on communication hardware. As an industrial application of CRSN, high pressure steam pipeline monitoring is illustrated in Figure 15.1, where

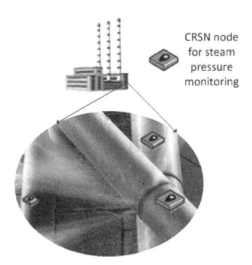

FIGURE 15.1
High pressure steam pipeline monitoring as a typical industrial application of CRSN.

sensor nodes are densely deployed on pipelines and exposed to shadowing due to surrounding pipelines. With the ability of cognitive radio, sensor nodes can adjust their communication parameters to overwhelm effects of dense deployment and fading with minimum energy expenditure [15].

- *Access to licensed spectrum bands*: Various wireless communication systems that use unlicensed bands may coexist, and different spectrum utilization patterns can be experienced in different areas of intelligent transport systems, smart grid, and health care applications. CRSN nodes can detect and use vacant licensed bands opportunistically without interfering with licensed users to fulfil industrial application specific QoS needs with respect to dynamic spectrum conditions. Therefore, not only adaptation to wireless link conditions is provided by cognitive radio capability, but also adaptation to different spectrum utilization patterns in different industrial environments is made possible.

- *Heterogeneous reliable and real-time communications*: Since taking the correct action on the event is highly depending on reliable estimation of event features, latency and losses can result in making the wrong decision and lead to malfunction of industrial equipment as well as hazardous situations for health. Thus, required number of sensing results should be reported to the sink within a desired interval. Requirement on number of collected sensing results and estimation interval, i.e., QoS requirements, may show heterogeneity according to industrial application, e.g., while healthcare applications may use delay-tolerant communication, a building automation may require delay-bounded communication. Real-time wireless monitoring and control requirements of the industrial applications can be fulfilled via spectrum adaptation to minimize delay and jitter. Multiple CRSNs in the same environment can share the spectrum to accomplish QoS needs of different industrial applications. CRSN nodes can collaboratively detect unused portions of the spectrum and access them to provide efficient and fair spectrum sharing between co-existing heterogeneous sensor networks. Cognitive radio empowers spectrum-efficient communication capability in sensor networks to assist achievement of multi-class QoS objectives for industrial applications.

15.3   CRSN Architecture for Industrial Applications

Cognitive radio sensor nodes form a wireless communication architecture for industrial applications over which the information obtained from the industrial equipments is conveyed to the actuators or monitoring center. Main duty of the sensor nodes is to perform sensing of the industrial equipment. CRSN nodes also perform spectrum sensing to detect vacant portions of the spectrum. In accordance with the spectrum opportunities, sensor nodes transmit

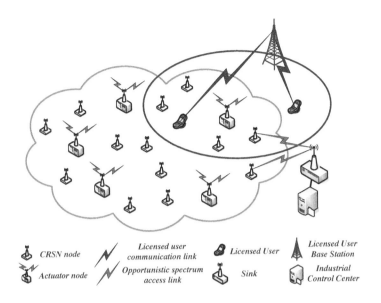

FIGURE 15.2
A typical CRSN topology coexisting with licensed users.

their event sensing results towards the sink. Sensors are also required to exchange information regarding spectrum conditions, and availability in industrial environments. Furthermore, industrial actuator units with high power resources, which act on the sensed event, may be part of the architecture as well [2]. These actor nodes may perform local spectrum management, or act as a spectrum broker. Therefore, they can be a vital part of the CRSN topology. A typical sensor field composed of resource-constrained CRSN nodes, actuators, and sink is illustrated in Figure 15.2.

CRSN node consists of sensing, processor, memory, power, and cognitive radio transceiver units as shown in Figure 15.3. In specific industrial applications, CRSN nodes can be equipped with localization and mobility functionalities as well. Limitations of WSN paradigm on energy, communication, processing and memory resources are inherited by CRSN nodes. Cognitive radio transceiver constitutes the primary distinction of a CRSN node against a WSN node. Cognitive radio unit provides adaptation to varying spectrum conditions via tuning its communication frequency, transmission power, and modulation technique.

In industrial applications, CRSN may exhibit different network topologies, and these topologies are as

- *Ad Hoc CRSN*: CRSN nodes deployed without any infrastructural element in industrial applications yields an ad hoc CRSN where nodes can send their

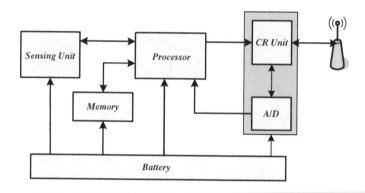

FIGURE 15.3
A systematic overview of CRSN node structure.

readings to the sink over multiple hops without any infrastructure. In ad hoc CRSN, spectrum sensing may be performed by each node individually or collaboratively. This topology brings reduced communication overhead for control data distribution for coordination of both event and spectrum sensing. Nevertheless, spectrum sensing results may be inaccurate due to hidden terminal problem, and performance degradation for the licensed user networks can be caused.

- *Clustered CRSN*: Coordination of spectrum sensing results, and spectrum allocation information among CRSN nodes can be performed via a common control channel. However, it may not be likely to find a network-wide common channel for coordination throughout the entire industrial application network. It has been shown in [25] that finding a common channel in a certain restricted area is highly possible due to the spatial correlation of channel availability. Therefore, a cluster-based network architecture with local common control channels is desirable for spectrum-aware clustering [24] in large-scale industrial applications such as intelligent transport systems and smart grid. Furthermore, industrial actor nodes may also be employed to handle additional tasks such as the collection and dissemination of spectrum availability information, and the local spectrum assignment. To this end, new cluster-head selection and cluster formation algorithms, which jointly addresses the resource limitations of CRSN nodes and the requirements of industrial applications, must be developed for CRSN.

- *Heterogeneous and Hierarchical CRSN*: Industrial actor nodes may be included in CRSN architecture [2]. These nodes can be used as relay nodes due to their longer transmission range. Therefore, a heterogeneous and multi-tier hierarchical topology composed of ordinary CRSN nodes, high power relay nodes, e.g., cognitive radio actor nodes, and the sink can formed in industrial applications distributed over a large geographical area such as intelligent transportation and smart grid. While the actor nodes in industrial applications may be exploited for effective OSA, this heterogeneity brings additional challenges. Sensor and actor deployment, increased communication overhead due to hierarchical coordination, and the cognitive radio challenges for the actor nodes are the main issues that should be addressed for realization of this architecture.

- *Mobile CRSN*: Sensor nodes, industrial actuator nodes, and even sink might be mobile depending on the industrial application and deployment scenario. Considering the mobility of the architectural elements of CRSN, dynamic topology challenge in industrial applications is further amplified with mobile CRSN nodes. Mobility should be incorporated into communication algorithms for CRSN in industrial applications. Furthermore, mobility-aware spectrum management schemes should be devised for resource-limited CRSN nodes, as well.

The CRSN node structure and wide range of CRSN architecture for industrial applications discussed above yield many open research issues outlined as

- *CRSN node development*: One of the fundamental issues for the realization of CRSN in industrial applications is the development of efficient, practical, and low-cost cognitive radio sensor nodes. Considering the basic design principles and objectives of sensor networks, as well as the inherent limitations of sensor nodes, hardware and software design for sensor nodes with cognitive radio capability must be extensively studied.

- *Topology control*: In cases where licensed user statistics for specific industrial sites are available, node deployment strategies for industrial applications considering spectrum availability and condition may provide considerable improvements on the life-time and communication efficiency of the network. Therefore, analysis of optimal node deployment for CRSN topologies in industrial applications should be performed, and, practical yet efficient deployment mechanisms should be investigated.

- *Clustering*: Clustering and forming hierarchy incur additional communication overhead for CRSN in industrial applications. This overhead may be amplified by the node mobility and spectrum handoff. Hence, for industrial applications requiring cluster-based and hierarchical CRSN topologies, dynamic spectrum-aware group formation [24] and maintenance techniques should be devised.

- *Collaborative spectrum access*: Spectrum sensing, spectrum decision, and spectrum handoff may be performed either individually or cooperatively by

CRSN nodes in industrial applications. A detailed efficiency analysis for the comparison of coordinated and uncoordinated schemes is required to assess CRSN deployment topologies for various industrial applications.

- *Network coverage*: In industrial applications, connectivity of CRSN is subject to changes, even in case of uniform deployment, since node failures and licensed user activities can disrupt event delivery to the sink. Therefore, to maintain maximum network coverage, certain nodes may be required to increase transmission power, which will increase their power expenditure. On the other hand, connectivity at longer transmission ranges in industrial applications may be achieved with lower carrier frequency employment, which may also help to reduce power consumption. An optimal network coverage analysis for CRSN is needed with respect to specific industrial application requirements. Trade-off between network life-time and communication coverage must be analyzed considering spectrum management challenges, and various CRSN deployment topologies for industrial applications.

15.4 Spectrum Management Requirements of CRSN in Industrial Applications

Minimization of environmental effects, adaptation to varying spectrum conditions, and overlay deployment of multiple sensor networks are some of the promising advantages of CRSN in industrial applications. However, the realization of CRSN in industrial applications mainly requires efficient spectrum management functionalities to dynamically manage the spectrum access of sensor nodes in challenging industrial communication environments. Requirements and research challenges for main three spectrum management functionalities of cognitive radio, i.e., spectrum sensing, spectrum decision, and spectrum mobility, are explored below for CRSN from the perspective of industrial applications.

15.4.1 Spectrum Sensing

To take advantage of spectrum sensing for CRSN, an efficient solution is needed considering jointly sensor network resource limitations, industrial application-specific sensing requirements, and OSA challenges. Considering high numbers of sensor nodes in large-scale industrial applications and low-cost requirements, it may not be feasible to equip sensor nodes with multiple radios and highly capable processors. Therefore, sophisticated spectrum sensing algorithms may not be used. Spectrum sensing should be performed with limited node hardware, possibly using single radio. Assuming that deployed sensor nodes in industrial applications have single radio due to their scalability and low-cost requirements, spectrum sensing durations should be minimized as much as possible with the consideration of possible communication activities and energy efficiency. There are various spectrum sensing methods, such as energy detection, feature detection, matched filter, and interference temperature [3]. Incorporating one or combination of these techniques, determination of dynamically changing noise components in industrial applications, and modelling of their interference with respect to time and space can be achieved. However, inherent sensor network limitations impose restrictions on the features of cognitive radio. For instance, CRSN nodes may only perform spectrum sensing over a limited band of the spectrum due to processing, power, and antenna size constraints.

In industrial applications, spectrum sensing algorithms that are designed considering the unique challenges posed OSA and by the resource constraints of CRSN nodes should address the following challenges:

- *Hardware limitations*: It is not feasible to equip CRSN nodes with highly capable processors and A/D units. Thus, complex detection algorithms cannot be used. Spectrum sensing must be performed with limited node hardware in industrial applications.

- *Minimum sensing duration*: Keeping the transceiver on even just for spectrum sensing will cause excessive power consumption of CRSN nodes. While detection accuracy increases for longer durations, spectrum sensing should be performed with minimum sensing duration.

- *Reliable sensing*: CRSN nodes can operate in licensed bands, unless they do not interfere with licensed users. Spectrum sensing algorithm should be devised to avoid interference on licensed users. Furthermore, event and spectrum sensing reliability should jointly be addressed [23].

Upon listed challenges for spectrum sensing, open research issues for spectrum sensing for realization of CRSN in industrial applications are as follows:

- *Hybrid sensing techniques*: A possible way to obtain spectrum information with minimum sensing duration and low computational complexity is to use hybrid sensing techniques, which is a balanced combination of the sensing approaches. For instance, energy detection may be used for a broader band to have an idea about which portions of the spectrum are possibly vacant. Based on this information, more accurate sensing methods can be performed over selected potentially vacant channels. Therefore, hybrid sensing techniques addressing the trade off between sensing accuracy and complexity must be investigated for employment of CRSN in industrial applications.

- *Cooperative sensing*: When nodes rely only on their own spectrum sensing results, they may not be able to detect the licensed user due to shadowing in industrial environment. Spectrum sensing duty can be performed cooperatively among CRSN nodes to increase sensing accuracy. Wideband sensing schemes can be employed via distributed CRSN nodes [22]. While cooperative sensing yields better sensing results, it also imposes additional complexity and communication overhead for CRSN. New cooperative sensing methods requiring minimum amount of coordination information exchange and having minimum impact on the sleep cycles of the node are needed for realization of CRSN in industrial applications.

- *Collaborative licensed user statistics*: With the channel usage statistics of the licensed users, efficiency of spectrum sensing methods can be significantly increased. Although licensed user statistics are not available on hand, sensors may collectively obtain these statistics by sharing their distributed spectrum sensing results. Intelligent and collaborative methods that estimate and then make use of licensed user channel usage statistics should be investigated for CRSN in industrial applications.

Overall, the benefits of OSA, such as lower packet collisions due to the capability of switching to the best available channel, less contention delay and more bandwidth, come with the additional energy expenditure for CRSN due to spectrum sensing functionality and distribution of spectrum sensing results. The trade off between energy efficiency and sensing accuracy for various spectrum methodologies should be addressed and a detailed analysis of cost vs. benefits for specific industrial applications should be performed.

15.4.2   Spectrum Decision

CRSN nodes must vacate accessed channels, when licensed user activity is detected. Selecting one radio frequency as network-wide cannot yield the expected performance gain due to spatio-temporally varying spectrum characteristics. Licensed users, such as TV and cellular phones in industrial environments, may use spectrum temporally, and hence, vacant portions of the spectrum also vary with time.

For efficient spectrum decision, parameter selection plays a vital role. To this end, possible parameter choices are ratio of spectrum sensing duration and data transmission duration, transmission power, expected duration to spend in a channel without spectrum handoff, predictive capacity and delay, energy-efficiency and error rate. Additionally, residual energy of CRSN nodes can be used for channel assignment [17]. For underlay approaches in industrial applications, the trade-off between spectrum handoff and adaptation to accessed spectrum band should be investigated. Licensed user protection and energy efficiency of sensor nodes should jointly be addressed [16]. To support industrial application QoS requirements, analytical models describing the relation between the spectrum decision parameters and performance indicators (reliability, delay, and energy expenditure) should be investigated.

Spectrum decision should be coordinated to prevent multi-user collision in crowded spectrum environments of the industrial applications [3]. Distributed carrier selection and power allocation should be investigated [10]. Furthermore, energy-adaptive power adaptation schemes can be employed for capacity maximization [12]. Different industrial applications may co-exist, and to satisfy their reliability and latency requirements, QoS-aware spectrum decision schemes are needed in industrial applications. An efficient spectrum decision scheme is required to address the QoS requirements of specific industrial applications and allocate CRSN resources in a spectrum-aware and fair manner.

Some of the main challenges for efficient spectrum decision schemes include time-synchronization, distributed spectrum decision coordination, and spectrum characterization for overall CRSN nodes. In distributed spectrum coordination approaches, sensor nodes sense the radio spectrum and communicate their results to overcome the spectrum decision problems caused by the limited knowledge of spectrum availability and network topology [3]. Energy efficient and scalable methods of spectrum decision mechanisms are yet to be investigated to efficiently realize the proposed CRSN in industrial applications.

15.4.3 Spectrum Mobility

When licensed user communication starts in currently occupied channel by CRSN nodes, it should be detected within a certain time through spectrum sensing, and spectrum handoff should immediately be performed to move a new vacant channel decided by an effective spectrum decision mechanism. These fundamental functionalities of cognitive radio create spectrum mobility. Determination of vacant channels, and selection of the new channel for spectrum handoff incur delays, and hence, buffer overflows which lead to packet losses, and degradation of reliability for industrial wireless monitoring and control applications. Considering the challenges posed by the inherent limitations of CRSN, analysis of the effects of spectrum handoff on industrial control and monitoring applications is required.

In CRSN, spectrum handoff can also be triggered by interference caused by other unlicensed wireless systems existing in industrial environments. In case of excessive interference and noise, ongoing communication should be carried onto an another channel selected by spectrum decision algorithm. To have effective spectrum mobility functionality, the trade-offs between communication parameters must be well understood. Since spectrum mobility brings interruptions to ongoing communication, buffer overflow prevention schemes to minimize the communication delay should be developed for reliable and real-time wireless monitoring and control applications in industrial environments. Based on variations on spectrum characteristics in time and space domains, spectrum handoff may be performed among heterogeneous channels, and hence, this will bring heterogeneous link conditions on the path to sink node. Furthermore, since industrial applications are usually placed in indoor environments, spectrum mobility functionality is also critical for adapting to time-varying fading challenges.

15.5 Communication Protocol Requirements of CRSN in Industrial Applications

Efficient operation of CRSN in industrial applications is tightly-coupled by the running communication protocol suite. In addition to the inherited WSN challenges, i.e., dense deployment, event-driven, and resource-limited nature of WSN, OSA and industrial environment challenges, i.e., spectrum efficiency, harsh environmental conditions, and variable link capacity, should be addressed to benefit from salient features of CRSN in industrial applications. In this section, industrial application specific challenges for communication layers of CRSN are investigated.

15.5.1 Cognitive Physical Layer

To overcome dynamic topology, fading, and licensed user interference in industrial environments, CRSN node's physical layer must be configurable in terms of transmission power, operating frequency, modulation technique, and channel coding. This configuration should be performed in accordance with spectrum sensing and decision functionalities. Due to resource-limited nature and cost constraints of CRSN nodes, implementing cognitive radio transceivers for CRSN nodes is a challenging task. CRSN physical layer should be capable of providing statistical information about channel conditions to upper layers to empower spectrum awareness. Therefore, efficient yet practical cognitive radio transceiver design for CRSN nodes is essential for the realization of CRSN in industrial applications. Existing open research issues for the physical layer of CRSN can be summarized as follows.

- In order to overwhelm temporally and spatially varying RF interference in industrial environments, adaptive power allocation schemes are essential for cognitive physical layer. Furthermore, power adaptation due to existing deployment of other wireless communication systems should also be considered. Devised physical layer solution should maximize energy usage efficiency.

- Adaptive modulation can be employed to maximize network life-time and meet industrial application-specific QoS requirements via reconfiguring parameters of the physical layer [11]. CRSN node's physical layer must be reconfigurable without hardware modification. Thus, software-defined radios providing efficient OSA should be devised for CRSN nodes.

- Cooperative transmission schemes should be investigated to benefit from transmitter diversity for CRSN nodes [21]. Cooperative relaying can help to realize energy-efficient communication in harsh RF interference and dynamic topology conditions of industrial environments.

- Statistical spectrum characterization methods should be designed for industrial environments. Considering limited processing capabilities and low-cost requirements of CRSN nodes, advanced yet practical signal processing algorithms should be devised at physical layer to enable effective spectrum management, and spectrum awareness in industrial applications.

15.5.2 Spectrum-Aware Collaborative Medium Access Control

The basic function of a medium access control (MAC) protocol is efficient coordination of the medium access among the nodes contending for communication. Therefore, it plays a vital role for CRSN to achieve high throughput in industrial applications, when varying wireless channel characteristics, licensed user activity, and dynamic topology in industrial environments are considered. In CRSN, MAC protocol should address OSA challenges as well as challenges inherited from WSN. Design challenges for MAC layer due to resource limitations, dense deployment, and industrial application-specific QoS requirements are further amplified by dynamic spectrum conditions of industrial environments. Since energy efficiency is still a crucial factor affecting design of MAC protocol, duty cycling, event reporting, and spectrum sensing of CRSN nodes should be optimized for industrial applications. The challenges for design of MAC layer of CRSN in industrial applications are outlined below.

- Common control channel should be used for coordination of spectrum decision before starting communication. Without use of dedicated common control channel, it is highly challenging to coordinate spectrum decision and negotiate the communication channel. Therefore, MAC protocol should be developed to efficiently address this issue to reduce the communication delay and energy expenditure.

- MAC design for CRSN should minimize the control packet exchange overhead to utilize spectrum opportunities efficiently. Although centralized solutions are highly efficient, they cannot be realized in CRSN due to dynamic topology and spatially varying wireless channel conditions in industrial applications. Therefore, distributed solutions that can provide coordination without central entity should be investigated for realization of CRSN in industrial applications.

- Joint design of spectrum sensing and duty cycling is required to achieve efficient utilization of spectrum and energy resources. Spectrum sensing and duty cycling algorithm should be devised with consideration of requirements of industrial application on sensed event. Furthermore, packet size optimization for CRSN nodes is essential [20] for energy efficiency in industrial applications.

MAC algorithms that address OSA challenges, dynamic topology, and indoor communications affected by spatially varying fading in industrial environments should be devised considering energy-limited CRSN regime. Performance of MAC protocols is highly depending on industrial application needs, hence, a flexible MAC protocol that can be adjusted to serve various industrial applications should be developed for employment of CRSN in industrial applications

15.5.3   Spectrum-Aware Event-Oriented Routing

Cognitive radio equipped sensor nodes amplifies the inherited energy-efficient and event-oriented routing challenges from WSN. In CRSN, the routing algorithm should be aware of spectrum opportunities to provide delivery of sensed event features satisfying industrial application-specific QoS requirements under the intermittently available licensed spectrum bands. Furthermore, indoor environmental conditions of industrial applications imposes challenges for routing due to spatially varying channels, node failures, and dynamic topology. The key design challenges for spectrum-aware event-oriented routing protocol development to realize employment of CRSN in industrial applications can be outlined as follows.

- Incorporating the spectrum parameters in the routing decisions will enable adaptation of event reporting flows toward sink to spectrum opportunities. This will yield lesser route failures in case of spectrum opportunity variations, and eventually lower energy expenditure and delay for CRSN.

- Dynamic topology and varying wireless links in industrial environments cause disruptions on event delivery. Thus, routing protocol should be able to distinguish the link failures either due to node failures, wireless errors, or spectrum mobility to efficiently perform route decision.

- The communication environment in industrial applications, i.e., vacant spectrum bands, dynamic topology, and harsh environmental conditions, is highly challenging and varying. It depends on the licensed user activity and the efficiency of spectrum management functionalities. Therefore, routing protocol should predict the spectrum and network dynamics to provide seamless QoS to industrial applications.

CRSN nodes communicating opportunistically make neighbor discovery and collaboration for route establishment quite challenging. Therefore, routing algorithms relying on excessive control messages for setting up and maintaining routes may not be practical for CRSN. This fact points out the need for design of a distributed routing algorithm based on local network state in challenging industrial environments. Various routing protocols [8, 9] are proposed for ad hoc cognitive radio networks to provide spectrum-aware routing decisions, however, they do not take the inherent energy constraints and event sensing requirements of CRSN into account. CRSN necessitates novel routing algorithms that incorporate sensed event signal and spectrum characteristics, such as channel access delay, interference, operating frequency, and bandwidth, in the routing decisions.

15.5.4   Reliable and Spectrum-Aware Event Transport

The union of cognitive radio and sensor networks makes reliability and congestion control an extremely challenging task in CRSN [5, 6]. Congestion control should be performed in accordance with spectrum management functionalities, and distributed rate control is needed for collaborative event reporting of sensor nodes on the same event. In industrial applications, rate control is a very challenging task due to dynamic spectrum conditions and spatially varying wireless links. Furthermore, a predictive congestion avoidance algorithm is needed as well. Proactive congestion mechanisms should be benefited using minimum coordination for energy conservation. Moreover, dynamic connectivity in industrial environments further amplifies the the reliable event transport problem. The challenges imposed on CRSN transport layer in industrial applications can be outlined as follows.

- There is no fixed channel path from the sensing node to the industrial actuator, which may cause significant variations on the communication parameters, e.g., link delay, bit error rate, capacity, over each hop.

- At the time of the spectrum handoff, buffer overflows and packet losses may occur due to delay until a suitable channel for communication is found. Furthermore, spectrum mobility during active communication or along the path to the sink may incur large variations and inaccuracy in estimation of end-to-end delay and packet loss rate, which makes providing reliability requirements of industrial applications challenging.

- CRSN nodes must sense spectrum periodically to detect licensed user activity. Since spectrum sensing sensor nodes cannot transmit and receive data, extra sensing delay and buffer overflows may result in additional packet losses. This may lead to loss of time-critical data, and cause malfunction of industrial control applications.

Industrial applications impose additional delay bounds on the reliable communication requirements. Delivery of queries, commands, and code updates may impose even tighter QoS requirements, which are challenging to cope with conventional fixed spectrum solutions due to large variations of channel characteristics over industrial environment. Open research issues for reliability and congestion control in industrial applications are summarized below.

- New reliability definitions, objectives, and metrics must be studied to incorporate the fundamental variables of dynamic spectrum access into industrial application specific QoS requirements. Analytical modelling of communication capacity, reliability, congestion, and energy consumption should be studied. Furthermore, queuing and network information theoretical analysis of reliable communication must be explored for employment of CRSN in industrial applications. Cross-layer interactions with spectrum management and congestion control mechanisms should be investigated to address large variations in channel characteristics over multi-hop paths to actuator.

- To provide reliable event transport under varying spectrum characteristics and spectrum sensing durations, probabilistic reporting schemes should be developed. Instead of aiming maximization of rate and maximum bandwidth utilization, rate control algorithm must aim to maximize reliability. Adaptive spectrum-aware transport solutions must be developed to address reliability requirements and application-specific QoS needs, as well as opportunistic spectrum access challenges for both sensor-to-actuator and actuator-to-sensor communication in industrial applications. New mechanisms to exploit the multiple channel availability towards reliable energy-efficient communication in CRSN must be developed.

- Industrial applications are required to take time-critical critical control actions on equipment. Therefore, reliable event transport functionality should be combined with real-time capability to support various latency and sample collection requirements at sink. Event-to-sink path characteristics, such as delay, and packet loss rate, should be considered for transport protocol design for employment of CRSN in industrial applications.

15.6 Conclusions

Recently, there has been an ongoing interest for low-cost and efficient wireless monitoring and control in industrial applications. Reliable communication in industrial environments is a challenging task for wireless monitoring applications, such as factory automation, smart grid, smart transportation, and healthcare applications, due to inherent sensor network energy limitations, spectrum scarcity, interference, equipment noise, dynamic topology, and fading. In this chapter, cognitive radio sensor networks (CRSN) in industrial applications is proposed to provide reliable and efficient communication for wireless monitoring and control in industrial applications. First, the potential advantages of CRSN in industrial applications are explored. Then, CRSN architectures in industrial applications are described, and their challenges challenges and requirements are discussed. Furthermore, spectrum management functionalities, i.e., spectrum sensing, spectrum decision, and spectrum mobility, are discussed from the perspective of CRSN and industrial applications. Moreover, the communication protocol suite is discussed for employment of CRSN in industrial applications as well. We have provided a contemporary perspective to the current state of the art for wireless monitoring and control via CRSN in industrial applications.

References

[1] O. B. Akan, O. Karli, and O. Ergul. Cognitive radio sensor networks. *Network, IEEE*, 23(4):34 –40, July-August 2009.

[2] I. F. Akyildiz and I. H. Kasimoglu. Wireless sensor and actor networks: research challenges. *Ad Hoc Networks*, 2(4):351 – 367, 2004.

[3] I. F. Akyildiz, W.-Y. Lee, and K. Chowdhury. Spectrum management in cognitive radio ad hoc networks. *Network, IEEE*, 23(4):6 –12, July-August 2009.

[4] I. F. Akyildiz, W. Su, Y. Sankarasubramaniam, and E. Cayirci. A survey on sensor networks. *Communications Magazine, IEEE*, 40(8):102 – 114, Aug 2002.

[5] A. O. Bicen and O. B. Akan. Reliability and congestion control in cognitive radio sensor networks. *Ad Hoc Networks*, 9(7):1154 – 1164, 2011.

[6] A. O. Bicen, V. C. Gungor, and O. B. Akan. Delay-sensitive and multimedia communication in cognitive radio sensor networks. *Ad Hoc Networks*, 10(5):816 – 830, 2012.

[7] A. O. Bicen, V. C. Gungor, and O. B. Akan. Spectrum-aware and cognitive sensor networks for smart grid applications. *Communications Magazine, IEEE*, 50(5):158 –165, May 2012.

[8] K. R. Chowdhury and I. F. Akyildiz. Crp: A routing protocol for cognitive radio ad hoc networks. *Selected Areas in Communications, IEEE Journal on*, 29(4):794 –804, April 2011.

[9] L. Ding, T. Melodia, S. N. Batalama, J. D. Matyjas, and M. J. Medley. Cross-layer routing and dynamic spectrum allocation in cognitive radio ad hoc networks. *Vehicular Technology, IEEE Transactions on*, 59(4):1969 –1979, May 2010.

[10] S. Gao, L. Qian, and D. R. Vaman. Distributed energy efficient spectrum access in wireless cognitive radio sensor networks. In *Wireless Communications and Networking Conference, 2008. WCNC 2008. IEEE*, pages 1442 –1447, April 2008.

[11] S. Gao, L. Qian, D. R. Vaman, and Q. Qu. Energy efficient adaptive modulation in wireless cognitive radio sensor networks. In *Communications, 2007. ICC '07. IEEE International Conference on*, pages 3980 –3986, June 2007.

[12] B. Gulbahar and O. Akan. Information theoretical optimization gains in energy adaptive data gathering and relaying in cognitive radio sensor networks. *Wireless Communications, IEEE Transactions on*, PP(99):1 –9, 2012.

[13] V. C. Gungor and G. P. Hancke. Industrial wireless sensor networks: Challenges, design principles, and technical approaches. *Industrial Electronics, IEEE Transactions on*, 56(10):4258 –4265, Oct. 2009.

[14] V. C. Gungor and F. C. Lambert. A survey on communication networks for electric system automation. *Computer Networks*, 50(7):877 – 897, 2006.

[15] G. Gur and F. Alagoz. Green wireless communications via cognitive dimension: an overview. *Network, IEEE*, 25(2):50 –56, March-April 2011.

[16] J. A. Han, W. S. Jeon, and D. G. Jeong. Energy-efficient channel management scheme for cognitive radio sensor networks. *Vehicular Technology, IEEE Transactions on*, 60(4):1905 –1910, May 2011.

[17] X. Li, D. Wang, J. McNair, and J. Chen. Residual energy aware channel assignment in cognitive radio sensor networks. In *Wireless Communications and Networking Conference (WCNC), 2011 IEEE*, pages 398 –403, March 2011.

[18] U.S. Department of Energy. Industrial wireless technology for the 21st century. *Office of Energy and Renewable Energy Rep.*, 2002.

[19] U.S. Department of Energy. Assessment study on sensors and automation in the industries of the future. *Office of Energy and Renewable Energy Rep.*, 2004.

[20] M. C. Oto and O. B. Akan. Energy-efficient packet size optimization for cognitive radio sensor networks. *Wireless Communications, IEEE Transactions on*, 11(4):1544 –1553, April 2012.

[21] J. Peng, J. Li, S. Li, and J. Li. Multi-relay cooperative mechanism with q-learning in cognitive radio multimedia sensor networks. In *Trust, Security and Privacy in Computing and Communications (TrustCom), 2011 IEEE 10th International Conference on*, pages 1624 –1629, Nov. 2011.

[22] H. Zhang, Z. Zhang, and Y. Chau. Distributed compressed wideband sensing in cognitive radio sensor networks. In *Computer Communications Workshops (INFOCOM WORKSHOPS), 2011 IEEE Conference on*, pages 13 –17, April 2011.

[23] H. Zhang, Z. Zhang, X. Chen, and R. Yin. Energy efficient joint source and channel sensing in cognitive radio sensor networks. In *Communications (ICC), 2011 IEEE International Conference on*, pages 1 –6, June 2011.

[24] H. Zhang, Z. Zhang, H. Dai, R. Yin, and X. Chen. Distributed spectrum-aware clustering in cognitive radio sensor networks. In *Global Telecommunications Conference (GLOBECOM 2011), 2011 IEEE*, pages 1 –6, Dec. 2011.

[25] J. Zhao, H. Zheng, and G.-H. Yang. Distributed coordination in dynamic spectrum allocation networks. In *New Frontiers in Dynamic Spectrum Access Networks, 2005. DySPAN 2005. 2005 First IEEE International Symposium on*, pages 259 –268, Nov. 2005.

16

Industrial WSN Standards

Tomas Lennvall, Krister Landernäs, Mikael Gidlund, and Johan
Åkerberg

ABB AB Corporate Research

CONTENTS

16.1   Introduction

The need for specific wireless communication standards in industrial automation has risen from the fact that most existing standards are targeting the
consumer industry. The consumer industry traditionally has very different re-

quirements compared to the industry. For consumers bandwidth and range is more important than real-time performance and reliability, while for the industry the opposite is usually true.

Another very important requirement from the industrial automation community is to have complete device interoperability. This means that a device in a wireless network must be replaceable with another device from another vendor, and still work flawlessly all the way up to the control system. This is only achievable using standards and the reason non-standardized solutions (e.g., proprietary) are only used in rare cases. However, it is not enough to only standardize the lower communication layers (e.g., physical, data-link, network, etc), because complete device interoperability also requires a standardized application layer. This means that wireless communication standards targeting industrial automation must include the application layer as well. This is how industrial wired communication standards (e.g., fieldbuses) are defined. The ZigBee Alliance [18] and Bluetooth SIG [2] are targeting this using profiles. However, there are no profiles for industrial automation.

To make things more complex, industrial automation is divided into three different application domains, each with different requirements: process automation, factory automation, and building automation. The latter domain, building automation, typically does not include residential homes, but rather malls, airports, apartment buildings, industrial buildings, etc.

All of this makes it difficult for the industry to use existing standards (e.g., ZigBee, Bluetooth) as is, thus there is a need for specific industrial automation standards. It is too early to determine whether recent development with IPv6 based wireless networks (e.g., 6LoWPAN, SNAP) will be suitable for industrial automation applications.

There are naturally different standards in each domain and in this chapter we will focus on standards for process automation. Although standards for communication in industrial automation have been around for decades it is only in recent years that wireless standards have emerged. The WirelessHART^{TM} specification was released in 2007 and back then it was the first wireless protocol developed specifically for process industry. Since then, both ISA100.11a and WIA-PA has been released. Wireless LAN (WLAN) technology [11, 17, 9] has always been interesting for industrial automation, and it is used in many places within the industry today. However, WLAN (or any other wireless technology) is currently not used for the most critical parts of an industrial plant due to lack of support for industrial real-time requirements [3].

This chapter will give an overview on all three protocols and discuss their strengths and weaknesses. A future outlook on industrial WSNs is also given.

16.2 History

Standardization has driven the development of industrial communication since the mid 1950s. From the introduction of digital communication in the '60s to the fieldbuses of today, standards have always been important to get interoperability between products and systems. Standardization of wireless protocols for industrial applications is relatively new. Bluetooth (1998) and ZigBee (2002) was introduced and marketed as protocols suited for industrial automation but had little impact on the industry. They simply did not meet the requirements in process industry and did not work well in an industrial setting. It was not until 2007 that a protocol tailored for process industry was released with the introduction of WirelessHART [8]. Soon after two more protocols followed; ISA100.11a [13] and WIA-PA [4].

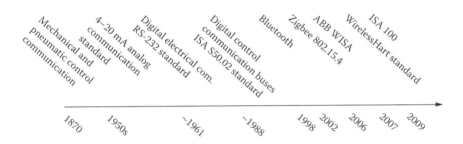

FIGURE 16.1
Industrial communication timeline.

16.3 Regulations and Standards

The process of getting an approved wireless standard is quite extensive and usually takes several years. Several bodies and groups are involved in the writing, review, and voting of the proposed standard. This section tries to give a high level overview on the process and the different bodies involved. A simplified chart of the standardization process is shown in Figure 16.2.

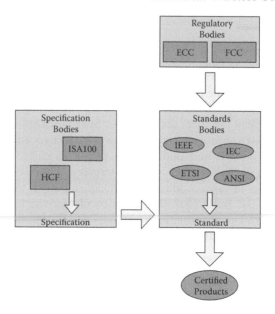

FIGURE 16.2
Standardization process.

16.3.1   Regulations

A regulation is a legislative text that constrains rights and allocates responsibility. The use of radio equipment is often permitted under different regulations. The main purpose of the regulation is to prevent people from injury and protect important radio frequency bands. These frequencies include bands used for public service radio, radar, and military communications. Regulations are different for different regions and countries around the world. In Europe, the ECC (Electronic Communications Committee) [5] develops regulations for the effective use and Europe-wide harmonization of the radio spectrum. The FCC (Federal Communications Commission) [7] has a similar task in the United States. Similar functions can be found in e.g., China (Ministry of Information Industry) [14] and Japan (Ministry of Internal Affairs and Communications) [15].

Regulations typically specify:

- Which frequencies can be used

- Maximum allowed transmission power

- Maximum duty cycle

16.3.2 Standards

A standard is a set of requirements that specifies, in this case, how the radio protocol works. Typically, the lower levels of the OSI model are specified in the standard, including modulation scheme, medium access, frame contents, coding, routing, and security. It is not unusual that the higher levels of the OSI model is not part of the standard (e.g., application layer). Standards bodies cooperate closely with the regulatory bodies in order to ensure that the finalized standard conforms to current regulations. A complete standard may also be a combination of different standards. WirelessHART is for example based on IEEE 802.15.4 [10], but it is also an IEC standard (IEC 62591) [12].

16.3.3 Specifications

Specifications are written by organizations like the HART Communication Foundation (HCF) and the International Society of Automation (ISA) to promote a wireless protocol. This is often done in working groups where members of the organization contribute to the specification. The work is quite extensive and often takes several years. Once the protocol has been finalized and approved by the members it can be put forward to a standards body. For example, the WirelessHART specification was approved in 2007 by the HCF members and then became an IEC standard (IEC 62591) in March 2010.

16.4 Industrial Requirements

One could argue that with several wireless standards available there would be no need for new industrial WSN standards like WirelessHART and ISA 100a. There are mainly two reasons why IWSNs are needed. First of all, requirements in process industry are quite different from the requirements in consumer market. Much tougher demands on real time performance and reliability in process industry have risen the need for protocols supporting frequency hopping, black listing, encryption, etc. Another important requirement in IWSNs is backward compatibility with legacy wired systems. Simple integration with existing DCS systems is a must. This has mainly implications on the design of the gateway which must support industrial protocols like Modbus, Profibus and ProfiNET.

All current wireless standards for process industry target mainly noncritical monitoring and control applications. This is application class 2-5 in the ISA definition shown in Figure 16.3.

Category	Class	Application	Description	
Safety	0	Emergency action	Always critical	
Control	1	Closed loop regulatory control	Often critical	Importance of message timeliness increase
	2	Closed loop supervisory control	Usually non-critical	
	3	Open loop control	Human in the loop	
	4	Alerting	Short term operational consequence	
Monitoring	5	Logging and downloading/ uploading	No immediate operational consequence	

FIGURE 16.3
ISA100 Application Classes.

16.5 IEEE 802.15.4

16.5.1 Introduction

The IEEE 802.15.4 is a wireless standard for low power, moderate speed communication networks. The standard specifies the lower layers (PHY and MAC) in the OSI model. Higher layers are specified by other standards that use the 802.15.4 as base. There are quite a few wireless products on the market that utilize the 802.15.4 standard, especially in consumer products. Zigbee is perhaps the most well known example. But also industrial protocols like WirelessHART and ISA 100.11a uses the IEEE standard. The main reasons for using 802.15.4 is that it operates in a license free band and there are low cost chipsets available from several vendors. The latest revision of the 802.15.4 standard was released in 2006.

16.5.2 Protocol Overview

IEEE 802.15.4 is specified for the three Industrial, Scientific and Medical (ISM) frequency bands:

- 886.0-868.6 MHz

- 902-928 MHz

- 2400-2483.5 MHz

The ISM band is a worldwide license free band. A maximum bit rate of 250 Kb/s is specified for the 2.4 GHz band while the bit rate is much less in the two other frequency bands. The medium access mechanisms include both carrier sense multiple access with collision avoidance (CSMA/CA) and guaranteed time slots (GTS). In the former method a device that wants to transmit will listen and check if the channel is idle before transmitting. If the channel is busy, the device will back off for a random period of time before retrying. This method can lead to collisions if two devices try to transmit simultaneously. Guaranteed time slots can be used to eliminate collisions. This is a way to give devices exclusive right to send at defined times. Seven GTS can be used in IEEE 802.15.4.

16.6   WirelessHART

WirelessHART is a part of the HART7 specification developed by the HART Communication Foundation [8]. It is an extension of the *Highway Accessible Remote Transducer* (HART) fieldbus protocol, which is the most widely used fieldbus protocol of today (30 million devices). WirelessHART was ratified in September 2007 and in 2010 it was approved as an IEC standard (IEC 62591).

WirelessHART is designed based on a set of fundamental industrial requirements: it must be simple (e.g., easy to use and deploy), robust (e.g., self-organizing and self-healing), flexible (e.g., support different applications), scalable (i.e., fit both small and large plants), reliable, secure, interoperable (use different vendors devices in the same network), and last but not least support existing HART technology (e.g., HART commands, configuration tools, etc).

One of WirelessHART's main advantages is that interoperability is enforced on all layers in the OSI model (including application layer). This makes it simple to mix and exchange devices from different vendors in the same network without any hassle. This is not enforced by ISA100.11a and WIA-PA.

16.6.1   Protocol Overview

WirelessHART is based on the PHY layer specified in the IEEE 802.15.4-2006 standard, but it only supports the 2.4 GHz spectrum. WirelessHART specifies new Data-link (including MAC), Network, Transport, and Application layers. WirelessHART uses channel hopping on a packet by packet basis and uses 15 of the 16 channels defined in IEEE 802.15.4-2006. Figure 16.4 shows a typical WirelessHART mesh network topology, in which any device can act as a router. Star topologies are also possible.

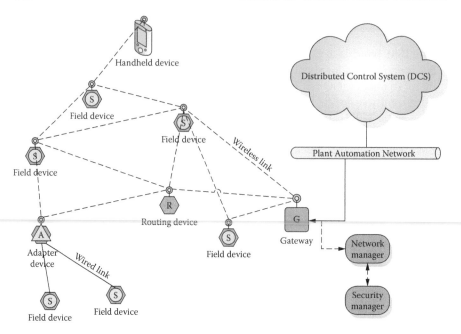

FIGURE 16.4
*Wireless*HART network topology.

A WirelessHART network consists of four main device types (see Figure 16.4):

Gateway: It connects the control system (via Ethernet, Profibus, etc.) to the wireless network. Device-to-device communication is not supported: all data must pass through the gateway;

Network Manager: This normally is part of the gateway and automatically builds the wireless network and manages its operations;

Field Device: These devices are connected to the process and consists of pressure, temperature, position, and other instruments but can also include adapters. All field devices are capable of routing packets on behalf of other devices in the network.

Adapter: The adapter enables wired HART field devices to connect to the WirelessHART network, i.e., it is intended to retrofit existing wired infrastructure.

Security Manager: This manages and distributes security encryption keys. It also holds the list of devices authorized to join the network.

Handheld: The handheld device is intended for commissioning, e.g., download security keys, set network id, as well as for configuration of field devices and adapters.

16.6.2   Protocol Features

In WirelessHART communications are based on *Time Division Multiple Access* (TDMA) and all devices are time synchronized and communicates in pre-scheduled fixed length time-slots. TDMA minimizes collisions and reduces the power consumption of the devices since they only have to have the radio on in time slots where they are active. Each time slot is of a fixed length, which is $10ms$ long. Devices are assigned to communicate within a slot, one device as the sender and one as the receiver. TDMA requires accurate internal slot timing in order for the sender and receiver to properly interact with each other during such a transaction. Time slots are grouped together to form what is called a *Superframe*, see Figure 16.5, which is continuously repeating and which can be activated or deactivated when necessary. It is possible to have multiple superframes of different length within a WirelessHART network and several of them can be active at the same time. The length of the superframe should correspond to the duty cycle requirements of the application, e.g., $500ms$, $1s$, $1hour$.

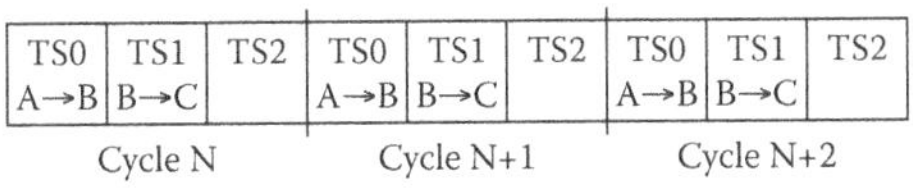

TS0	TS1	TS2	TS0	TS1	TS2	TS0	TS1	TS2
A→B	B→C		A→B	B→C		A→B	B→C	

Cycle N   Cycle N+1   Cycle N+2

FIGURE 16.5
*Wireless*HART superframe.

WirelessHART uses channel hopping to further increase reliability. Each time slot uses a new channel for communication, and the hopping pattern is determined using a pseudo-random algorithm.

All WirelessHART devices must support routing, i.e., there are no simple end/leaf devices. Since all devices can be treated equally in terms of networking capability, installation, formation, and expansion of a WirelessHART network becomes simple. Two different mechanisms are provided for message routing. Graph routing uses predetermined paths to route a message from a source to a destination device. Graph routes include redundant paths. This is the preferred way of routing messages both up- and downstream in a WirelessHART network. Source routing uses ad-hoc created routes for the messages without providing any path diversity, and the id's of the devices in the source route is stored in the packet which reduces the available payload size. WirelessHART also supports data aggregation in the source node, and a solution for how packet aggregation can be performed is proposed in [16].

WirelessHART is a secure and reliable protocol, which uses advanced encryption standard (AES) with 128 bit block ciphers. A counter with CBC MAC mode (CCM) is used to encrypt messages and calculate the message integrity code (MIC). Security in WirelessHART is mandatory and there are no selectable levels of security, i.e., there is only one security configuration available. End-to-end security is provided on the network layer, while the data link layer provides per-hop security between the two neighboring devices.

HART and WirelessHART shares the same OSI application layer, which specifies exactly what and how a device should provide measurement values, parameters, configuration settings, etc.

16.7   ISA100.11a

ISA-100.11a, approved by the ISA Standards & Practices Board as an official ISA standard in September 2009. The focus of the ISA100 committee [13] is that ISA100 shall be a family of industrial wireless standards covering different applications such as process applications, asset tracking and identification, and so on. ISA100.11a is the first standard released in the ISA100 family with the specification for process automation.

ISA100.11a is developed to have a broad coverage of process automation networks and is aimed at converging existing networks and assimilate devices communicating using different protocols. This flexibility has the drawbacks of increased complexity (compared to WirelessHART) and that full device interoperability is not possible.

16.7.1   Protocol Overview

ISA100.11a use the PHY and MAC layers defined in IEEE 802.15.4-2006, and also operates on all the defined channels (11 - 26, where 26 is optional). The ISA100.11a specification defines the data-link, network, transport, and application layers. ISA100.11a supports a wide range of network topologies: star, mesh, and star-mesh. Figure 16.6 shows a star-mesh network topology, which is typical for ISA100.11a networks. The figure also shows a backbone network connecting all the backbone routers, however the backbone network is not part of the ISA100.11a specification so no details are available.

An ISA100.11a network consists of several different devices:

Field Device Field devices are connected to the process and consists of pressure, temperature, position, and other instruments and actuators. It is optional for field devices to support routing.

Backbone router The backbone router routes data packets from a subnet over the backbone network to its destination e.g., another subnet connected to the backbone or the distributed control system.

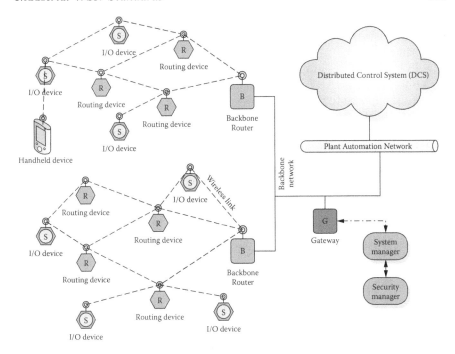

FIGURE 16.6
ISA100.11a network topology.

Gateway The Gateway acts as an interface between the ISA100.11a field network and the plant network or the distributed control system.

System Manager The System Manager is the administrator of the whole ISA100.11a network. It monitors the network and is in charge of system management, device management, network run-time control, and communication configuration (resource and scheduling).

Security Manager The Security Manager is responsible for providing security services based on the security policies defined. It performs security keys management, and guarantees secure system operation.

As the specified network architecture in Figure 16.6, the Gateway and System Manager must include a backbone router in order to get access to the backbone network.

16.7.2 Protocol Features

ISA100.11 uses a TDMA based communication with time slots and superframes similar to WirelessHART. However, in ISA100.11a time slots have a variable length of $10-12$ ms per superframe. This flexibility allows ISA100.11a to send packets to several receivers and receive multiple acknowledgments in one time slot.

ISA100.11a uses graph and source routing in the subnets, the main difference to WirelessHART being that the routing is performed in the data-link layer (as opposed to the network layer). It is possible to create routes that start in one subnet, going through the backbone network and backbone routers, to another subnet. Above the subnet (on the network layer) ISA100.11a uses IPv6 addressing and routing.

ISA100.11a supports three Channel Hopping schemes: slotted hopping is where each time slot uses a new channel for communication. Slow hopping is a mechanism where several time slots use the same channel before it is changed, and hybrid hopping is a combination of both, see Figure 16.7. ISA100.11 defines five default hopping patterns which must be supported by all ISA100.11a devices. It is possible to add custom hopping patterns.

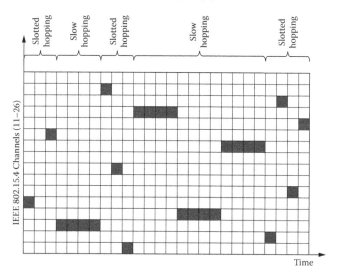

FIGURE 16.7
ISA100.11a hybrid channel hopping.

ISA100 has pushed the traditional OSI model network layer down to the data-link layer and replaced it with 6LoWPAN [1]. 6LoWPAN provides the use of IPv6 over IEEE 802.15.4 networks. Since the Maximum Transmission Unit (MTU) of IPv6 packets is 1280 octets, which is far too large compared to IEEE 802.15.4 packets, both header compression and fragmentation is handled by 6LoWPAN. Still this can have a fairly negative impact on the available application layer payload of ISA100.11 packets.

Security in ISA100.11a is optional and is provided on both the data-link and transport layers. The data-link layer provides hop-to-hop security while the transport layer provides end-to-end security. Symmetric keys are used for both message integrity (MIC) as well as confidentiality (encryption), and optionally asymmetric keys can be used for device joins.

ISA100.11a provides a generic object-oriented application layer which has the intention of providing base support for any type of protocol being implemented on top. This is in contrast to WirelessHART which provides a very specific application layer protocol (i.e., the HART protocol).

16.8 WIA-PA

Wireless network for Industrial Automation - Process Automation (WIA-PA) is an industrial WSN standard developed by the *Chinese Industrial Wireless Alliance* (CIWA) for the Chinese market. WIA-PA became a IEC Publicly Available Specification (PAS) in 2008 (IEC/PAS 62601) and in 2011 it was approved as an IEC standard.

16.8.1 Protocol Overview

WIA-PA is based on the *IEEE 802.15.4-2006* standard and supports two types of network topologies; a hierarchical topology which combines star and mesh, and a pure star topology. The hierarchical network topology that combines star and mesh (both based on IEEE 802.15.4-2006) is illustrated in Figure 16.8. The first level of the network uses a mesh topology, where routing devices and gateway devices are deployed. The second level of the network is in star topology, where routing devices, field devices, and handheld devices are deployed.

A WIA-PA network consists of four main device types (see Figure 16.8):

Host computer An interface through which users can interact with the WIA-PA network.

Gateway device The Gateway handles protocol-translation and data-mapping between the WIA-PA network and other networks.

Routing device A device which forwards packets between different devices.

Field device The field devices are connected to the process and consists of pressure, temperature, position, and other instruments.

Handheld device A portable device used for configuration of the network devices and monitoring of network performance.

In addition there are two important logical roles which must be present in the network (usually in the Gateway device):

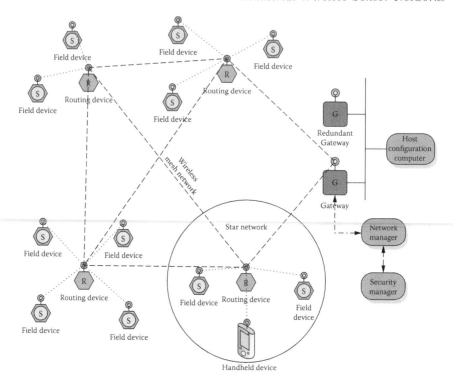

FIGURE 16.8
WIA-PA hierarchical network topology.

Network Manager The Network Manager manages and monitors the entire network, and there should only be one network manager per WIA-PA network.

Security Manager The Security Manager deals with security key management and security authentication of gateway devices, routing devices, field devices, and handheld devices.

The Network Manager and Security Manager that are used for system management should reside in a gateway device. One physical device may perform the functions of several logical roles. In the hierarchical network that combines star and mesh, a gateway device may perform the logical roles of gateway, Network and Security Managers, and cluster head. A routing device should act as a cluster head. A field device/handheld device should only act as a cluster member.

16.8.2 Protocol Features

WIA-PA provides time slotted communication using beacons specified in IEEE 802.15.4 [10]. The NM generates superframes (see Figure 16.9) for each routing device, and the superframe length is set as the lowest data update rate of all the members in a cluster. The time slot types of WIA-PA includes shared time slots and dedicated time slots, where shared time slots are used for transmission of aperiodic data, and dedicated time slots are used for intra- and inter-cluster transmission of periodic data.

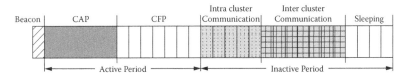

FIGURE 16.9
Details of the WIA-PA superframe.

Figure 16.9 shows what a WIA-PA superframe looks like. It consists of a beacon, an active, and inactive periods (as defined in [10]). Within the active period there are two sub periods; the Contention Access Period (CAP) and the Contention Free Period (CFP). The CAP is used for device joins, intra-cluster management, and retries, and the CFP is used for handheld to cluster head communication. The inactive period is used for intra-cluster communication, inter-cluster communication, and sleeping time. WIA-PA uses static routing which is configured by the Network Manager based on information on devices neighbors.

WIA-PA also supports frequency hopping and uses three different mechanisms for this. Which mechanism is used depends on which type of communication it is used for, i.e., if its in the active or inactive periods. *Adaptive Frequency Switch* (AFS) is used to change the channel used for the Beacon, CAP, and CFP for each superframe period. The channel is only changed if the conditions are bad (i.e., high packet loss ratio). *Adaptive Frequency Hopping* (AFH) changes channel per time slot if the channel condition is bad (i.e., number of retries are high) and is used for intra-cluster communication. *Timeslot Hopping* (TH) always changes the channel each time slot and is used for inter-cluster communication.

There are two types of aggregation used in WIA-PA: data and packet. Data aggregation is used when the source of data combines several data packets it intends to send into one packet, and packet aggregation is used by routing devices to combine several data packets from different field devices.

WIA-PA provides a flexible way to use the security measures defined in the IEEE 802.15.4 standard. Both data integrity (*Message Integrity Code* (MIC)) and confidentiality (encryption) is provided on the application and data-link layers, i.e., end-to-end and per hop security. Security is optional in WIA-PA, and it is possible to configure the use of either application or data-link layer security separately as well as MIC or encryption only protection.

16.9 Conclusions

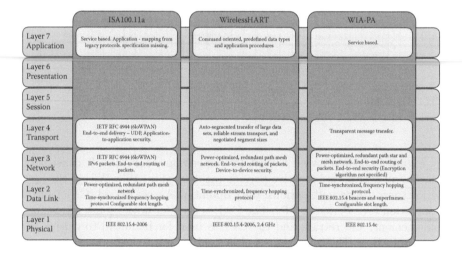

FIGURE 16.10
Comparison of the three industrial WSN standards.

In recent years opinions have been raised advocating a convergence of the standards, since all three target the same applications. There are some ongoing initiatives looking into merging WirelessHART, ISA100.11a, and WIA-PA. Progress is, however, slow and it is still to early to predict the outcome of this work.

Figure 16.10 shows how the three different specifications are structured according to the OSI layers. As can be seen they all build on top of IEEE 802.15.4 with only minor modifications, i.e., all three standards provides similar performance (e.g., 250 kbps). On the higher layers the standards diverge making convergence a challenge. WirelessHART is the only standard which completely specifies the application layer protocol. The advantage of this is full device interoperability. Interoperability is the ability for like devices from different manufacturers to work together in a system and be substituted one for another without loss of functionality at the host system level.

Work on enhancing the IEEE 802.15.4 standard is also progressing. The IEEE 802.15 task group 4e is currently working on enhancing the MAC layer of the 802.15.4-2006 standard. This new version aims for a better support for the industrial markets and better compatibility with modifications being proposed within the Chinese WPAN.

Another concern in the industry is the increasing problem with co-existence. In general, more and more technologies utilize the ISM frequency band. The more systems that operate in the same frequency band the more likely interference is. This has caused standards bodies, like ETSI [6], to act. In 2011, ETSI passed EN300-328 that forces any radio operating in the 2.4GHz band to stay below 10 mW unless it has a back-off mechanism similar to WLAN.

A general trend in industrial communication is the adoption of Ethernet based communication. This is for example the case in wired communication with ProfiNET, which replaces the non-deterministic TCP/IP layers with a deterministic real-time communication layer. The same approach has been taken by for example WirelessHART, which adds new layers on top of the IEEE 802.15.4 layers.

Within the WSN area there is a trend to use the IPv6 protocol. Both ZigBee Smart Energy 2.0 and 6LoWPAN are IP based protocols under development. These are protocols are primarily intended for home automation and not industrial applications. The main question with IP based wireless protocols is if the large protocol overhead can be justified in an industry setting.

It is clear that wireless standards tailored for the process industry are needed. The market share for industrial WSNs is, however, small compared to the WSN market in the consumer and residential area. Industrial WSNs could therefore be forced to follow the same path and become IP based in the future. Another possibility is that the industrial sector move away from the 2.4GHz ISM band. This is already the case in the medical area.

16.10   Glossary

AES: Advanced Encryption Standard.

BA: Building Automation.

CIWA: Chinese Industrial Wireless Alliance.

CSMA: Carrier Sense Multiple Access.

DCS: Distributed Control System.

ECC: Electronic Communications Committee.

FCC: Federal Communications Commission.

ETSI: European Telecommunications Standards Institute.

FA: Factory Automation.

GTS: Guaranteed Time Slot.

HART: Highway Accessible Remote Transducer.

HART7: The HART standard which includes Wireless HART.

HCF: HART Communication Foundation.

IEC: International Electrotechnical Commission.

IEEE: Institute of Electrical and Electronics Engineers.

IP: Internet Protocol.

ISA: The International Society of Automation.

ISA100.11a: An ISA developed IWSN specification for process automation.

IWSN: Industrial Wireless Sensor Network.

OSI: Open Systems Interconnection.

PA: Process Automation.

TDMA: Time Division Multiple Access.

WIA-PA: Wireless network for Industrial Automation - Process Automation.

WLAN: Wireless Local Area Network (IEEE 802.11 family).

WPAN: Wireless Personal Area Network (IEEE 802.15 family).

WSN: Wireless Sensor Network.

6LoWPAN: IPv6 over Low power Wireless Personal Area Networks.

References

[1] 6LoWPAN - IPv6 over Low-Power Wireless Personal Area Networks. http://tools.ietf.org/html/rfc4919.

[2] Bluetooth Special Interest Group (SIG). http://bluetooth.org.

[3] G. Cena, L. Seno, A. Valenzano, and C. Zunino. On the Performance of IEEE 802.11e Wireless Infrastructures for Soft-Real-Time Industrial Applications. *IEEE Transactions on Industrial Informatics*, 6(3):425–437, 2010.

[4] Chinese Industrial Wireless Alliance. http://www.industrialwireless.cn/en/06.asp.

[5] Electronic Communication Committee. http://www.cept.org/ecc/.

[6] European Telecommunications Standards Institute. www.etsi.org.

[7] Federal Communications Commission. http://www.fcc.gov.

[8] HART Communication Foundation. http://www.hartcomm.org.

[9] Hirschmann Industrial Wireless LAN. http://www.hirschmann.com.

[10] IEEE 802.15.4. http://www.ieee802.org/15/pub/TG4.html.

[11] IEEE Wireless Local Area Networks. http://grouper.ieee.org/groups/802/11/.

[12] International Electrotechnical Commission. http://www.iec.ch.

[13] International Society of Automation. www.isa.org/isa100.

[14] Ministry of Information Industry. http://www.gov.cn/english/2005-10/02/content_74175.htm.

[15] Ministry of Internal Affairs and Communications. http://www.soumu.go.jp/english.

[16] J. Neander, T. Lennvall, and M. Gidlund. Prolonging Wireless HART Network Lifetime Using Data Aggregation. *IEEE International Symposium on Industrial Electronics (ISIE), 2011*, June 2011.

[17] Siemens Industrial Wireless LAN. http://www.automation.siemens.com/.

[18] The ZigBee Alliance. http://www.zigbee.org.

Index